RENEWABLE ENERGY

ECONOMICS, EMERGING TECHNOLOGIES AND GLOBAL PRACTICES

Renewable Energy: Research, Development and Policies

Additional books in this series can be found on Nova's website under the Series tab.

Additional e-books in this series can be found on Nova's website under the e-book tab.

Energy Science, Engineering and Technology

Additional books in this series can be found on Nova's website under the Series tab.

Additional e-books in this series can be found on Nova's website under the e-book tab.

RENEWABLE ENERGY: RESEARCH, DEVELOPMENT AND POLICIES

RENEWABLE ENERGY

ECONOMICS, EMERGING TECHNOLOGIES AND GLOBAL PRACTICES

ANDREAS POULLIKKAS
EDITOR

nova publishers
New York

For permission to use material from this book please contact us:
Telephone 631-231-7269; Fax 631-231-8175
Web Site: http://www.novapublishers.com

NOTICE TO THE READER

Additional color graphics may be available in the e-book version of this book.

Library of Congress Cataloging-in-Publication Data

ISBN: 978-1-62618-231-8

Published by Nova Science Publishers, Inc. † New York

CONTENTS

PREFACE

Renewable Energy: Economics, Emerging Technologies and Global Practices provides an insight to the current trends in the field of renewable energy which are expected to play an important role in the future sustainable energy systems. The current edited book collection is divided into eleven chapters as follows:

Chapter 1 - *What lessons have been learned from renewable energy policy and R&D Activities failures in Japan*: Japan had succession of the failures of policies and projects to promote renewable energies to society. This chapter examines the background of the failures in terms of policy implementation structure in Japan through a case study of policies for promoting biomass utilization. The result shows that one of the reasons of the failure is that Japan has not put much importance on the progress and effectiveness of policy implementation. The R&D program of JST-RISTEX in the area of environment and energy, which is one of limited examples of program implementation in Japan, is also examined. As a result, it is pointed out that the feedback system of the program to the next program cycle or policy level are still challenges to be tackled, as well as increasing the understanding of participants of the program through R&D proposed projects and building.

Chapter 2 - *Supporting decision makers in assessing environmental policy instruments*: The European Union (EU) has formulated ambitious environmental policy targets aimed at forming the basis for a sustainable European growth with increasing prosperity and reduced pressure on natural resources and the environment. A major framework for this is the EU Sustainable Development Strategy (SDS) and the EU 2020 strategy, in particular, with its flagship initiative for a resource efficient Europe. More specifically, targets include the reduction of greenhouse gas emissions, improvement of energy efficiency, and increasing the use of renewable energy technologies. In order to achieve these targets, a broad package of Policy Instruments (PIs) to promote Sustainable Energy Technologies (SETs) will be required. In this context, the aim of this chapter is to present a decision support tool to assess how specific PIs could affect the implementation of SETs and provide policy recommendations, based on a Strategy Roadmap. In particular, the tool explores the consequences of the implementation of a given policy with regards to environmental, economic and social sustainability. Moreover, the tool contributes to the selection of the most suitable SETs for implementation, so as to direct national sustainable development priorities. The current approach lays emphasis on the formulation of the process so as to be relatively straightforward to incorporate direct experts' preferences, as well as trying to deal with the needs of flexibility and validity. The types of impacts and conclusions, as well as the tool's evaluation are discussed.

Chapter 3 - *Energy mix, intermittency of renewables and electricity export drivers*: In this chapter, the electricity export drivers are analyzed for a panel of 16 European Union countries, from 1994 to 2009. Exporting can be an apt solution to cope with the characteristic of intermittency of renewables sources. The characteristic of non-constant generation of electricity, which is conditional to the promptitude of natural resources (intermittency) could provoke either deficit or excess of electricity generation when compared to domestic demand. Exports can therefore be a precious tool in optimizing the electricity supply. Exported electricity is evaluated contingent on an energy mix, by using two approaches: i) *capacities approach* - installed capacities of the various energy sources; and ii) *generation approach* - electricity actually generated by these sources. Results show that both hydropower and wind contribute to electricity exports, unlike the solar photovoltaic source. When considering fossil sources, two effects are observed. On the one hand, coal and oil show a positive relationship with exported electricity, suggesting that they are generating electricity above the mean value of domestic consumption needs. This result could even be a sign of a precautionary function to avoid sharp shortages in the system, namely under an energy mix increasingly accommodating renewables. Coal and oil are thus corroborating their main role of baseload energy generation. On the other hand, the combined cycle gas turbine plants are commonly used to back up renewables and, as expected, gas restrains exports. Indeed, the use of gas is mainly used to back up the volatility in the electricity generation from renewables, and not to support exports. Overall, by using the actual generated electricity, by source, rather than the installed capacity by source, the robustness of the results is enhanced. This can be explained by the lag between installed capacity and the actual utilization of that capacity. In other words, diversifying the energy mix may be creating excess capacity, and if the literature is focused on installed capacity, then erroneous results could arise.

Chapter 4 - *Hydrogen based energy storage*: The transformation of the energy system towards a power generation based on renewable energies necessitates the implementation of storage technologies. A number of short term storage options exist and are already in use. However, large scale, long term storage can only be performed on hydrogen based chemical storage systems. Nevertheless, the low storage densities of hydrogen constitute a severe constraint of these concepts. To overcome this limitation hydrogen can be liquified, compressed or chemically transformed. A major focus of recent research activities has been the chemical transformation. A conversion into a carbon based fuel by Fischer-Tropsch or related processes is one possibility for this. The existing infrastructure and the high storage densities for this kind of fuels make this option very interesting even though the overall efficiencies are quite poor. A crucial requirement for the production of these fuels is the availability of a concentrated carbon oxide stream. A second option is the reversible storage of hydrogen on an organic carrier compound. Therefore an unsaturated compound is hydrogenated and later on dehydrogenated again, when the energy is needed. Certain constraints should be considered during the selection of the carrier material. For example, liquid compounds are easier to handle than solid or gaseous substances. Since the melting points of the hydrogenated and dehydrogenated forms often significantly vary, a close look on the phase change behavior is necessary. A big advantage of organic hydrogen carriers compared to other forms of chemical conversions of hydrogen is the fact that the hydrogen is released again as hydrogen. Therefore, the efficient power generation systems like fuel cells that are available for hydrogen can be used. The overall efficiency thus can be significantly increased compared to the conversion into fuels such as Fischer-Tropsch-gasoline.

Chapter 5 - *Green nanotechnology for renewable energy*: The amount of the world's primary energy supply provided by renewable energy technologies is required urgently because the current fossil fuel usage is unsustainable and associated with greenhouse gas production. Moreover, it is well known that the greatest challenge of nanotechnology today is the understanding the long-term impact on the renewable energy and the related environment of structures that cannot be seen even with the most sophisticated optical microscopes and structures that can interact with the basic components of life. Therefore, the relevant technologies such as hydrogen fuel, solar cell, biotechnology based on nanotechnology suitable to the friendly environment are reviewed aim to provide the vital information about the growing field of green nanotechnology for renewable energy to minimize the potential environmental risks. It is pointed out that the significantly feasible world's eco-energy for the foreseeable future should not only be realized, but also methods for using the current energy and their by-products more efficiently should be found correspondingly, alongside technologies that will ensure minimal environmental impact.

Chapter 6 -*Stochastic solutions for the unit commitment of power systems with wind power generation*: This chapter focuses on the comparison of two main classes of stochastic unit commitment formulations in the presence of uncertainty due to wind power generation: the scenario based stochastic unit commitment and the robust stochastic unit commitment. Both methods have an explicit representation of volatility / stochasticity of wind power production. The advantages and disadvantages of each method will be discussed with an emphasis on their practical implementation for the TSO or the market operator. The analysis focuses on uncertainty characterization when wind power generation is to be considered and a discussion on the choice of objectives and constraints and their implications on problem dimension and data requirements. An evaluation of the stochastic solutions and their implications on a day-ahead market will be provided.

Chapter 7 - *Evaluation methodologies and input data for offshore wind power plant electric topology assessment*: Offshore wind power is attracting the interest of experts in a range of disciplines such as electricity, electronics, mechanics, materials, control, project management and financing. The objectives are bold. On the one hand, with the fossil era showing slight signs of decay, renewable sources of electricity are seen as a realistic supply of energy for societies of the XXI century. Marine renewable energies in general, and offshore wind energy in particular, are one of the most promising alternatives in this slow but global shift to renewable sources of electricity. On the other hand, during the last decade onshore wind has proven itself as a powerful economic drive for the countries leading its development. And that drive continues worldwide in areas with vast land and wind resources. In contrast, appropriate onshore sites are dwindling in some countries and governments, companies and academia are focusing on offshore wind.

Chapter 8 - *Exploitation of olive solid waste for energy purposes*: The promotion of biomass for energy purposes consists one of the main initiatives of the European energy policy. The Mediterranean region possesses a very promising source of biomass, olive solid waste (OSW) that can potentially be exploited for energy purposes. In the recent past, many studies have been conducted for the investigation of the performance of OSW as an energy source. Although a lot of knowledge has already been gained with regard to the appropriate conditions for an efficient and "clean" OSW energy conversion, this type of biomass still remains unexploited, resulting to a significant loss of green energy on an annual basis. This chapter aims to present the current research status on the exploitation of OSW for energy purposes. Research efforts conducted in the field of OSW direct combustion, pyrolysis and gasification are presented and discussed. European standards in the field of olive solid waste quality classes and specifications are also presented and discussed. The most promising technology solutions that could potentially be applied are identified and further analyzed. Emphasis is given on the gas emissions resulting from the OSW energy conversion as well as the necessary fuel pretreatment conditions.

Chapter 9 - *Electrification of Rural Areas using Renewable Energy Sources; review of best practices around the world*: 1.6 billion people live in areas where the electricity grid is not available. One of the electrification options is the use of Renewable Energy Sources (RES) based off-grid systems. This chapter describes the progress in RES technologies utilized for this purpose focusing on small scale RES. A general outline of the financing mechanism is provided in order to understand potential financing ways. A review of Best practices for rural electrification via RES in various, usually remote areas of the world, is provided. This review shows that off-grid electrification via RES is a viable solution. The examples may show sometimes innovative financing, marketing and technical solutions and try to give the paradigm for imitations.

Chapter 10 - *Technical, economic, and mitigation performance of small-scale (≤1 kW_e) hydroelectric and photovoltaic water pumping systems relative to network electricity in rural areas in the southwest of Western Australia*: This work describes rural hydroelectric and photovoltaic (PV) water pumping systems technical performance simulations and economic and greenhouse gas mitigation modelling research. The pico-hydroelectric (for electricity service provision) and PV water pumping (for stock water supply) simulations were selected for a detailed analysis due to their high sensitivity to water resource availability in rural areas. Both observed and projected changes in rainfall and runoff in the in the predominantly agricultural southwest (SW) of Western Australia (WA) will heavily influence land, water, and energy supply decision-making at all levels, including landholders. The hydroelectric systems are also contrasted against similar technical alternatives in the region (≤3 kWe grid-connected PV and wind turbine systems). The results suggest the technical performance of small-scale grid-connected (≤1 kW_e) hydroelectric systems have the higher technical potential of all the small-scale grid-connected technical alternatives over the short-to-medium term, despite current drying conditions in the SW of WA. Similarly, the small-scale stand-alone PV water pumping system could not compete against grid-connected electricity when a connection was available, although the systems would be an effective market adaptation and market mitigation measure to supply stock water in most cases. This is due to stock watering requirements being generally off-grid, and a stand-alone PV pumping system NPV would likely be greater than a grid-connected pumping system that requires electricity network extension. Furthermore, a farm water supply pipe network extension may be economically less attractive than a stand-alone PV pumping system, although this would require further analysis of water resource quality and availability at the water point, and would require cost comparisons between the PV option and the costs associated with extension of the pipe network, including the actual pipe gauge costs, the pipe network extension distance, and whether the pipe was buried or otherwise (etc). However, all of the renewable energy systems were expensive adaptation and mitigation options and were unable to compete with the low cost of network electricity when a connection is available in terms of the NPV against 2010 market prices, subsidies, and costs.

Chapter 11 - *Modeling the carbonization process to develop a cost-effective, smokeless, continuous, down-draft rice husk carbonier suitable for poor rice growing regions*: This work discusses modeling of the operational processes occurring in a small-scale, down-draft, continuous rice husk (40 kg h^{-1}) carbonizer suitable for application in poor rice growing regions. The model was used as a tool to optimize the performance of a constructed carbonizer using material and heat balances. The carbonizer technology operating principles are discussed in terms of four operational "zones" and the possible reactions occurring in each zone. The material balance model was used to determine the amounts of each participating material at each zone, and the energy balance was generated using the material balance solutions. The final output of the model for O_2, CO, and CO_2 was reconciled with testing performance of the constructed carbonizer. The results suggested that 99.2% (weight basis) of the total CO produced during carbonization was burnt at the ignition chamber zone, resulting in only 0.8% CO emission from the chimney. The energy balance determined there was a high potential for the carbonizer to produce useful heat, for rice farm activities, with flue gasses calculated at 724°C. The material and heat balance models were successfully verified by prototype testing.

Renewable Energy: Economics, Emerging Technologies and Global Practices is not by any means exhaustive, nor is it intended to be, but to provide current advancements in the field.

Dr. Andreas Poullikkas
Visiting Associate Professor
College of Engineering
American University of Sharjah

In: Renewable Energy
Editor: Andreas Poullikkas
ISBN: 978-1-62618-231-8

Chapter 1

WHAT LESSONS HAVE BEEN LEARNED FROM RENEWABLE ENERGY POLICY AND R&D ACTIVITIES FAILURES IN JAPAN

Sawako Shigeto*

Associate Fellow, Research Institute of Science and Technology for Society (RISTEX), Japan Science and Technology Agency (JST), Tokyo, Japan

ABSTRACT

Japan had a succession of the failures of policies and projects to promote renewable energies to society. This chapter examined the background of the failures in terms of policy implementation structure in Japan through a case of the policy implementation for promoting biomass utilization.

The result shows that one of the reasons of the failures is that Japan has not put much importance on the progress and effectiveness of the policy implementation process. The Research and Development (R&D) program of JST-RISTEX in the area of environment and energy, which is one of the limited examples of "program" implementation in Japan, is also examined and the effectiveness is discussed in light of the ROAMEF cycle established in the UK.

The lack of a feedback system as the organization to the policy improvement cycle and the next program cycle was also pointed out. Although the government is under the serious consideration to introduce "program" as an effective tool of policy implementation, based on few experiences of program implementation and the low awareness about the significance in academia, there would be still a long way to overcome the past policy failures.

Keywords: Renewable energy, Policy failure, R&D program implementation

* Corresponding author address: K's Gobancho 9F, 7, Gobancho, Chiyoda-ku, Tokyo 102-0076, Japan. E-mail: sawakoshigeto@gmail.com.

INTRODUCTION

The Great East Japan Earthquake on March 11th 2011 (3.11) and the following Fukushima nuclear accidents brought an abrupt change in the circumstances surrounding actions against global warming and fossil-fuel-dominant society. The Japanese government had been recognized renewable energy as a key element to achieve greenhouse gas (GHG) emission reduction, and the huge amount of budget has been allocated to research and development (R&D) activities to promote renewable energy to society. However, the contribution of renewable energy is accounted for only 3.5 % of the total electricity output in 2010. Ministry of Internal Affairs and Communications evaluated the Biomass Nippon Strategy and the projects undertaken by the Strategy between 2003 and 2008, and resulted that almost all the projects failed not only to promote biomass utilization in society but also to achieve CO_2 reduction. Thus, before March 11th in 2011, Japan had a succession of the failures of policies and projects to promote renewable energy to society. There are two significant aspects of the failures. Firstly, the developing renewable energy utilization system requires humane and social approaches working with local people and adopting communicative, institutional and economic methods, particularly in the case of biomass energy resources. However, the greater part of the R&D and promotion projects had been still "technology push" or "initial investment" approach. Secondly, Japan had not worked seriously on effective policy and program or project design and the appraisal and evaluation cycle development not only for achieving the specific objectives and goals but also for contributing to the social well-being, whilst other developed countries for example European countries and the United States have developed the unique appraisal and evaluation process, such as ROAMEF (Rationale, Objectives, Appraisal, Monitoring, Evaluation and Feedback) cycle in the UK and the PART (Program Assessment Rating Tool) in the US, learned from the past failures. In Japan the introduction of program to R&D as a tool of policy implementation is finally under the consideration responding to the drastic change of the Basic Plan for Science and Technology from the 3rd of specific area focused to the 4th (Cabinet decision, Augst 2011) of the problem solution based. This change was stemed from the critical views in the general public if the previous science and technology policy was really helpful and beneficial for society, which was enhanced further after 3.11. It is also an important aspect that there was a lack of the viewpoint to propose a new social-economic system in the program and project design. The significant increase of renewable energy in society means the reform from the fossil-fuel-dominant and large-scale centralized social system to the sustainable and de-centralized social system, particularly after 3.11. However, most of R&D activities were conducted as alternatives on the premise of the conventional fossil-fuel-dominant system within the classical social structure. This chapter aims to examine the background of the failures of renewable energy policies and R&D activities particularly in terms of policy implementation structure in Japan through an analysis of policies for promoting biomass utilization. Firstly the renewable energy policies in Japan are overviewed. Then the failure of the past renewable energy policies and R&D activities in Japan particularly for biomass utilization are examined from the two aspects mentioned above. The third, the R&D program of "Community-Based Actions, against Global Warming and Environmental Degradation (FY2008-2013)", which is one of the limited examples of "program" implementation in Japan, is examined, and the effectiveness is discussed in light of the ROAMEF cycle.

Overview of Renewable Energy Policies in Japan

1970s – 1990s

The political interest to renewable energies in Japan was initiated by the first oil cricis in 1973. The "Sunshine Project", which is a big narional project, was launched in 1974 to develop and commercialize new energy technologies, including renewables. Just after the second oil crisis (in 1980), Energy and Industrial Technology Development Organization (NEDO) was established to implement the project, and the sunshine project was updated to the "New Sunshine Program" in 1993. The New Program was characterized to address the rationale of the program for the contribution to the CO_2 reducton against grobal warming, whilst the sunshine project was focused on the development of the particular alternative energies instead of oil (i.e., solar, geothermal, coal and hydrogen energy technologies). As to the implementation of renewable energy technologies to society, subsidies for residential photovoltaic systems were applied during 1994-2003, which contributed the diffusion of photovoltaic power generation in Japan. Subsidies for entities to introduce wind power systems were also applied in 1994 (till 2010). The hosting COP3 and the ratification of the Kyoto Protocol accelerated the government's budget increase for renewable energy. However, policies in 1990s were characterized as "technology push" or "initial investment" approach (Maruyama, et al., 2007).

2000s – Before 3.11, 2011

In the late 1990s, the introduction of market-pull policy such as feed-in tariffs (FIT) was started discussing according to the introduction of FIT in Germany and Spain. "Federation of Diet Members for Promoting National Energy" was launched. With the cooperation of an environmental NGO, the Federation took the action to enact a law to promote renewable energy based on FIT. However, with the strong opposition of the Ministry of Economy, Trade and Industry (METI) and power companies, the law never passed in 2000. After the election of House of Representatives, the initiatives to enact a law for promoting renewable energy was placed to the METI, and the New and Renewable Portfolio Standard (RPS) Law based on a fixed quota system was enacted in 2002 and came into effect in 2003. The RPS law legally required electric power suppliers to sell a certain percentage of electyricity generated from the new and renewable sources. This policy framework was not effective to promote renewable energy drastically because the target was quite low and fixed, that is only 1.35 % of the total electric power supply in 2010. Thus the policy was not designed to promote renewable energy over the targeted amounts, rather resulted in discouraging entities to participate in the renewable energy businesses. Moreover, the end of the subsidies for residential photovoltaic systems was announced in 2003 (abolished in the end of the FY 2005 and restored 2009). As to biomass energy, "Biomass Nippon Strategy" was approved at the Cabinet meeting in 2002, followed by the implementation of 214 projects between 2003 and 2008 to develop biomass utilization technology, local planning, technical assistance/extension, and biomass conversion facilities in local communities. However, the policy evaluation concluded that those were not effective to promote biomass utilization (Ministry of Internal Affaires and Communications, 2011). Although a variety of renewable energy policy measures were applied in 2000s, the renewable energy market was not established in Japan because those were still mostly "technology push" or "initial investment" subsidies without the certain commercial viability. In many local communities, renewable energy technologies or facilities were symbolically introduced by the use of those subsidies, and never established as a business. The turning point was in 2009 when FIT was introduced although only for solar surplus power. After the political regime change in the same year, the political discussion to broaden the coverage of the FIT for the all renewable energy was accelerated.

After 3.11

The energy policy was shifting to the nuclear dependent reflected by the intention of policy makers and the industry sector before 3.11. However, the severe nuclear disaster at Fukushima Daiichi nuclear power plant brought a tremendous shock to the government and Japanese society. The Japanese government has undertaken the re-examination of the Basic Energy Plan established in 2010 with a clean slate. As the result, the Innovative Strategy for Energy and Environment of Japanese government (Cabinet decision, September 2012) indicates the development of renewable energy of over 300 billion kWh by 2030 and 110 billion kWh in 2010 (three times) (The Energy and Environment Council, Government of Japan, 2012). Meanwhile, the FIT for all renewable energy was implemented in July 2012.

Policy Implementation for Promoting Biomass Utilization and the Evaluation

"Biomass Nippon Strategy" and Policy Implementation

"Biomass Nippon Strategy" (Cabinet dicision, December 2002) is the first national strategy for Japan to utilize biomass as valuable resources. The formation of the strategy was led by Japan's Ministry of Agriculture, Forestry and fisheries (MAFF) with the cooperation of Ministry of Education, Culture, Sports, Science and Technology (MEXT), METI, Ministry of Land, Infrastructure, Transport and Tourism (MLIT) and Ministry of Environment (MOT). The rationale of the government intervention through the strategy was: the prevention of climate change; the creation of a recycling-oriented society; fostering new strategic industries; and activation of agriculture, forestry, and fishery as well as associated rural communities. Thus, the strategy was explained as the first national comprehensive strategy to promote biomass utilization including the viewpoints of technology, social system and economy (Kuzuhara, 2005). As goals and objectives, three types of target rate of biomass utilization was set through use of manure, power generation and gas production etc. until 2010, i.e. technological, regional and national with the action plans. Concretely, as to nationwide perspective, the strategy had a goal to utilize more than 80% of waste biomass and more than 25% unused biomass in terms of their carbon equivalent, and to start the commercial use of energy crops until 2010. Similarly, as to technological perspective, the goal of the energy conversion efficiency was also set as follows: 20% in terms of electricity, and 80% in terms of heat. Meanwhile, in the regional perspective, it was targeted the launch of about 300 Biomass Towns that have implementation plans of more than 40% of unused biomass or 90% of waste biomass, in terms of their carbon equivalent. The strategy was once revised based on the change of the situation such as the adoption of Kyoto Protocol (in February, 2005). The new strategy clearly uphold promoting to use domestic agricultural products for transportation fuel and unutilized resoures such as thinned lumbers left in forest areas.

To achive the goals, the forenamed five ministries and Ministry of Internal Affaires and Communications (MIC) implemented 214 biomass promotion projects which cost about 6 trillion yen (US$75 million when 1$=80yen) between 2003-2008. The projects consisted of R &D, technology demonstration tests, facility implementations in local areas, and dissemination and rising awareness of them, etc.

The Policy Evaluation of "Biomass Nippon Strategy"

Article 19 of the Government Policy Evaluation Act (formulated in 2001 and effective from 2002) requires the Japanese government to create an annual report on the status on policy evaluation to reflect in policy revision, and the Ministry of Internal Affairs and Communications (Administrative Evaluation Breau) are responsible for planning and coordinating of policy evaluation. The policy evaluation was conducted over the policy level (i.e. "Biomass Nippon Strategy") and the individual projects level (i.e. implemented 214 biomass promotion projects), and the report was leleased to the public in Feb 2011. The report revealed the serious consequences that only 35 projects (i.e.,16.4% of the total) were effective but stilll far from the expected level, and only 8 facilities out of 77 biomass related facilities have the effect of CO_2 emission reducaion, whilst the others have no effect or rather increaseing the emission. With regards to Biomass Towns, the report also shows the negative perspective that the progress and feasibility seems quite poor after the planning although 286 towns (at the end of Jan 2011) were announced their biomass implementation plan.

The duprication of the similar projects over several ministries were also pointed out as the inefficiency of policy implementation. Following the critical evaluation, the policy recommendations were made in Feb. 2011 to MIC, MEXT, MAFF, METI, MLIT, and MOT.

Why the Policy Implementation Failed?

"Technology Push" and "Initial Investment" Approach

The policy evaluation of "Biomass Nippon Strategy" categorized 214 projects into "R&D", "Technology demonstration test", "Facility implementations", "Dissemination and rising awareness" and "New standard setting". As seen in Figure 1, projects categorized into "R&D" and "Technology demonstration test", that are "technology puch" type of approach, account for 54 % of the total projects. Whilst those categorized into "Facility implement-tations", that are “initial investment approach”, accounts for 24.8% of the total, in fact 81.6% of the total budget were allocated into it.

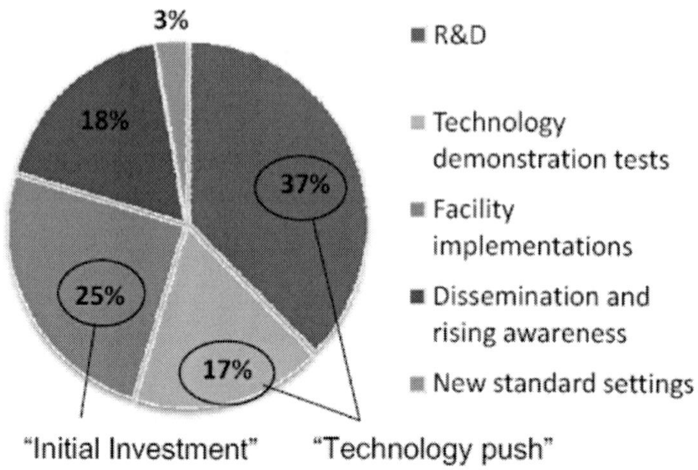

Figure 1. The type of the implemented projects under the “Biomass Nippon Strategy”.

Thus, the great part of the R&D and promotion projects under the “Biomass Nippon Strategy” was still “technology push” and “initial investment approach” although biomass energy utilization requires developing humane and social approaches working with local people and adopting communicative, institutional and economic methods, particularly among the renewable energy resources.

The Way of Policy Implementation

The other important aspect of the policy failure is the way of policy implementation. The most case of the policy implementation has been through projects under the policy implementation plan (see Figure 2). The ministries have tended to focus on the budget allocation for the policy implementation, and not much on the progress or effectiveness of the projects as the policy implementation.

Although the concept of PDCA (Plan-Do-Check-Action) cycle has introduced into policy evaluation system by the Government Policy Evaluation Act, the reality is that it has been neglected for a long time at the policy implementation level.

Ministries have recently started applying their own project evaluation systems, but they were no more than administrative check if the projects are conducted as planed. It has come to the little effect as shown in the previously mentioned “Biomass Nippon Strategy”.

Beside, there is also the startling fact that the accounts of about 40% of the projects were not specified even by the related ministries.

On the other hand, other developed countries for example in the United States and European countries have often set “programs” over the individual project to have responsibility for achieving the expected policy goals with developing the strict program evaluation system based on the PDCA cycle.

Currently in Japan, the introduction of program particularly to R&D is under the consideration for more effective policy implementation responding to the drastic change of the Basic Plan for Science and Technology from the 3rd of the specific area focused to the 4th of the problem solution based (Working Group on system for evaluating government funded R&D, Cabinet Office of Japan, 2012). “Working group on system for evaluating government funded R&D ” was organized under the Cabinet Office.In the working group, ROAMEF (Rationale, Objectives, Appraisal, Monitoring, Evaluation and Feedback) cycle in the UK was introduced as an oversea example developed to achieve policy goals effectively.In the UK, the Treasury published “The Green Book” which sets out a ‘policy cycle’ to represent the desired policy process as seen in Figure 3.In the ROAMEF cycle, each stage follows on rationally from the previous one, so that a rationale is developed, then objectives are set, then options are appraised (HM Treasury, 2003).Virtually, this cycle is a form of ‘policy cycle’, and recently has judged being divorced from reality by most academics and practioners (Rutter, et al., 2011). However, it is still valuable as a reference to improve the way of policy implementation in Japan.

Japan has to work seriously on the following aspects to improve the way of policy implementation to avoid the futher policy failures like “Biomass Nippon Strategy”even if ‘program’ was introduced as policy implementation: one is how to manage ‘program’ in the effective PDCA cycle, and the other is how to build the ‘policy cycle’ in the continuous vision to achieve policy goals in Japanese government which has neglected while the UK has tackled for a long time.

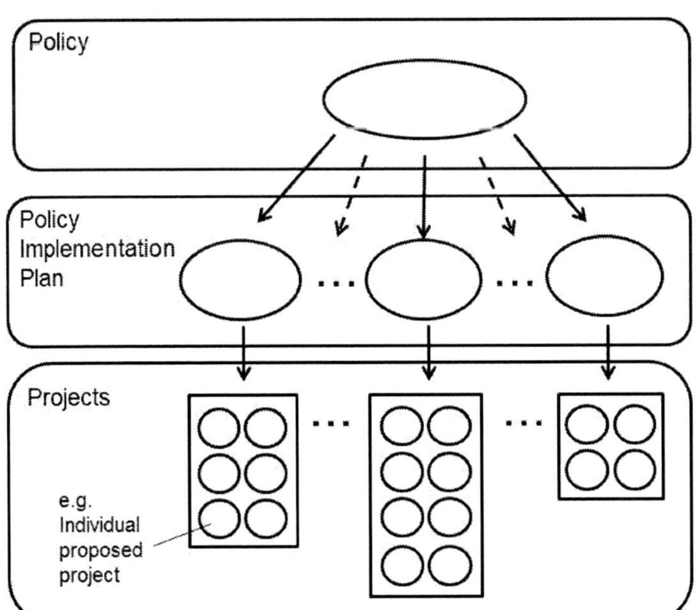

Figure 2. The Current Policy Implementation Flow.

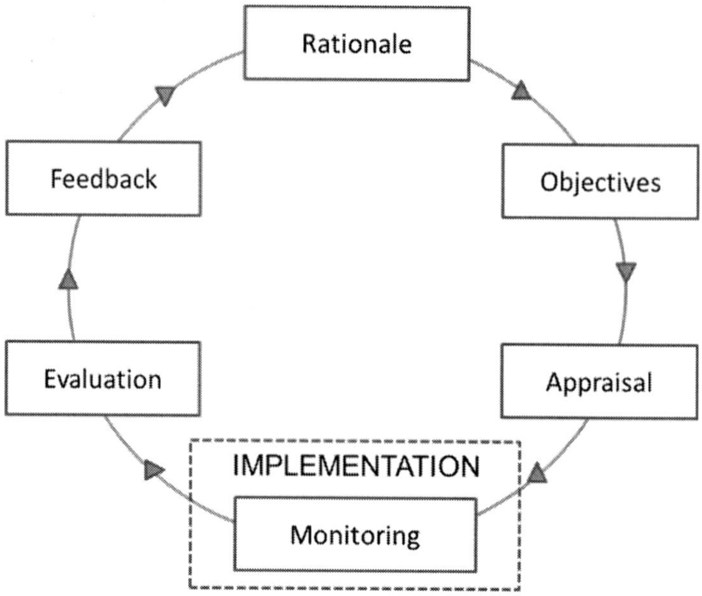

Figure 3. ROAMEF Cycle (HM Treasury, 2003).

Program Design and Management in the Area of Energy and Environment: A Case in Japan

Research Institute of Science and Technology for Society (RISTEX) of Japan Science and Technology Agency (JST) is a unique organization that has introduced R&D programs to contribute to the specific social problem solving. The research institute has developed responding to internationally increasing the awareness to the "science in society and science for society" that was declared as the role of science in the 21st century" at the 1999 World Conference on Science in Budapest, jointly hosted by UNESCO and the International Council for Science (ICSU). For establishing a program, the RISTEX firstly identifies current social issues to be solved or that can be expected to occur in the future, and invites relevant experts and stakeholders to survey and investigate the respective issues then to narrow down specific issues. After the establishment of the specific goals and objectives relating to the specific soial problem solving, the program starts inviting R&D project proposals and selecting 10-20 R&D projects as a tool of program implementation. To reach the objectives or goals of the project, the PDCA cycle is applied both in the program management level and the individual project level. In this passage, the effectiveness of the implementation of the R&D program in the area of environment and energy is examined in light of the ROAMEF cycle.

Program Design: Setting Program Objectives and Its Appraisal

Similar to other OECD member countries, Japan has a number of research funds in the area of energy and environment for tackling climate change and environment degradation. However, those were often "technology push" and "initial investment approach" founded mostly on scientific/engineering scenarios developed by the experts including researchers as seen the overview of renewable policies in Japan as well as the case of "Biomass Nippon Strategy". However, not only for introducing technologies, but also for changing society toward low-CO_2-emissioned system, it is necessary to develop humane and social approaches working with people and adopting communicative, institutional and economic methods. The deficiencies associated with the classical R&D approaches should be overcome by adopting the concepts of "science and technology for society" and "on-the-ground perspectives" and by accumulating the collaborative relationship with local people. The R&D program "Community-Based Actions against Global Warming and Environmental Degradation" (FY2008-2013) sets the objectives as below to tacke climate change and to achieve 80% GHGs reduction by 2050 in the combination of technical scenario and socio-economic scenario with the quantitive evidence (JST-RISTEX, 2011).

1 Research into ways to eliminate global warming and achieve coexistence with the environment is framed in terms of activities to achieve sustainable social systems, based on new concepts that are both comprehensive and laterally integrated. Efforts to scientifically verify the results will be conducted in regional frontline areas. An effort will be made to generalize and systematize research achievements to enable them to be used effectively both at home and abroad.
2 To promote active local communities through redefining the issue in terms of actions against global warming and those for environmental coexistence, existing activities, policies, administrative systems, programs and so on are being organized and analyzed and endogenous and trans-disciplinary plans, techniques for identifying, evaluating and disseminating new values of local communities are developed.

Corresponding to the objectives, the characteristics of the program was defined: identification of problems from trans-disciplinary, comprehensive, endogenous and productive view points to turn local communities into complex sustainable systems; appropriate cooperation among researchers of humanities and natural science, industry, government, academia and the general public; formulation and implementation of a methodology for nurturing the people who will bear the construction of sustainable and autonomous local communities; development of diverse venues in which local residents and stakeholders can share a vision of the community's future.

After the program was appraised as ppropriate about rationale of the RISTEX's intervention into the targeted area, goals and objectives and the program implementation plan including the selection of program officer (PO) and advisers as a management team by the outside appraisal committee, then the program was officially established, and started a call for R&D project proposals and selecting cycle (for three years from the FY2008).

Program Management (Program Implementation and Monitoring)

For program implementation, it is necessary to design the appraisal endpoints of proposed R&D projects. The appraisal endpoints define the program characteristics as well as effectiveness for achieving the program objectives and expected outcomes.

The deficiencies associated with the classical R&D approaches should be overcome (see Table 1): by showing convincing needs, novelty, uniqueness of the project and its approach; by defining clearly the goals and objectives with the targeted amounts of the GHGs reduction; by clearly targeting the contribution to the problem solving at the regional level; by building realistic scenario associated with the medium-term scenario and having appropriate PDCA cycle application; by developing problem solving and GHGs reduction scenario in the combination of technical scenario and socio-economic scenario with the quantitative evidence; by collaborating diverse local stakeholders and entities for implementation of the project.

Through the first R&D project proposals and selecting cycle, it was revealed that there was a difficulty for applicants to design the R&D project based on the combination of the technical scenario and socio-economic scenario with the quantitive evidence. The majority of the proposals were still techno-centric or regional activities oriented with few R&D element. Therefore, the program guideline was reinforced by adding the scenario building guide (see Figure 4) and providing the quantitative check sheet to be filled in by the applicants.

Table 1. Appraisal endpoints of the RISTEX's energy and environment program to overcome the conventional R and D approach in the area

	The classical R&D approach	Proposed R&D projects in the RISTX's energy-environment program
Rationale	Not so clear	The needs, novelty, uniqueness of the project and its approach should be convincing.
Goals and objectives	Not clearly defined	Goals and objectives should be clearly defined with the targeted amounts of the GHGs reduction.
Contribution for problem solving	Not much focus on actual contribution to communities and society	The contribution to the problem solving at the regional level shouled be clearly targeted.
Scenario building	Not much associated with the medium or long term scenario	The clear relationship has to be shown between medium-term scenario and project implementation plan with appropriate PDCA cycles application.
	Not realistic due to too much technocentric nature	The problem solving and GHGs reduction scenario should be developed in the combination of technical scenario and socio-economic scenario with the quantitive evidence.
Implementation system	Often not appropriate for achieving the objectives	Diverse stakeholders and entities in the communities should be participated, accumulating the collaborative relationship with local people.
Continuity after the project period	Not much considered	The project implementation plan shoud also include the plan for continuous contribution to building sustainable society in the regions after the project period.

Note: The author reorganized the table 1 in (Shigeto and Horio, 2011).

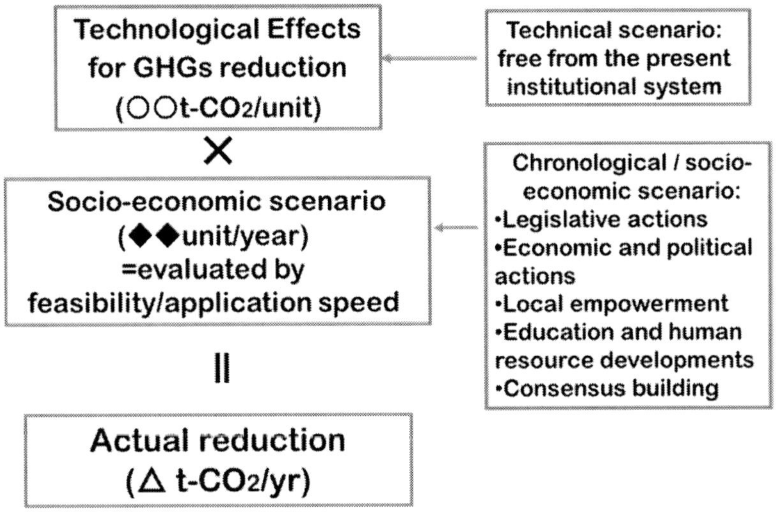

Figure 4. The guide to quantify and structure R&D project.

Thus, the PDCA cycle has also applied to the program management so that it improves to achieve objectives of the program effectively.

The program currently has 12 projects underway (4 projects already completed) in Hokkaido, the Tohoku region, the Kanto region, the Chubu region, the Kansai region, the Chugoku region, and on the islands of Shikoku and Kyushu, which generally covers the entire country.

These projects are highly motivated by a variety of local-problem-solving that include: urban and rural development through renewable energy utilization and the shift from the fossil-fuel-dominant society; rural remigration through the establishment of rules for self-government and the creation of local industries; revitalization of forestry by urban-rural collaboration; development of new logistics and financial systems to promote the transition to a low CO_2 emission society; revitalization and creation of local industries by these means.

With regard to the monitoring at the R&D project level, the program management team checks the progress through the review of the annual plan and report as well as visiting project sites, meetings and events.

When there are concerns over the progress, the management team sometimes takes some direct actions to improve under the program director decision.

The Further Challenge (Evaluation and the Feedback)

There was a mid-term evaluation to the program and projects having more than 3 year duration as an outside evaluation in 2010. While the program policy, management and intervention for improving projects were evaluated positive, it was pointed out that the insufficiency of the understanding and motivation for the individual R&D project to challenge the goals and objectives set by the program.

It stems from the long history of R&D funds that researchers could utilize them without much obligation and constraints once they were allocated. At the policy level, the importance of introducing 'program' for effective policy implementation has been increasingly recognized, but for researchers and local actors it still tends to be one of sources for continuity of their research and their local activities, respectively.

This result implies that how to share the program goals and objectives and guide them is still the challenge in Japan which has few experiences of program implementation. The end of the program period, the outside evaluation will be applied to the program and projects. The evaluation result should give feedback to the policy level or the next program cycle according to the ROAMEF cycle.

However, the RISTEX does not have a feedback system as an organization because the program is established through the own process of identifying important social issues. The related policy and implementation plans tend to come across the several ministries, and it is difficult to access them only by the program own efforts.

Beside, currently there has little continuity as a program within the RISTEX because the RISTEX does not necessarily run the program under the mid-term or long-term plans in the related area. Therefore, it also currently relys on the program own effort to bridge the outcome diffusion to society.

CONCLUSION

Japan had a succession of the failures of policies and projects to promote renewable energies to society. This book chapter examined the background of the failures in terms of policy implementation structure in Japan through a case study of policies for promoting biomass utilization.

In Japan the introduction of program to R&D as a policy implementation is under the consideration responding to the drastic change of the Basic Plan for Science and Technology from the 3rd of specific area focused to the 4th of the problem solution based.

This paper firstly examined the evaluation results of biomass utilization promotion policies which are implemented from 2003 to 2008 based on the "Biomass Nippon Strategy". The result shows that the policy failed because Japan has not put much importance on the progress and effectiveness of the policy implementation process compared to such as the ROAMEF cycle established as a policy making cycle in the UK.

Research Institute of Science and Technology for Society (RISTEX) of Japan Science and Technology Agency (JST) has introduced R&D programs to contribute to the specific social problem solving. Therefore, the next, the effectiveness of program implementation of the R&D program of JST-RISTEX in the area of environment and energy is also examined in light of the ROAMEF cycle. The result shows that there is a lack of the feedback system as the organization to the policy improvement cycle and the next program cycle.

It is also pointed out as a problem that the meaning and significance of the program implementation has not still understood well by the individual R&D project participants because they have been hardly monitored the R&D progress under the PDCA cycle in the past in Japan.

As a result, there is still a risk to continue policy failures after the program implementation if it is not faithfully learned from the ROAMEF cycle and dose not value on the PDCA program management, although there is an urgent need to introduce program as an

effective policy implementation tool in Japan to overcome the past policy failures particularly in the area of environment and energy.

ACKNOWLEDGMENTS

This book chapter is addressed as a part of R&D program management of "Community-Based Actions against Global Warming and Environmental Degradation (FY2008-2013)" of Japan Science and Technology Agency (JST), Research Inst. of Science and Technology for Society (RISTEX) that the author engages in. However, the view indicated in this chapter is personal not representative of the view of the organization.

REFERENCES

HM Treasury, The Green Book: Appraisal and Evaluation in Central Government, (2003). http://www.hm-treasury.gov.uk/d/green_book_ complete.pdf (last date accessed: 15 Oct. 2012)

JST-RISTEX., (2011). Research Institute of Science and Technology for Sociery (RISTEX) 2011-2012. http://www.ristex.jp/public/ristexnews/pdf/11 pamph_en.pdf (last date accessed: 18 Oct 2012)

Kuzuhara, Y. Biomass Nippon Strategy-Why "Biomass Nippon" now?. *Biomass and Bioenergy* 29 (5), 2005, 331-335.

Maruyama, Y., Nishikido, M., Iida, T., (2007). The rise of community wind power in Japan: Enhanced acceptance through social innovation. *Energy Polocy* 35, 2761-2769.

Ministry of Internal Affaires and Communications, (2011). Report on evaluation of Biomass Utilization Policy. (in Japanese)

Rutter, J., Parker, S. and Sworth, M., (2011). Policy making in the real world: evidence and analysis. Institute for Government. http://www.instituteforgovernment.org.uk/sites /default /files /publications/Policy%20making%20in%20the%20real%20world.pdf (last date accessed: 18 Oct 2012)

Shigeto, S. and Horio, M. (2011). Adapting the concepts of "science and technology for society" and "on-the-ground perspectives" for restoration of Tohoku area and the society suffering serious damage from the Great East Japan Earthquake. *Proceedings of the 26th annual meeting of The Japan Society for Science Policy and Research Management*. 15-16 Oct. Yamaguchi University. (in Japanese)

The Energy and Environment Council, Government of Japan, (2012). Inovative Strategy for Energy and the Environment.

Working Group on system for evaluating government funded R&D, Cabinet Office of Japan. (2012). Report on the examination of R&D evaluation system development.

In: Renewable Energy
Editor: Andreas Poullikkas

ISBN: 978-1-62618-231-8

Chapter 2

SUPPORTING DECISION MAKERS IN ASSESSING ENVIRONMENTAL POLICY INSTRUMENTS

Charikleia Karakosta*, Vangelis Marinakis and John Psarras
National Technical University of Athens, School of Electrical and Computer Engineering, Management and Decision Support Systems Laboratory (EPU-NTUA), Prof. John Psarras, Athens, Greece

ABSTRACT

The European Union (EU) has formulated ambitious environmental policy targets aimed at forming the basis for a sustainable European growth with increasing prosperity and reduced pressure on natural resources and the environment. A major framework for this is the EU Sustainable Development Strategy (SDS) and the EU 2020 strategy, in particular, with its flagship initiative for a resource efficient Europe. More specifically, targets include the reduction of greenhouse gas emissions, improvement of energy efficiency, and increasing the use of renewable energy technologies. In order to achieve these targets, a broad package of Policy Instruments (PIs) to promote Sustainable Energy Technologies (SETs) will be required. In this context, the aim of this chapter is to present a decision support tool to assess how specific PIs could affect the implementation of SETs and provide policy recommendations, based on a Strategy Roadmap. In particular, the tool explores the consequences of the implementation of a given policy with regards to environmental, economic and social sustainability. Moreover, the tool contributes to the selection of the most suitable SETs for implementation, so as to direct national sustainable development priorities. The current approach lays emphasis on the formulation of the process so as to be relatively straightforward to incorporate direct experts' preferences, as well as trying to deal with the needs of flexibility and validity. The types of impacts and conclusions, as well as the tool's evaluation are discussed.

Keywords: Sustainable Development; Policy Instruments; Decision Support

---

* Corresponding Author address: Email: chkara@epu.ntua.gr.

INTRODUCTION

Nowadays, there is growing emphasis on the two-way relationship between Sustainable Development (SD) and climate change mitigation. In this context, the European Union (EU) has formulated ambitious environmental policy targets in a variety of sectors aimed at forming the basis for a sustainable European growth with increasing prosperity and reduced pressure on natural resources and the environment. Indeed, six themes have been addressed by the Environmental Action Programme of the European Commision (EC), namely acidification and air quality, waste prevention and recycling, marine environment, soil, pesticides, natural resources and urban environment. A major framework for these challenges is the EU Sustainable Development Strategy (SDS) and the EU 2020 strategy, in particular, with its flagship initiative for a resource efficient Europe [1]. In the area of energy and climate change the EU has set a 20/20/20 target to reduce Greenhouse Gas (GHG) emissions, increase the use of renewable energy technologies, as well as increase energy efficiency by 20% by 2020 through its Energy and Climate Package and subsequent directives [2]. In order to achieve these targets, a broad package of Policy Instruments (PIs) will be required to promote Sustainable Energy Technologies (SETs) which varies from market-based instruments to government-imposed legislation. These policies and technologies need to function together in an efficient way whereby mutual reinforcement needs to be sought and counter productivities avoided. Each EU Member State should monitor, assess and report the effectiveness of its policy for climate change mitigation. For this purpose, a list of the EU SD indicators has been developed to provide a picture of progress towards the goals and objectives of the EU's SD strategy [3, 4]. Moreover, a number of SD indicators have been presented by the relevant studies and reports of the International Atomic Energy Agency (IAEA) [5], the Organisation for Economic Co-operation and Development (OECD) [6] and the United Nations (UN) [7, 8].

A number of studies exist in the international literature, which mainly focuses on the assessment of energy technologies based on SD indicators [9-13]. The studies of Olsen and Fenhann [14], Sutter and Parreño [15] presented an overview of SD criteria and respective indicators to assess the SD benefits of all the CDM projects. It is thus necessary to incorporate environmental policies into a broader sustainability strategy aiming at reducing the use of natural resources required by the socio-economic system. While environmental policies are increasingly harmonised at the EU level, national policies still are fragmented and may be creating hurdles to or impeding achievement of EU-level policy goals, including through increased costs and unsustainable policy outcomes. Currently available information on policies and technologies often enables only qualitative assessments, which limits the analysis of opportunities for further refinement and tuning of EU and national policies. To remedy this, more thorough evaluations of their effects are needed. In this context, the aim of this chapter is to present a decision support tool to assess how specific PIs could affect the promotion of SETs and provide policy recommendations that lead to a Strategy Roadmap. In particular, the tool explores the consequences of the implementation of a given policy with regards to environmental, economic and social sustainability. Moreover, the tool contributes to the selection of the most suitable SETs for implementation, so as to direct national sustainable development priorities. The types of impacts and conclusions, as well as the tool's evaluation are discussed.

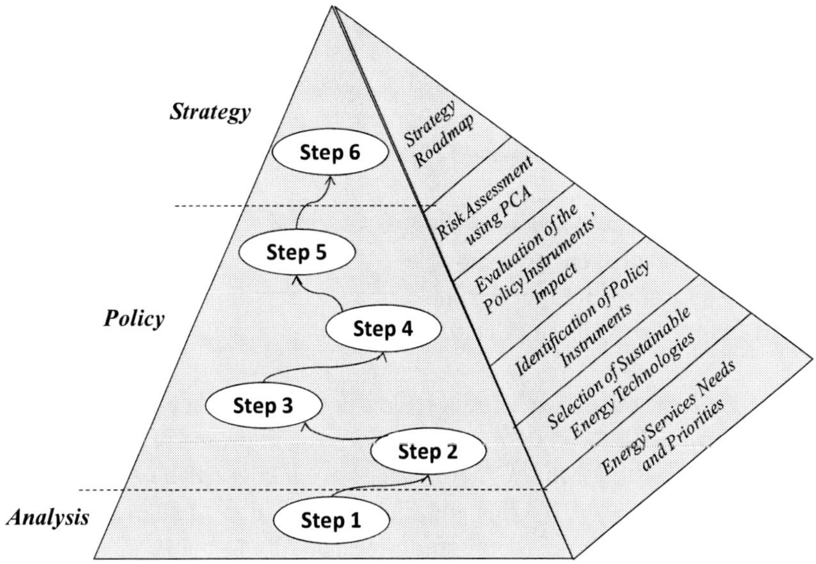

Figure 1. The General Philoshophy of the Approach Adopted.

Apart from the introduction, the chapter is structured along three sections. The main steps of the approach proposed are presented in the second section. The third section is devoted to a critical analysis of the proposed tool's impacts and its evaluation. Finally, the last section is summarizing the key issues that have arisen in this paper.

Methodological Procedure

The Approach

The current approach lays emphasis on the formulation of the process so as to be relatively straightforward to incorporate direct experts' preferences, as well as trying to deal with the needs of flexibility and validity. The general philosophy of the adopted approach is presented in Figure 1.

More specifically, the proposed tool includes the following steps:

- *Step 1 – Energy Services Needs and Priorities:* Before answering on "how" the country promotes diffusion of suitable SETs, the need is evident to turn the selection of the technology implemented towards the country's specific needs and priorities [16]. The assessment of the country's energy needs and priorities is targeted as a country driven exercise aiming at the effective involvement of the widest possible groups of stakeholders for in-depth consultations. Securing adequate stakeholder participation requires a clear, structured and ongoing framework that involves all relevant parties at an early stage, makes the impacts upon them and their responsibilities clear, and continues to engage with all stakeholders throughout the assessment and implementation process. In this context, a broad range of energy services have been identified as alternatives, such as electricity for industry, agriculture, households and service sector, heat for industry, households and service sector, etc. The adopted approach uses a Linguistic Decision Making method, based on the Linguistic Ordered Weighted Averaging (LOWA) operator, to assess stakeholders' preferences, in terms of assessing the country's energy services and identify the best alternatives. The criteria to be used for the assessment of the priority energy services are the following [17]:

- Need for increased access to energy;
- Need for reliable energy supply;
- Need for affordable energy supply;
- Need for energy decentralization.

- *Step 2 - Selection of SETs:* This step aims at the selection of the most suitable SETs in accordance with the country's energy needs and priorities, as well as PIs through the use of an existing Multi Criteria Decision Making (MCDM) method, namely the ELECTRE Tri Method. To this end, the identification of alternative SETs has been realized. The six criteria incorporating the benefits in the economic, environmental and social dimension are the following [19-21]:

  - Accordance with strategic/developmental planning;
  - Local and regional economic development;
  - CO_2 emissions reduction;
  - Minimization of the negative effects on the natural environment at national - regional level;
  - Contribution to the employment;
  - Contribution to energy independence.

  In this context, stakeholders and key energy actors were interviewed so as to elicit all the required input. Stakeholders for the interview procedure were selected so as to represent cross section groups in the countries' energy market.

  The stakeholders selected included among others governmental organizations, non-governmental organizations (NGOs), energy companies (products and services), industrial sectors, industrial associations, legal and financial bodies, researchers and developers. The experts determined the weights foe each criterion as well as evaluated the alternatives over each one of the criteria.

- *Step 3 - Identification of PIs:* Member States have a number of different options that they can use to promote SETs. Often, a mix of PIs is the key to their success (e.g. regulatory policies, fiscal incentives, public financing). In this context, a number of PIs have been identified, such as feed-in tariff, capital subsidies, grants, rebates, tradable renewable energy certificates, energy production payments or tax credits, energy tax, excise tax, or VAT reduction, public investment, loans, or financing, etc.

- *Step 4 – Evaluation of the PIs' Impact:* This step includes the evaluation of the impact that PIs have in a broad range of environmental, social and economic sectors upon criteria. The tool incorporates six general categories of criteria, as follows [18]:

- *Effectiveness:* Static and dynamic effectiveness, environmental integrity, security of supply, degree of coercion and prescription, rebound effect, etc;
- *Efficiency:* Static and dynamic efficiency, compliance costs, administration costs, transaction costs (search, negotiation, approval, monitoring, insurance), institutional compatibility, etc;
- *Impacts on energy and market prices:* Prices of electricity and fossil fuels;
- *Impacts on society:* Equity and fairness for directly affected parties or indirectly affected parties (e.g consumers), employment, business opportunities and trade, governmental revenues, political acceptability, etc;
- *Innovation:* Innovation of efficient technologies, diffusion of new efficient technologies, etc;
- *Market competition:* Liberalization of market, market transparency, entry barriers, product market distortions, tax distortions, etc.

- *Step 5 - Risk Assessment using Principal Component Analysis (PCA):* This step includes the risk assessment of the chosen SETs combined with the appropriate PI using PCA, as follows:

  - Identification of the PCA's criteria;
  - Assessment of each alternative technology and criterion;
  - Determination of PCA's variation (Pearson or Covariance matrix);
  - PCA's results (score of alternative technologies and criteria).

- *Step 6 - Strategy Roadmap:* Through this step the transition from technology-oriented considerations to policy and strategy considerations/ roadmap will be elaborated. The Strategy Roadmap will generate the integrated package of strategies needed to support the transfers through identification of the blockages and opportunities. The development of the Strategy Roadmap is based on the following actions:

  - Strategy Roadmap draft version;
  - Strategy Roadmap enhanced version;
  - Stakeholders' mobilization on the consultation process;
  - Online public consultation of the enhanced roadmap;
  - Strategy Roadmap's finalization.

Sources and Processing of Data

A desk study analysis is proposed to be elaborated for the identification of the different PIs that countries can use to promote SETs, as well as the alternative SETs.

An overview of the current status of the already known policies and technologies is necessary. Analytical information could be also collected from international sources and publications on the existing PIs used and examined technologies in energy systems in various regions.

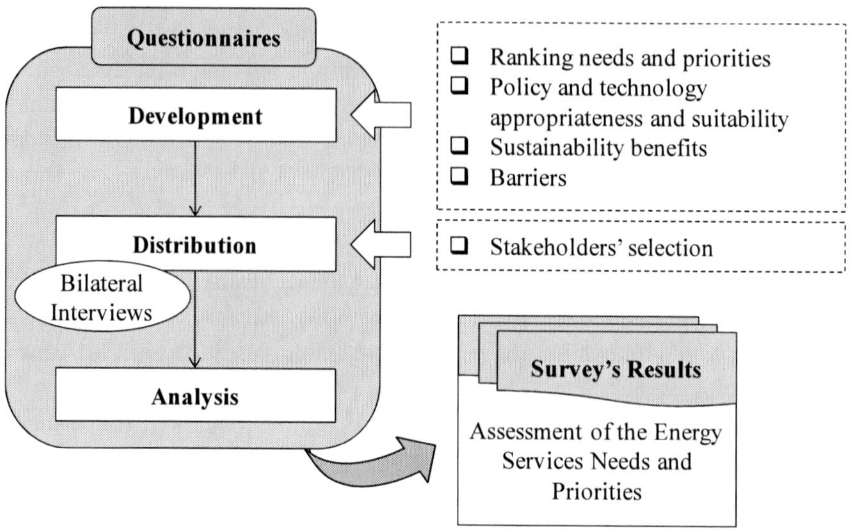

Figure 2. Questionaires' Distribution

Finally, an estimation of growing potential of the technologies with a view to Research and Development (RandD), as well as the availability of space and resources and the social acceptability, could be conducted. Moreover, an appropriate questionnaire could be used in bilateral interviews with energy sector, research and policy stakeholders (Figure 2). In particular, questionnaires' careful development is essential to ensure that questions will elicit all the required information and that the questions are clear and unambiguous. It is essential to clarify the specific items of information required that can be reasonably obtained through a questionnaire approach. A number of interested stakeholders will be asked to evaluate (e.g. on a Lickert scale from 1 to 7) the countries' energy services needs and priorities. Stakeholders will be selected from governmental departments with responsibility for energy, environment and development issues, representative national and international companies or bodies in other GHG intensive sectors (e.g. energy intensive industry). There are also stakeholders from financial institutions involved in the manufacture, import and sale of environmentally sound technologies, international organizations and donors, local organizations, as well as NGOs involved in the promotion of environmental and social objectives. Regarding the evaluation of energy services, a linguistic term set (insignificant, low, medium, high, perfect) is used for the required data acquisition by policy stakeholders. For the assessment of energy technologies, the performances will be assigned by the stakeholders in an ordered scale from 0 (not relevant for the country) to 5 (very high priority).

Critical Analysis of the Proposed Methodology

Types of Impacts and Conclusions

In our approach, a range of energy services, policies, technologies and criteria for the assessment were defined. Moreover, related stakeholders were also involved; so as to identify the country's needs and priorities, policies and technologies fulfilling the country needs that are most applicable to the country's SD. The research focuses on the provision of a small but clearly understood set of evaluation criteria, which can form a sound basis for the comparison of the examined energy options in terms of their contribution to SD objectives. More specifically, the evaluation criteria incorporate the benefits in the economic, environmental and social dimension. For the evaluation of the PI's impact a broad range of criteria were identified, including effectiveness, efficiency, impacts on energy and market prices, impacts on society, innovation and market competition. In addition, for the assessment of the energy technologies a number of sustainable development objectives had to be incorporated in the analysis, such as social impacts, local and regional economic development, GHG reductions, as well as environmental impacts, based on the specific characteristics, development needs and perspectives of each country.

The adopted approach leads to a Strategy Roadmap, through the integration of outcomes reached by the relevant evaluations. This Roadmap will contribute to the definition of a technology strategy by displaying interaction between technological priorities and socio-economic conditions over time, taking into account a series of considerations on the relevant PIs. In particular, the Roadmap will involve exploration of the following:

- Policy vision within a specific time horizon, suitably tailored for the country context.
- Strategies (country specific) based on the analysis of the research and testing outputs;

Evaluation

The proposed tool addresses the question of what the key priorities and needs with respect to a country's SD are. Nonetheless, this process assists only in describing a country's development needs and priorities before moving to the identification, evaluation and prioritization of policy and technological means (Table 1). Furthermore, the tool leads to technologies that are based on the specific countries' large indigenous energy resources that can, in most cases, be considered by the government of high priority for introduction in their energy sector. In addition, the adopted approach secures adequate stakeholder participation by using a clear, structured and ongoing framework that involves all relevant parties at an early stage, while simplifying process input and data elaboration provided by stakeholders so as to result in concrete outcomes. In contrast to other models, the tool does not lay emphasis to the evaluation of alternative scenarios by comparing their energy requirements, their social costs and benefits and their environmental impacts. The methodology faces the problem of imprecise and uncertain data, since stakeholders information may often be qualitative or quantitative information that is not stated precisely. PIs often interact, as many instruments also aim and claim to achieve certain co-benefits. Due to a lack of knowledge and data, as well as methodological know-how, however, current impact assessments do not adequately take into account these interactions at the sectoral, national and European level. Indeed, it is not possible to discern differences in effectiveness among policies given the diverse environmental needs and objectives of a particular country. A perspective for further research is to incorporate a more comprehensive understanding of the policy systems at the sectoral level, the national level, and a pan-European level. Including interrelationships and trade offs within each level on different scales of observation is of crucial importance when designing efficient policy mixes that can lead Europe to more significant sustainable pathways.

Table 1. Strengths and Shortcomings of the Tool

Strengths	Shortcomings
Identification of a country's energy development needs and priorities	Evaluation of alternative scenarios
Evaluation of the PI's impacts	Data uncertainty
Provision of specific actions and/or strategies	Interrelationships and trade offs within each level on different scales of observation when designing efficient policy mixes
Stakeholders engagement	Consideration of different sectors (e.g. air, land, water, health, etc)
Process simplicity and flexibility	

European and national policymakers need a more holistic understanding of these relationships in order to design efficient environmental policies. Except from climate change and energy, the proposed approach could consider a range of different sectors in order to gain a broader picture of the impacts of environmental policies on the three dimension of sustainability. This includes incorporation of interactions between policies in different sectors, as follows:

- Environmental performance, such as impacts on climate, air, land, water, bio-diversity;
- Social performance, such as impacts on health, safety, income distribution;
- Economic performance such as impacts on investments, employment, trade balance, public budget, value added, security of (energy) supply.

Conclusion

Environmental problems go beyond climate change and many of them, global as well as local, have their root causes not just in the increasing use of energy, but in the rapidly increasing use of natural resources as a whole. In this context, a broad package of PIs will be required to promote SETs which varies from market-based instruments to government-imposed legislation. These policies and technologies need to function together in an efficient way whereby mutual reinforcement needs to be sought and counter productivities avoided. It is thus necessary to incorporate environmental policies into a broader sustainability strategy aiming at reducing the use of natural resources required by the socio-economic system. This chapter presented a decision support tool to assess how specific PIs could affect the promotion of particular SETs and provide policy recommendations, through a Strategy Roadmap. For the evaluation of the PI's impact and the energy technologies a number of sustainable development criteria are incorporated in the analysis. The Strategy Roadmap through the integration of outcomes reached by the proposed approach will contribute to the definition of a technology strategy by displaying interaction between technological priorities and socio-economic conditions over time, taking into account a series of considerations on the relevant PIs. The adopted approach secured adequate stakeholder participation by using a clear, structured and ongoing framework that involves all relevant parties at an early stage, while simplifying process input and data elaboration provided by stakeholders so as to result in concrete outcomes. The approach can assist policymakers in reducing inefficiencies in policy design and to create win-win situations whereby economic strength goes hand in hand with environmental protection and efficient use of natural resources. It should be clearly stated that this approach does not intent to replace the decision maker. With appropriate use, it could be proved to be a useful decision support tool by structuring the decision process and assisting the decision makers to preassess the most suitable PIs to promote SETs.

ACKNOWLEDGMENTS

The main motivation for the elaboration of this study was provided through the research conducted within the "Assessment of Policy Impacts on Sustainability in Europe (APRAISE)" FP7 project (Grant agreement no: 283121), funded by the European Commission. The content of the paper is the sole responsibility of its authors and does not necessary reflect the views of the EC. In addition, Mrs. Charikleia Karakosta wishes to acknowledge with gratitude the Alexander S. Onassis Public Benefit Foundation for supporting her PhD research.

REFERENCES

[1] European Comission, (2009). Communication from the Commission to the European Parliament, the Council, the European Economic and Social Committee and the Committee of the Regions : Mainstreaming Sustainable Development into EU policies: 2009 *Review of the European Union Strategy for Sustainable Development*. Brussels, Belgium. COM(2009) 400 final.

[2] European Comission, (2010). Communication from the Commission Europe 2020: A Strategy for Smart, *Sustainable and Inclusive Growth*, Brussels, Belgium. COM(2010) 2020 final.

[3] European Commission, (2011). Europe 2020 indicators: Headline indicators. Eurostat Statistics, Luxemburg. Available at: http://epp.eurostat.ec.europa.eu/portal/page/portal/globalisation/indicators online database on European Commission > Eurostat > Globalisation Indicators.

[4] European Commission, (2007). Measuring Progress towards a more Sustainable Europe – 2007 Monitoring Report of the EU Sustainable Development Strategy, ed. Ledoux L., Lock G., Wolff P., Hauschild W.: Eurostat - Statistical books, *ISBN* 978-92-79-05043-5.

[5] International Atomic Energy Agency, (2005). Energy Indicators for Sustainable Development: Guidelines and Methodologies. Vienna, Austria: IAEA Publishing, *ISBN* 92–0–116204–9.

[6] Organisation for Economic Co-operation and Development, (2011). OECD Factbook 2011-2012: Economic, Environmental and Social Statistics, Paris, France, OECD Publishing, *ISBN* 9789264124189.

[7] United Nations (United Nations Department of Economic and Social Affairs), (2007). Indicators of Sustainable Development: Guidelines and Methodologies - Third Edition. New York, USA: United Nations Publication, *ISBN* 978-92-1-104577-2.

[8] United Nations, 2011. The Millennium Development Goals Report 2011. New York, USA: United Nations Publication, *ISBN* 978-92-1-101244-6.

[9] Carrera, D.G., Mack, A., (2010). Sustainability Assessment of Energy Technologies via Social Indicators: Results of a Survey among European Energy Experts. *Energy Policy*, 38(2), 1030-39.

[10] Evans, A., Strezov, V., Evans, T.J., (2009). Assessment of Sustainability Indicators for Renewable Energy Technologies. *Renewable and Sustainable Energy Reviews*, 13(5), 1082-88.

[11] Svensson, N., Roth, L., Eklund, M., Martensson, A., (2005). Environmental Relevance and use of Energy Indicators in Environmental Management and Research. *Journal of Cleaner Production*, 14(2), 134-45.

[12] Krajnc, D., Glavic, P., (2003). Indicators of Sustainable Production. *Clean Technologies and Environmental Policy*, 5(3-4), 279-88.

[13] Veleva, V., Ellenbecker, M., (2001). Indicators of Sustainable Production: Framework and Methodology. *Journal of Cleaner Production*, 9(6), 519–549.

[14] Olsen, K.H., Fenhann, J., (2008). Sustainable Development Benefits of Clean Development Mechanism Projects: A New Methodology for Sustainability Assessment based on Text Analysis of the Project Design Documents submitted for Validation. *Energy Policy*, 36(8), 2819-2830.

[15] Sutter, C., Parreño, J.C., (2007). Does the Current Clean Development Mechanism (CDM) Deliver its Sustainable Development Claim? An Analysis of Officially Registered CDM Projects. *Climatic Change*, 84(1), 75-90.

[16] Karakosta, C., Askounis, D., (2010). Developing Countries' Energy Needs and Priorities under a Sustainable Development Perspective: A Linguistic Decision Support Approach. *Energy for Sustainable Development*, 14(4), 330-338.

[17] European Comission, (2008). Communication from the Commission to the Parliament, the Council, the European Economic and Social Committee and the Committee of the Regions: 20 20 by 2020, *Europe's climate change opportunity*. Brussels, Belgium. COM (2008) 30 final.

[18] Karakosta, C., Doukas, H., Psarras, J., (2009). Sustainable Energy Technologies in Israel under the CDM: Needs and Prospects. *Renewable Energy*, 34(5), 1399-406.

[19] Oikonomou, V., Jepma, C.J., (2008). *A Framework on Interactions of Climate and Energy Policy Instruments. Mitigation and Adaptation Strategies for Global Change*, 13, 131-56.

[20] Karakosta, C., Doukas, H., Psarras, J., (2009). Directing Clean Development Mechanism towards Developing Countries' Sustainable Development Priorities. *Energy for Sustainable Development*, 13, 77-84.

[21] Doukas, H., Karakosta, C., Psarras, J., (2010). Computing with Words to Assess the Sustainability of Renewable Energy Options. *Expert Systems with Applications*, 37, 5491-97.

In: Renewable Energy
Editor: Andreas Poullikkas

ISBN: 978-1-62618-231-8

Chapter 3

ENERGY MIX, INTERMITTENCY OF RENEWABLES AND ELECTRICITY EXPORT DRIVERS

António Cardoso Marques* and José Alberto Fuinhas
NECE, University of Beira Interior,
Management and Economics Department, Covilhã, Portugal

ABSTRACT

In this chapter, the electricity export drivers are analyzed for a panel of 16 European Union countries, from 1994 to 2009. Exporting can be an apt solution to cope with the characteristic of intermittency of renewables sources. The characteristic of non-constant generation of electricity, which is conditional to the promptitude of natural resources (intermittency) could provoke either deficit or excess of electricity generation when compared to domestic demand. Exports can therefore be a precious tool in optimizing the electricity supply. Exported electricity is evaluated contingent on an energy mix, by using two approaches: i) capacities approach - installed capacities of the various energy sources; and ii) generation approach - electricity actually generated by these sources. Results show that both hydropower and wind contribute to electricity exports, unlike the solar photovoltaic source. When considering fossil sources, two effects are observed. On the one hand, coal and oil show a positive relationship with exported electricity, suggesting that they are generating electricity above the mean value of domestic consumption needs. This result could even be a sign of a precautionary function to avoid sharp shortages in the system, namely under an energy mix increasingly accommodating renewables. Coal and oil are thus corroborating their main role of baseload energy generation. On the other hand, the combined cycle gas turbine plants are commonly used to back up renewables and, as expected, gas restrains exports. Indeed, the use of gas is mainly used to back up the volatility in the electricity generation from renewables, and not to support exports. Overall, by using the actual generated electricity, by source, rather than the installed capacity by source, the robustness of the results is enhanced. This can be explained by the lag between installed capacity and the actual utilization of that

* Corresponding Author address: Email: amarques@ubi.pt; acardosomarques@gmail.com.

capacity. In other words, diversifying the energy mix may be creating excess capacity, and if the literature is focused on installed capacity, then erroneous results could arise.

Keywords: Electricity export drivers; idle capacity; energy mix; and renewables intermittency

INTRODUCTION

The growing use of renewable energy leads to new challenges for energy policymakers, including how to properly incorporate increasing amounts of electricity, which results in much of the intermittent generation sources and inaccuracies in the forecasting of natural resource availability, such as solar power and wind power. These characteristics require solutions both to meet periods of scarce generation, as well as to avoid wastage of electricity at times of excess generation. Exported electricity arises as a possible solution. In the enormousness of international trade, and in particular among countries of the European Union (EU), energy and, especially, electricity has become increasingly important. Furthermore, the EU has assumed a leading role in the diversification of the energy mix, by setting ambitious targets for use of renewable sources in its Member States. Renewables are growing faster than the global energy market (Varun et al., 2009), and the reasons for the development of renewables are studied, for example, by Carley (2009), Sovacool (2009), and Marques and Fuinhas (2011). In the long term, an even greater use of renewable technologies is expected, including wind, solar, geothermal and hydro (Nikolakakis and Fthenakis, 2011). According to the predictions of Varun et al. (2009), by the mid-21st century, the demand for renewable energy could reach 50% of total energy demand, through appropriate policies and technological developments. In the last two decades, the use of solar and wind increased so rapidly that these two technologies are considered by Zahedi (2011) as "success stories", because beyond ensuring environmental sustainability, they further promote technological innovation. The energy obtained from renewable sources is replacing other sources of electricity generation, thereby lowering costs of fuel and greenhouse gas emissions (Luickx et al., 2010), though the real consequences of the increased use of renewable sources on CO_2 emissions are far from consensual in the literature. However, the fact is that their variability and unpredictability limit the potential of these sources, because they depend on uncontrollable factors like the weather (sun and wind), making electricity generation more variable and less predictable (Welle and Joode, 2011).

From the developments in technology, not only those linked directly to renewable sources, but also from the evolution of technology that fosters more rational use of electricity, much will depend on the development of both renewable sources and the whole energy mix of a country. High electricity prices, which are artificially granted through public policies, may be causing adverse effects on the economic growth of countries, as Marques and Fuinhas (2012a) alerted. In short, the promotion of Research and Development (RandD) activities to lower generation costs, the incentive for patenting and to increase competitive environment in the market for renewable energy will together play a crucial role in the full affirmation of renewables as competitive sources in the near future. The re-composition of the energy mix and the pursuit of energy independence may actually be causing a new dependence. In fact, despite the increasing development of technology, intermittency and unpredictability remain

key features of these sources, including the sun and wind. Consequently, they require the presence of alternative sources, namely fossil sources capable of supplying the periods in which those resources are not available, in what is known as backup power. Generating plants compatible with storage of resources, such as coal, gas, oil or water will remain compulsory in the energy portfolio composition. The role played by these traditional sources, predominantly fossil, may thus be undergoing change. From a baseload energy role, they are increasingly assuming the role of backup, which means idle capacity in large periods of the day.

In fact, as pointed out by Jacobson and Zvingilte (2010) or Purvins et al. (2012), the growth of renewables, together with the need to maintain the capacity of fossils, may be influencing electricity exports. The investment in renewables can transform what are normally energy-importing countries into export countries, after ensuring that domestic demand is fulfilled (Supersberger and Fuhrer, 2011). As noted by Kost et al. (2011), the electricity generated can be used either to meet demand, or to export to other countries. Both intermittency and the inaccuracies in resource availability forecasting for some renewables, notably wind, can cause two major effects. On the one hand, there may be shortages during periods of insufficient wind, so there will be a need to use reserve power plants (backup) or import electricity from neighboring countries. On the other hand, over-generation may occur, and exports become a solution for this electricity that goes beyond the domestic demand (e.g. Purvins et al., 2012; and Jacobson and Zvingilte, 2010). The correlation between wind and exported electricity also merited some attention in the literature, such as Mignard et al. (2007). These authors show that the generation of electricity from wind actually contributes to electricity exports in Western Denmark, but the local combined heat and power plants contributed more to the net exports than wind. It therefore seems imperative to study electricity exports, controlling not for a single source of energy individually, but instead for all the sources within the energy mix.

The location of the electric grid, its size and its interconnection with countries with diverse characteristics (e.g. weather) may influence the incorporation of more intermittent sources in the energy mix of a country. Therefore, the central countries, with easier interconnection with other grids, may incorporate higher levels of this type of electricity generation than in isolated systems, which will have greater difficulty in circumventing problems like shortage or excess generation. Focused on the electricity generation costs in the isolated system of the Canary Islands, Marrero and Ramos-Real (2010) conclude that there is an increase in the cost of electricity and in uncertainty, due to the isolation of the network. The characteristics of irregularity and discontinuation of renewable source generation, together with the isolation of the economy, may make it hard for renewables to penetrate these isolated systems. Generally, as pointed out by Weisser (2004), such an increase in the costs is mainly due to higher transport costs resulting from the distance between sources of supply. In systems that do not suffer these constraints, it is much easier to manage the supply of renewable electricity, and consequently extend is use. Despite some literature that is focused on the analysis of the main incentives and constraints to the development of renewables, the study of the relationship between the mix of electricity generation sources and electricity exports remains scarce. Focused on a panel of EU countries, this chapter seeks to contribute empirically to the study of the relationships between the various sources of generation and electricity exports, while controlling for other potential drivers, such as the price of electricity, the size of the electric grid and the dynamics of RandD. There are two central questions which it seeks to answer: (i) are renewable and fossil sources playing an identical role in exported electricity?; and (ii) taking into account the characteristic of the intermittency of renewables, and the shutdown periods of the fossil backup, a question arises. Is there any difference between putting the focus on the installed capacities and focusing on actual electricity generation, i.e., on the used capacities? Several panel data econometric techniques are applied in the analysis, both as the appropriate response to specific characteristics of the data, and in the search for econometrically robust results.

Analysis

This section analyzes the data, and thereafter it presents and discusses the results. The main objective of this chapter is to assess the effectiveness of the energy mix on electricity exports. Furthermore, the effects of drivers on exports, as indicated by the literature are controlled for. To meet the main objective, two approaches were followed to assess the potential drivers of exports. A first approach focuses on electricity generated, by source, as explanatory factors of the exports. The second approach focuses on the installed capacity, instead of on the electricity generated. Hereafter, we will name the first approach the *generation approach*, and the second the *capacities approach*. Our choice of approach belongs to the recent and open debate in the literature about the lag between installed capacity and capacity effectively used (e.g. Boccard, 2009). This lag is provoking serious problems of inefficiency in economies, given that in some periods there is substantial idle capacity. That idle capacity is verified both in some fossils and in renewables: for fossil sources, mainly as a consequence of the required backup power, and for renewables, when the natural resources are absence, namely wind and sun.

This option for two approaches has three main advantages: (i) it allows the effects of using different sources of energy generation mix in the electricity exports to be evaluated; (ii) it allows the possible high correlations to be tackled, which could be present between installed capacity and electricity generated, when jointly estimated; and (iii) it allows observation of whether the results are quite different when considering installed capacity or actual electricity generation. The panel consists of 16 EU countries, namely Belgium, Czech Republic, Denmark, Germany, Ireland, Greece, Spain, France, Italy, the Netherlands, Austria, Poland, Portugal, Finland, Sweden and the UK, for the time span covering the years 1994-2009. The selection criterion of this balanced panel was the availability of data for all variables, for the longest period of time. Table 1 shows the definition of the variables, sources, and summary statistics. The dependent variable is the exported electricity (*ELETEXP)*. The exports reveal erratic behavior, mostly after the early 2000s. This fact can be directly related to the development of renewable energy sources in electricity generation. Germany has the highest average exported electricity, while Ireland is the country with the lowest average. The explanatory variables are taken in accordance with the literature. It is worthwhile noting that in a prior phase of this research, the inclusion of several other variables such as CO_2 as factors behind exported electricity were tested, but proved not to be statistically significant and were therefore excluded from the analysis.

Table 1. Data: definition, summary statistics and source

Variable	Definition [Mean; Standard Deviation]	Source
$ELETEXP_{ct}$	Electricity export (GWh) – total for EU-27 [10522.48; 9528.19]	Eurostat, April 2012, EC
$ELETWIND_{ct}$	Electricity from wind power (Mtoe) [240.94; 583.34]	idem
$ELETSUN_{ct}$	Electricity from solar power (Mtoe) [10.39; 58.65]	idem
$ELETGAS_{ct}$	Electricity from gas power (Mtoe) [30564.59; 42525.76]	idem
$ELETGEO_{ct}$	Electricity from geothermal power (Mtoe) [299.33; 1152.08]	idem
$ELETOIL_{ct}$	Electricity from oil power (Mtoe) [9569.52; 19921.68]	idem
$ELETCOAL_{ct}$	Electricity from coal power (Mtoe) [55792.51; 74940.14]	idem
$ELETNUCL_{ct}$	Electricity from nuclear power (Mtoe) [54925.38; 103321.2]	idem
$WINDCAP_{ct}$	Wind installed capacity (MWh) [1658.06; 4021.25]	idem
$SUNCAP_{ct}$	Solar photovoltaic installed capacity (MWh) [151.77; 816.94]	idem
$MUNCAP_{ct}$	Municipal waste installed capacity (MWh) [190.17; 264.40]	idem
$HYDRCAP_{ct}$	Hydropower installed capacity (MWh) [7596.54; 7916.52]	idem
$NUCLCAP_{ct}$	Nuclear installed capacity (MWh) [7786.16; 15295.59]	idem
$POPDENS_{ct}$	Population density (pop/km^2) [154.29; 117.18]	World Bank, World Development Indicators Database
$ELETPRIC_{ct}$	Electricity prices by Medium size households (€/kWh) [10.11; 2.70]	Eurostat, Mar 2012, EC
$RDEXPEND_{ct}$	Gross domestic expenditure on RandD (% of GDP) [1.80; 0.87]	OECD, Factbook 2011

Additionally, for other variables, namely those related to installed capacities in fossil sources, data were not available for the entire time span. We then list the variables used and briefly discuss the rationale behind their inclusion in our analysis.

- *Electricity generated from wind power and wind installed capacity (ELETWIND, WINDCAP).* Wind energy has been growing in recent years (Varun et al., 2009) and it is characterized by its inherent intermittency. During periods with favorable weather conditions for the generation of electricity from wind, two effects may occur. On one hand, there could be some electricity not absorbed by domestic demand, and therefore the effect on exports is positive. On the other hand, the additional electricity could be fully absorbed by domestic demand or even take off the backup from fossils. In this case, a null effect of both electricity from wind and installed capacity in exports is anticipated.
- *Electricity generated from solar photovoltaic (PV) power and solar PV installed capacity (ELETSUN, SUNCAP).* When compared to the former variable, in this case there may be a peculiarity, which may influence the overall effect. Indeed, although there is also intermittent generation, the predictability of this resource is larger. Therefore, it may result in an easier accommodation of the electricity generated by this source in domestic demand. Two possible effects can be anticipated: a negative or a null effect. In other words, the management of the required backup is more

predictable for solar than for wind. As such, generation from the sun can be more easily compatible with domestic consumption and hence the net effect of these variables on exports could be neutral or even negative.

- *Electricity generated from nuclear power and nuclear installed capacity (ELETNUCL, NUCLCAP).* The expected effect of nuclear power on electricity exports may be positive, null or negative. The effect will be positive if, in order to enjoy the benefits of economies of scale, countries generate more electricity from nuclear than required domestically, and then they export the remainder. This effect will be not statistically significant if nuclear power is scaled just to meet the domestic needs of each country and thus not materialized in exports. The predictability of generation from this source may also reduce incentives for the use of a more diverse energy mix that could lead to over-generation. In fact, it may well have a negative effect on exports. It should be noted that, as for geothermal, some countries do not have electricity generated from nuclear power.
- *Electricity generated from fossil sources, gas, oil, and coal (ELETGAS, ELETOIL, ELETCOAL).* The effect of these sources on electricity exports may be different depending on the source and the role it plays in the energy mix. Indeed, oil is considered mainly as baseload power. In turn, coal and gas have been largely used as backup power, mostly in combined cycle power plants to support the development of renewables. Actually, such as noted in the literature (e.g. Schaber et al., 2012; and Marques and Fuinhas, 2012a) both natural gas and coal could be used as backup power for renewables. However, coal is a source of cheap energy and therefore there may be generation for exports. Thus, for oil a positive effect on electricity exports is expected, while both for coal and gas, the dominant effect could be positive or negative.
- *Municipal waste installed capacity (MUNCAP).* Unlike most of renewable sources, the use of this source of electricity generation is not dependent on weather conditions. The expected sign of this source on exported electricity is positive. Higher *MUNCAP* could mean more available resources to generate electricity. Largely, the generation of electricity from this source does not depend on the local electricity demand but the installed capacity and the availability of fuel waste. Hence, excess generation ought to be exported.
- *Electricity generated from geothermal power (ELETGEO).* This energy source is not available in all countries. Electricity generated using geothermal allows countries to decrease their energy dependence, while it can deliver electricity for export. The expected effect on electricity exports is therefore positive, because its generation may occur continuously during periods of low local electricity consumption. This excess can thus be exported.
- *Hydric installed capacity (HYDRCAP).* Hydropower is seen as essential in pursuing sustainable development (e.g. Varun et al., 2009), while it has the added benefit of enabling storage and thereby adjusting supply to the demand for electricity. For this reason, it is expected that this feature will both satisfy domestic demand, as well as being generated for export, depending on weather conditions. Hence, the expected sign of this variable is neutral or positive.

- *Population density (POPDENS).* Traditionally, population density is used as a proxy of the length of the grid. Marrero and Ramos-Real (2010) study the interconnection between grid characteristics and energy dependence. Increased population density requires a grid with higher generation capacity and may thus lead to export during periods of lower demand, so as to drain the excess generation. In short, we control for the population density, expecting a positive effect on exported electricity.
- *Electricity prices by medium size households (ELETPRIC).* The literature (e.g. Ochoa and Ackere, 2009) takes into account the price of electricity on exports. It is expected that higher prices of electricity promote electricity exports. However, considering that most national electricity generation systems may be operating at the maximum of their capacity, the effect of the price of electricity could be ineffective.
- *Gross domestic expenditure on RandD (RDEXPEND).* Most of the literature focused on exports, not only on electricity export, considers RandD expenditure as a driver of exports (e.g. Yang and Chen, 2012). We control for gross domestic expenditure on RandD as a percentage of GDP, expecting a positive effect on electricity exports. Higher levels of RandD can enable higher levels of productivity and competitiveness, as well as more dynamism within countries that encourage electricity exports. In fact, data on patents for all sources of energy are not available for all countries and for the time span under analysis. As a consequence, this variable was used as a proxy one. Overall, it is expected that the higher RandD expenditure is, the greater electricity exports will be. This effect may be a consequence, on the one hand, of the fact that these expenses could be directly incurred in the search for technologies for efficient generation of electricity. On the other hand, they may be a mirror of the most dynamic economic activity, with more robust and surplus electricity generating systems.

Method

In panel data analysis, working upon countries with specific characteristics, with variables of a different nature, and used simultaneously, it might be possible to expect the presence of heteroskedasticity, panel autocorrelation, and contemporaneous correlation phenomena. The presence of these phenomena on the data should therefore be adequately appraised, otherwise inefficiencies in coefficient estimation and biasedness in the estimation of standard errors could arise. In particular, cross sectional dependency must be assessed, in order to avoid common Fixed Effects and Random Effects panel data estimators being plagued by these inefficiencies and bias. In the presence of panel-level heteroskedasticity and contemporaneous correlation of observations among panels, two estimators could be used: the Panel Corrected Standard Errors (PCSE) and the Feasible Generalized Least Squares (FGLS). The PCSE estimator generally performs better in finite samples than the asymptotically efficient FGLS estimator (Beck and Katz, 1995). The econometric analysis was performed using Stata 12.1.

The general model is:

$$ELETEXP_{ct} = \alpha + \sum_{k=1}^{k} \beta_k X_{kct} + d_c + d_t + \mu_{ct}. \tag{1}$$

The error term is $\mu_{ct} = \rho_c \mu_{c,t-1} + \gamma_{ct}$, where γ_{ct} is serially uncorrelated but correlated over countries. Following Cameron and Triverdi (2009), the PCSE estimator allows the error term to be correlated over the countries, allows the use of first-order autoregressive model for over time and allows to be heteroskedastic. $ELETEXP_{ct}$ is the electricity export for each country c for each t year, and X_{kct} is a vector of the k explanatory variables. The dummies d_c and d_t refer to country and time, respectively. For both the *generation approach* and for the *capacities approach* the correlations matrixes, in the appendix (see tables A.1 and A.2), are useful to assess the possible existence of multicollinearity. In this first inspection of the data, no indicia of multicollinearity are evident, despite the moderate correlation coefficient between solar and wind. Estimations are run, using these variables separately and simultaneously. No substantial changes are observed. Nonetheless, to cast aside any doubt about potential collinearity, the variance inflation factor (VIF) for multicollinearity test was conducted.

Table 2. Specification tests

	Pooled	Random Effects	Fixed Effects
Generation approach			
Modified Wald test (χ^2)			1040.86***
Wooldridge test F(N(0,1))	18.390***		
Pesaran's test		2.822***	2.738***
Frees' test		1.501***	1.068***
Friedman's test		18.950	23.80*
Capacities approach			
Modified Wald test (χ^2)			538.48***
Wooldridge test F(N(0,1))	7.757**		
Pesaran's test		1.420	1.996**
Frees' test		1.677***	1.299***
Friedman's test		10.229	8.958

Notes: The Modified Wald test has $\chi 2$ distribution and tests the null hypothesis of: $\sigma_c^2 = \sigma^2$, for $c = 1,\ldots,N$; The Wooldridge test is normally distributed N(0,1) and tests the null hypothesis of no serial correlation; Pesaran's test is a parametric testing procedure and follows a standard normal distribution; Frees' test uses Frees' Q-distribution; Friedman's test is a non-parametric test based on the Spearman's rank correlation coefficient; Pesaran, Frees and Friedman's tests test the null hypothesis of cross-section independence. xtcsd Stata's command was used (De Hoyos and Sarafidis, 2006); ***,**,* denotes 1%, 5%, and 10% significance level, respectively.

For both approaches, the Mean VIF is far from the rule of thumb 10 (2.55 and 2.32 for *generation* and *capacities* approaches, respectively), as shown in tables A.1 and A.2 in the appendix. It is advisable to precede panel estimation procedures with the study of the data properties. Therefore, moving forward, we test the presence of serial autocorrelation, heteroskedasticity, and contemporaneous correlation phenomena. Table 2 synthesizes the battery of tests conducted to appraise those phenomena. The modified Wald test was provided to test for groupwise heteroskedasticity in the residuals of a fixed effect regression model (Baum, 2001). It is concluded that the errors exhibit groupwise heteroskedasticity, given the strong rejection of the null hypothesis. As far as the serial correlation is concerned, the Wooldridge test was provided. The null hypothesis of no first-order autocorrelation is strongly rejected. With regard to testing the presence of contemporaneous correlation phenomenon, three distinct tests were carried out, both for Fixed Effects and Random Effects: (i) Pesaran's test which is a parametric testing procedure (Pesaran, 2004); (ii) Frees' test, which is a semi-parametric test (Frees, 1995, 2004); and (iii) Friedman's test, which is a non-parametric test (Friedman, 1937). In the *generation approach*, in general, the tests suggest that residuals are not spatially independent, though in the Friedman's test the null hypothesis of cross-sectional independence is rejected in the model under Fixed Effects assumption at 10% significance level. In its turn, in the *capacities approach*, only the Frees' test strongly rejects the null of cross-sectional independence. In short, from the specification tests it is concluded that there is evidence of: (i) heteroskedasticity, probably because of heterogeneity in the countries with regard to *ELETEXP*; (ii) first-order autocorrelation; and (iii) contemporaneous correlation across the countries, though with stronger evidence on the *generation approach*.

The presence of contemporaneous correlation may be a consequence of the high degree of interconnection policies of the countries in analysis, given that they belong to the same supra-national organization, the European Union. With respect to heteroskedasticity, it may result from differences among the countries, such as the existence of geothermal potential or even the use of nuclear power. In short, table 2: (i) induces the appropriateness of using the PCSE estimator in the *generation approach*; and (ii) suggests the suitability of use and compares both the traditional panel data estimators and PCSE estimator for the *capacities approach*, taking into consideration the weak evidence of the existence of the contemporaneous correlation phenomenon. In accordance with the results of the specification tests, in the *generation approach* the PCSE estimator was mainly used, while the common panel data estimators were also used, under a procedure of checking for the robustness of results. In this way, for the *generation approach*, our option was to reveal results from the PCSE estimator in a separate table (table 3) and the traditional estimators are presented in the appendix (table A.3). In contrast, for the *capacities approach*, the results from the traditional estimators and PCSE are both revealed in the same table (table 4). In order to cope with the characteristics of the data, the models are estimated with four options: correlation over countries and no autocorrelation - model I; country-level heteroskedastic errors and common first-order autoregressive error (AR1) – model II; correlation over countries and autocorrelation AR(1) – model III; and correlation over countries and autocorrelation country-specific AR(1) – model IV[1].

[1] The subscripts "G" and "C" in tables 3 and 4 denote generation approach and capacities approach, respectively.

Table 3. Results of electricity generation, by source – *generation approach*

Independent variables	Dependent variable $ELETEXP_{ct}$			
	PCSE			
	Model I_G	Model II_G	Model III_G	Model IV_G
$ELETWIND_{ct}$	4.1483*** (0.6886)	3.4039*** (1.1647)	3.4039*** (1.2267)	3.6465*** (1.2867)
$ELETSUN_{ct}$	-17.2778*** (6.2526)	-9.3940 (6.9966)	-9.3940 (7.8851)	-10.0701 (8.2900)
$ELETGAS_{ct}$	-0.0506*** (0.0075)	-0.0442*** (0.0167)	-0.0442*** (0.0149)	-0.0593*** (0.0115)
$ELETGEO_{ct}$	2.3125*** (0.2768)	2.2780*** (0.5528)	2.2780*** (0.5593)	2.7037*** (0.4625)
$ELETOIL_{ct}$	0.0501*** (0.0147)	0.0639*** (0.0227)	0.0639*** (0.0204)	0.0631*** (0.0218)
$ELETCOAL_{ct}$	0.0724*** (0.0046)	0.0713*** (0.0090)	0.0713*** (0.0097)	0.0826*** (0.0105)
$ELETNUCL_{ct}$	-0.0081*** (0.0010)	-0.0063 (0.0042)	-0.0063** (0.0026)	-0.0062*** (0.0020)
$POPDENS_{ct}$	35.9585*** (2.6197)	35.4586*** (5.5902)	35.4586*** (4.9155)	39.6020*** (3.7284)
$ELETPRIC_{ct}$	348.4220*** (129.1517)	137.1927 (172.2022)	137.1927 (159.2615)	6.2645 (160.2376)
$RDEXPEND_{ct}$	3510.4947*** (217.0689)	3316.8079*** (599.3993)	3316.8079*** (431.0225)	3315.254*** (441.0598)
CONS	-8111.6347*** (1212.6881)	-6186.6179*** (1707.0313)	-6186.6179*** (1442.2765)	-5428.399*** (1875.79)
Observations	190	190	190	190
R^2/Pseudo R^2	0.7517	0.5767	0.5767	0.7810
Wald (χ^2)	5392.32***	394.52***	844.30***	959.35***

The assumptions about variances across panels and serial correlations in the PCSE estimators remain within the two approaches. These various assumptions also allow the robustness of the results to be checked.

Generation Approach

Results from the PCSE estimators, under the *generation approach*, are shown in table 3. Overall, the results from the several assumptions about variances across panels and serial correlations in the estimators reveal strong consistency. Our choice of the unbundling of the energy sources, rather than considering them on an aggregate basis, proves to be relevant and appropriate. Indeed, both for fossils and for renewables, the effects are dissimilar depending on the specific source. Parameter values and significances are very similar and no signal change is observed. The core results could be synthetized as follows:

- A positive and strongly statistically significant relationship between *ELETWIND* and exported electricity. This effect is robust and is verified for all estimations.

- The effect of *ELETSUN* on exported electricity is not always statistically significant. In the model I_G, the effect is statistically siginifcant and, unlike wind, it is negative.
- The parameter measuring the effect of *ELETGAS*, is negative and statistically significant throughout the models.
- Larger generation of electricity from *ELETGEO* improves exported electricity. This effect is highly significant and consistent throughout all models.
- The effects of fossil sources *ELETOIL* and *ELETCOAL* on exported electricity are positive and highly significant for all models.
- Although not statistically significant in all models, greater generation of electricity from nuclear source (*ELETNUCL*) hampers exported electricity.
- The *POPDENS* effect on exported electricity is positive, as expected, and is statistically significant in all models.
- The effect of *ELETPRIC* on exported electricity is not always statistically significant.
- Expenditure on RandD is driving exported electricity. This effect is highly significant, consistent, and reveals similar magnitudes throughout the models.

As stated earlier, in order to check the robustness, results from traditional panel data estimators of Random Effects and Fixed Effects are revealed in table A.3, in the appendix. As in Marques and Fuinhas (2012b), we are testing robustness in a reverse way to what is standard. Indeed, results from PCSE will be robust if the results from the other estimators are somewhat dissimilar, given that the former estimators suffer from violations of the traditional hypothesis. In other words, considering the results of the specification tests (cf. table 2), and given that the PCSE estimator are able to handle these phenomena, then obvious differences between those traditional estimators and the PCSE are expected. These differences are, in fact, detected.

The battery of diagnostic tests shown on table 2 strongly support the three options used for the residuals: (i) Conventional Standard Errors - CSE; (ii) Robust Standard to achieve heteroskedasticity-robust estimates - RSE; and (iii) First-order autoregressive errors to remove serially correlated errors - Corr(AR1). By providing a Lagrange Multiplier (LM) test, the existence of individual effects was tested, and the null of non-relevance of individual effects was strongly rejected. The Hausman test leads to the rejection at 5% significance level of the null of unobservable individual effects not correlated with the explanatory variables. Additionally, pursuing additional proof of robustness, the model was estimated with robust regression with robust weight functions, which is an M estimation with Huber and Tukey bisquare weight functions. When comparing the results from tables 3 and A.3, and keeping in mind the reverse robustness check explained above, there is evidence of inefficiencies in coefficient estimation and biasedness of standard errors. This outcome was expected seeing that classical assumptions of the usual panel data estimators are violated, as documented in table A.3.

These inefficiencies and bias cause losses of statistical significance in some variables, such as *ELETWIND*, *ELETGEO*, and *RDEXPEND*. Notwithstanding, there is no signal change in parameters and significances, revealing consistency. Bearing in mind the results from the robust regression, the stability and reliability of the estimators are once again proved robust.

Capacities Approach

Results focused on the installed capacities as explanatory variables are shown in table 4. As stated before, in view of the specification tests from table 2, under this approach the option was to reveal both the results from common panel data estimators (Fixed Effects and Random Effects) and the results from PCSE estimator. The comparison among them allows a robustness check. Once again, the LM test suggests that there is evidence of relevance of individual effects by country. In its turn, from the Hausman test comparing Fixed and Random estimates, there is no rejection of the null that Random Effects provide consistent estimates. In general, in this approach, on the one hand there is higher variability in the levels of statistical significance of PCSE estimates. On the other hand, the results achieved with the Random Effects estimator do not differ substantially from those of PCSE. Actually, this constitutes no surprise considering the specification tests, namely upon contemporaneous correlation, as discussed before. The results to highlight are:

- When statistically significant, the effect of *WINDCAP* on exported electricity remains negative. However, this effect is significant only in some models of the PCSE estimator.
- The effect of *SUNCAP* on exported electricity reinforces the negative effect achieved when considering electricity generated.
- There is a consistently positive and statistically significant relationship between *MUNCAP* and exported electricity.
- The positive effects of *POPDENS* and *RDEXPEND* are borne out, despite some variability in the significance level.

For the panel under analysis, exported electricity experienced, on average, a growth of 26%, while in some countries exported electricity more than doubled. Moreover, the large increase in installed capacity of renewables in recent years is a real fact, as noted by Varun et al. (2009). We focused, therefore, on noticing the impact of renewables on electricity exports. We control for this effect, by testing both the installed capacity and the electricity generated for wind. Under both approaches, there is a booster effect of the wind on electricity exports, in line with Mignard et al. (2007). It is worthwhile noting, however, that the results under the *generation approach* are much more robust. Actually, when compared with the *capacity approach*, the quality of panel regressions under the *generation approach* is visibly improved. One reason for this is certainly the fact that the problem of intermittency causes long periods in which the installed capacity is not harnessed. Boccard (2009), for instance, points to the capacity factor realized in Europe of around 21%. It follows that for the design of public policies these two dimensions must be accommodated. First, it is quite different to focus on installed capacity rather than on generation. The installed capacity may be hidden - a problem of economic inefficiency - considering the fact that there is idle capacity. Second, in order to encourage electricity exports, those public policies should not just stimulate the installation of more capacity. Rather, they should promote greater levels of efficiency in the generation of the capacity already installed, in the sense that reducing idle capacity means increasing exports.

Table 4. Results with installed capacity – *capacities approach*

Independent variables	OLS	Random Effects	Fixed Effects	PCSE			
				Model I_C	Model II_C	Model III_C	Model IV_C
$WINDCAP_{ct}$	0.8082 (0.6195)	0.3802 (0.3385)	0.3635 (0.2745)	0.8082*** (0.2044)	0.6493** (0.2647)	0.6493*** (0.2436)	0.4429 (0.2873)
$SUNCAP_{ct}$	-2.4672 (1.6430)	-2.2033*** (0.5414)	-1.7318*** (0.2363)	-2.4672** (1.1826)	-1.5912* (0.9451)	-1.5912* (0.8945)	-1.0910 (0.7550)
$MUNCAP_{ct}$	16.3994** (7.5599)	8.6415*** (2.9854)	8.4191*** (2.6510)	16.3994*** (2.1991)	12.9136*** (3.0994)	12.9136*** (2.7832)	7.9221*** (3.0460)
$HYDRCAP_{ct}$	0.2607 (0.1810)	0.3235* (0.1677)	-1.7862 (1.1913)	0.2607*** (0.0506)	0.2863*** (0.0893)	0.2863*** (0.1039)	0.2520* (0.1484)
$NUCLCAP_{ct}$	-0.1772** (0.0751)	-0.1284 (0.1389)	1.2429 (0.8179)	-0.1772*** (0.0233)	-0.1395*** (0.0467)	-0.1395*** (0.0430)	-0.0921* (0.0558)
$POPDENS_{ct}$	30.4362*** (9.4669)	39.8285*** (9.7439)	-12.9486 (102.1801)	30.4362*** (2.5750)	32.7227*** (6.4057)	32.7227*** (6.1523)	30.6391*** (5.2355)
$RDEXPEND_{ct}$	1004.296 (831.9162)	2092.971** (1012.443)	3927.22* (1972.06)	1004.296*** (239.4894)	931.5091 (634.1159)	931.5091** (472.8136)	740.0474** (355.7451)
CONS	-605.1542 (2254.317)	-2448.593 (1912.062)	7887.245 (16393.2)	-605.1542 (629.8957)	-953.1903 (1320.979)	-953.1903 (1394.091)	132.7924 (1487.783)
Observations	199	199	199	199	199	199	199
R^2/Pseudo R^2	0.6874		0.2882	0.6874	0.4753	0.4753	0.5101
F (N(0,1))	116.93***		77.80***				
Hausman (χ^2)			9.48				
LM (χ^2)		315.70***					
Wald (χ^2)		164.70***		1336.76***	149.55***	204.58***	216.03***

Notes: The F-test is normally distributed N(0,1) and tests the null hypothesis of non-significance as a whole of the estimated parameters; the LM test has $\chi 2$ distribution and tests the H_0 of non-relevance of individual effects; the Hausman test has $\chi 2$ distribution and tests the H_0 that unobservable individual effects are not correlated with the explanatory variables; The Wald test has $\chi 2$ distribution and tests the null hypothesis of non-significance of all coefficients of explanatory variables; Asymptotic robust standard deviations are reported in brackets for Pooled OLS, Random Effects and Fixed Effects; ***, **, *, denote significance at 1, 5 and 10% significance levels, respectively.

This is in line with the indications in recent literature (e.g. Frondel et al., 2010; and Marques and Fuinhas, 2012a) suggesting that public policy should rethink its trajectory. In other words, public policies should evolve from policy-driven instruments, which have led to massive installation of capacity, to market-driven mechanisms, which seek to increase the efficiency of the technology. The main aims should be aiming to increase both the productivity and the generation times, accommodating various climatic conditions in which the capacity can be used. As a final, but highly relevant outcome, the cost of backup power will be reduced. With respect to the effect of the renewable source sun, we observe that the results achieved in the two approaches are relatively similar, thus indicating that for this source, working either upon the installed capacity or on actual generation causes no significant differences. This outcome was not at all expected, because there is less variability in predicting sun than wind. At the same time, periods when the sun is available coincide with periods of high electricity demand.

The dominant effect is negative. In fact, the effect of using the solar PV is contributing to restraining electricity exports, unlike wind, as discussed above. Controlling the required backup is more predictable under solar PV than for wind. Consequently, the results suggest that the backup procedures designed to fulfill interruptions of solar PV are much more predictable and programmable, thus not generating electricity in excess to export. Regarding the contribution of the main fossil fuels for electricity export, it appears that the effects are not homogeneous. Coal and oil have contributed to the increase in exports, while gas has restricted them. The full explanation for these phenomena implies analyzing both the main role of each source within the energy mix, as well as its relative cost. The role of oil is essentially baseload, while the coal thermal power plant and the combined cycle gas turbine plants are commonly used to back up renewables. This means that they are used mainly as an energy reserve, and not on an ongoing basis allowing exports to increase. It should, however, be noted that coal is the fuel with the lowest relative price of the three sources. As a result, coal has been used on a large scale to meet not only the periods of lack of renewable electricity, but also as baseload. We even think that in the wake of the findings of Gomez et al. (2011) and Marques and Fuinhas (2012a), coal is being used to allow lower average electricity prices, which have been improved by renewables. This explains the results obtained for the three fossil fuels. It could be concluded that, in general, the price of electricity is not a main driver of exports. This variable is statistically significant only in the generation approach and only occasionally. This suggests that the market mechanism is not a dominant force in electricity exports, but rather the effective capacity of generation and eventual oversupply resulting from the energy mix. This oversupply may result from the activity of backup sources, namely wind. In its turn, RandD expenditure stimulates electricity exports, which in themselves are also a sign that exports are predominantly "technical", in the sense that they appear to be largely a consequence of the technical characteristics of the energy mix. Indeed, as expected, the higher the level of expenditure on RandD, the greater the level of productivity and competitiveness should be, and these factors together should be encouraging electricity exports.

Regarding the size of the electric grid, which is analyzed in this chapter by the proxy *POPDENS*, this proves to be statistically significant only in the capacities approach, which is in accordance with expectations. In fact, this variable, which is representative of infrastructure, proves to be statistically significant in the approach that is focused precisely on installed capacities.

As anticipated, it is observed that large population density levels require a grid with higher generation capacity, thus leading to exports during periods of lower demand as a way to drain excess generation. A final note to highlight is that, as was expected, the variables that revealed a larger discrepancy between installed capacity and used capacity (electricity generated) are contributing to degrading the quality of the results for *capacities approach.* Instead, when the lag between installed capacity and capacity used is small, then it is expected that the results obtained with the two approaches will get closer. This actually occurs, for example, with the nuclear source.

CONCLUSION

The drivers of electricity exports within a context of diversification of energy mix have not been broadly studied in the literature. Working upon a panel of European countries, this chapter seeks to analyze empirically the nature of the individual effect of each source of electricity on electricity exports. Considering, in particular, the problem of intermittency in electricity generation from renewables and the interruptions imposed on backup power, there is often some potential for untapped generation, i.e., the installed capacity from various sources may not coincide with the actual electricity generated. Thus, the chapter is focused on that subject, using two approaches: capacities and generation. The results are generally robust and they suggest the appropriateness of studying this subject focused on the approach of electricity actually generated, rather than on the installed capacity approach. Installed capacity proves to be a poor explainer of electricity exports, which makes all sense considering that incentives towards renewables have eminently encouraged more capacity.

Moreover, it appears appropriate to treat the variables in a disaggregated way in the explanation of exports. The effects and the significances achieved are consistent with what the literature indicates as the main role of variables. Specifically, the effects found in this chapter support the role of baseload for oil and the main role of backup for gas, for example. The results also suggest that a role beyond backup has been reserved for coal. The use of this source has been a way of lowering the overall generation cost, since it is the cheapest source of energy in relative terms. These results are further evidence of the robustness of the analysis, not only from the standpoint of econometric technique, but also from an economic perspective. This analysis may signal that economic agents have been unfocused on the target of exports. In fact, installed capacity, on its own, does not guarantee such exports. A country with large installed capacity can export less than one that makes better use of this capacity. Further research is needed to fully understand the role that electricity exports can play in the financial rationality of renewables projects, as well as in the closely associated power backup. The public energy policy should be focused on stimulating generation, avoiding the inefficiencies associated with idle capacity. In particular, the regulatory authorities may play a crucial role, such as promoting highly electricity-intensive events during off-peak periods, ensuring the rationalization of energy consumption within an electric grid of interconnected countries.

APPENDIX

Table A.1. Correlation Matrix – *Generation Approach*

	$ELETEXP_{ct}$	$ELETWIND_{ct}$	$ELETSUN_{ct}$	$ELETGAS_{ct}$	$ELETGEO_{ct}$	$ELETOIL_{ct}$	$ELETCOAL_{ct}$	$ELETNUCL_{ct}$	$POPDENS_{ct}$	$ELETPRIC_{ct}$	$RDEXPEND_{ct}$
$ELETEXP_{ct}$	1										
$ELETWIND_{ct}$	0.4695	1									
$ELETSUN_{ct}$	0.3083	0.7968	1								
$ELETGAS_{ct}$	0.4095	0.3067	0.2283	1							
$ELETGEO_{ct}$	0.2238	-0.0633	-0.0421	0.4501	1						
$ELETOIL_{ct}$	0.2120	0.0681	0.0369	0.4082	0.8554	1					
$ELETCOAL_{ct}$	0.6468	0.5457	0.3683	0.4050	-0.0528	0.0151	1				
$ELETNUCL_{ct}$	0.0721	0.1527	0.1023	0.0454	-0.1386	-0.0571	0.2322	1			
$POPDENS_{ct}$	0.4992	0.0510	0.0504	0.4315	0.0725	0.0231	0.2155	-0.0295	1		
$ELETPRIC_{ct}$	0.4461	0.2633	0.2129	0.4279	0.4764	0.4418	0.1793	-0.0354	0.3596	1	
$RDEXPEND_{ct}$	0.1684	0.0854	0.0816	-0.1059	-0.2167	-0.3104	-0.0329	0.2265	-0.1006	-0.0929	1
VIF		3.86	2.81	2.03	4.96	4.48	1.75	1.16	1.49	1.70	1.25
Mean VIF	2.55										

Table A.2. Correlation Matrix – *Capacities Approach*

	$ELETEXP_{ct}$	$WINDCAP_{ct}$	$SUNCAP_{ct}$	$MUNCAP_{ct}$	$HYDRCAP_{ct}$	$NUCLCAP_{ct}$	$POPDENS_{ct}$	$RDEXPEND_{ct}$
$ELETEXP_{ct}$	1							
$WINDCAP_{ct}$	0.5497	1						
$SUNCAP_{ct}$	0.3567	0.7862	1					
$MUNCAP_{ct}$	0.7289	0.6772	0.5539	1				
$HYDRCAP_{ct}$	0.1527	0.1957	0.1111	0.3180	1			
$NUCLCAP_{ct}$	0.0532	0.1628	0.1080	0.3469	0.6073	1		
$POPDENS_{ct}$	0.5012	0.0913	0.0863	0.3581	-0.2567	-0.0338	1	
$RDEXPEND_{ct}$	0.1869	0.1224	0.1097	0.3252	0.1039	0.2322	-0.0904	1
VIF		3.57	2.65	3.33	2.02	1.73	1.66	1.30
Mean VIF	2.32							

Table A.3. Results from usual panel data random effects and fixed effects estimators - *Generation Approach*

	Dependent variable $ELETEXP_{ct}$								
	OLS		*Random Effects*			*Fixed Effects*			*Robust Regression*
Independent variables	*CSE* *Model V*	*RSE* *Model VI*	*CSE* *Model VII*	*RSE* *Model VIII*	*AR(1)* *Model IX*	*CSE* *Model X*	*RSE* *Model XI*	*AR(1)* *Model XII*	*Model XIII*
$ELETWIND_{ct}$	4.1483*** (1.0936)	4.1483** (1.8557)	4.7752*** (1.1178)	4.7752*** (1.4182)	3.0957** (1.3808)	6.0403*** (1.3287)	6.0403*** (1.9666)	0.2611 (2.0121)	3.8088*** (0.9482)
$ELETSUN_{ct}$	-17.2778* (9.1247)	-17.2778 (11.8094)	-20.6280*** (6.3702)	-20.6280** (9.7328)	-9.7424 (7.1617)	-25.1391*** (7.8611)	-25.1391* (12.1312)	-17.6778** (8.3729)	-14.1382* (7.9113)
$ELETGAS_{ct}$	-0.0506*** (0.0117)	-0.0506*** (0.0163)	-0.0930*** (0.0268)	-0.0930*** (0.0222)	-0.0437* (0.0258)	-0.1710*** (0.0367)	-0.1710*** (0.0417)	-0.0524 (0.0497)	-0.0549*** (0.0102)
$ELETGEO_{ct}$	2.3125*** (0.7152)	2.3125*** (0.7153)	4.4073*** (1.2050)	4.4073*** (0.5335)	2.6881** (1.1419)	8.4426** (3.7557)	8.4426*** (2.2326)	4.8063 (3.9655)	2.6700*** (0.6201)
$ELETOIL_{ct}$	0.0501 (0.0428)	0.0501 (0.04548)	0.0075 (0.0397)	0.0075 (0.0234)	0.0626 (0.0466)	-0.0026 (0.0934)	-0.0026 (0.0810)	0.0520 (0.0886)	0.0437 (0.0371)
$ELETCOAL_{ct}$	0.0724*** (0.0063)	0.0724*** (0.0154)	0.0622*** (0.0181)	0.0622*** (0.0161)	0.0674*** (0.0141)	-0.0211 (0.0463)	-0.0211 (0.0519)	-0.0366 (0.0444)	0.0800*** (0.0055)
$ELETNUCL_{ct}$	-0.0081** (0.0036)	-0.0081 (0.0057)	0.0040 (0.0127)	0.0040 (0.0105)	-0.0078 (0.0098)	0.0269 (0.0308)	0.0269 (0.0161)	-0.1349*** (0.0456)	-0.0054* (0.0032)
$POPDENS_{ct}$	35.9585*** (3.7026)	35.9585*** (5.3416)	50.5504*** (12.8903)	50.5504*** (6.3760)	37.6299*** (9.4680)	320.9636*** (88.6830)	320.9636** (149.4489)	14.4545 (108.1013)	39.8939*** (3.2103)
$ELETPRIC_{ct}$	348.4220* (180.9972)	348.4220 (312.2399)	-34.2451 (157.2043)	-34.2451 (123.9300)	-98.5419 (206.2852)	-381.7282 (195.5602)	-381.7282 (269.7689)	-216.3911 (256.3805)	120.9457 (156.9293)
$RDEXPEND_{ct}$	3510.495*** (472.4128)	3510.495*** (1065.809)	3848.41*** (1073.024)	3848.41* (1996.721)	3742.492*** (1016.628)	4054.043*** (1338.817)	4054.043 (2876.561)	5048.007*** (1802.878)	2479.964*** (409.5943)
CONS	-8111.635*** (1913.969)	-8111.635** (3579.847)	-6256.527* (3256.442)	-6256.527** (2962.947)	-4592.782 (3015.029)	-41502.06*** (13537.28)	-41502.06* (22168.22)	13005.94** (6104.031)	-5462.293*** (1659.462)
Observations	190	190	190	190	190	190	190	174	190
R^2/Pseudo R^2	0.7517	0.7517				0.2897	0.2897	0.1322	0.8112
F (N(0,1))	54.20***	231.49***				6.69***	14.55***	2.25**	76.91
Hausman (χ^2)						15.65**			
LM (χ^2)			352.01***						
Wald (χ^2)			80.08***	427.57***	76.12***				

Notes: The F-test is normally distributed N(0,1) and tests the null hypothesis of non-significance as a whole of the estimated parameters; the Hausman test has $\chi2$ distribution and tests the H_0 that unobservable individual effects are not correlated with the explanatory variables; the LM test has $\chi2$ distribution and tests the H_0 of non-relevance of individual effects; the Wald test has $\chi2$ distribution and tests the H_0 of non-significance of all coefficients of explanatory variables; Standard errors are reported in brackets as follows: CSE – Conventional Standard Errors; RSE – Robust Standard Errors; Corr(AR1) – first-order autoregressive error. REE means Random Effects Estimator, and FEE means Fixed Effects Estimator. ***, **, *, denote significance at 1, 5 and 10% significance levels, respectively.

References

Baum, C. (2001). Residual diagnostics for cross-section time series regression models. *The Stata Journal, 1*, 101-104.

Beck, N., and Katz, J. (1995). What to do (and not to do) with time-series cross-section data. *American Political Science Review, 89*, 634-647.

Boccard, N. (2009). Capacity factor of wind power realized values vs. Estimates. *Energy Policy, 37*, 2679-2688.

Cameron A., and Triverdi P. (2009). *Microeconometrics using Stata.* Texas, StataCorp, Stata Press.

Carley, S. (2009). State renewable energy electricity policies: an empirical evaluation of effectiveness. *Energy Policy, 37*, 3071-3081.

De Hoyos, R. E., and Sarafidis, V. (2006). Testing for cross-sectional dependence in panel-data models. *The Stata Journal, 6*, 482-496.

Frees, E. (1995). Assessing cross-sectional correlation in panel data. *Journal of Econometrics, 69*, 393-414.

Frees, E. (2004). *Longitudinal and Panel Data: Analysis and Applications in the Social Sciences*. Cambridge University Press.

Friedman, M. (1937). The use of ranks to avoid the assumption of normality implicit in the analysis of variance. *Journal of the American Statistical Association, 32*, 675-701.

Frondel, M., Ritter, N., Schmidt, C., and Vance, C. (2010). Economic impacts from the promotion of renewable energy technologies: the German experience. *Energy Policy, 38*, 4048-4056.

Gómez, A., Zubizarreta, J., Dopazo, C., and Fueyo, N. (2011). Spanish energy roadmap to 2020: Socioeconomic implications of renewable targets. *Energy, 36*, 1973-1978.

Jacobson, H., and Zvingilte, E. (2010). Reducing the market impact of large shares of intermittent energy in Denmark. *Energy Policy, 38*, 3403-3413.

Kost, C., Pfluger, B., Eichhammer, W., and Ragwitz, M. (2011). Integration of renewable energies and nuclear power into North African Energy Systems: An analysis of energy import and export effects. *Energy Policy, 39*, 7136-7145.

Luickx, P., Delarue, E., and D´haeseleer, W. (2010). Impact of large amounts of wind power on the operation of an electricity generation system: Belgian case study. *Renewable and Sustainable Energy Reviews, 14*, 2019-2028.

Marques, A. C., and Fuinhas, J. A. (2011). Drivers promoting renewable energy: A dynamic panel approach. *Renewable and Sustainable Energy Reviews, 15*, 1601-1608.

Marques, A. C., and Fuinhas, J. A. (2012a). Is renewable energy effective in promoting growth? *Energy Policy, 46*, 434-442.

Marques, A. C., and Fuinhas, J. A. (2012b). Are public policies towards renewables successful? Evidence from European Countries. *Renewable Energy, 44*, 109-118.

Marrero, G., and Ramos-Real, F. (2010). Electricity generation cost in isolated system: The complementarities of natural gas and renewables in the Canary Islands. *Renewable and Sustainable Energy Reviews, 14*, 2808-2818.

Mignard, D., Harrison, G., and Pritchard, C. (2007). Contribution of wind power and CHP to exports from Western Denmark during 2000 – 2004. *Renewable Energy, 32*, 2516-2528.

Nikolakakis, T., and Fthenakis, V. (2011). The optimum mix of electricity from wind- and solar-sources in conventional power systems: Evaluating the case for New York State. *Energy Policy, 39*, 6972-6980.

Ochoa, P., and Ackere, A. van (2009). Policy changes and the dynamics of capacity expansion in the Swiss electricity market. *Energy Policy, 37*, 1983-1998.

Pesaran, M. H. (2004). General diagnostic tests for cross section dependence in panels. *Cambridge Working Papers in Economics*. Faculty of Economics, University of Cambridge, 0435.

Purvins, A., Papaioannou, I., Oleinikova, I., and Tzimas, E. (2012). Effects of variable renewable power on a country-scale electricity system: High penetration of hydro power plants and wind farms in electricity generation. *Energy, 43*, 225-236.

Schaber, K., Steinke, F., Mühlich, P., and Hamacher, T. (2012). Parametric study of variable renewable energy integration in Europe: advantages and costs of transmission grid extensions. *Energy Policy, 42*, 498-508.

Sovacool, B. (2009). Rejecting renewables: the socio-technical impediments to renewable electricity in the United States. *Energy Policy, 37*, 4500-4513.

Supersberger, N., Fuhrer, L., (2011). Integration of renewable energies and nuclear power into North African Energy Systems: An analysis of energy import and export effects. *Energy Policy*. 39, 4458-4465.

Varun, Prakash, R., Bhat, I. (2009). Energy, Economics and environmental impacts of renewable energy systems. *Renewable and Sustainable Energy Reviews, 13*, 2716-2721.

Weisser, D. (2004). On the economics of electricity consumption in small island developing states: a role for renewable energy technologies? *Energy Policy, 32*, 127-140.

Welle, A., and Joode, J. (2011). Regulatory road maps for the integration of intermittent electricity generation: Methodology development and the case of The Netherlands. *Energy Policy, 39*, 5829-5839.

Yang, C-H, and Chen, Y-H (2012). RandD, productivity, and exports: Plant-level evidence from Indonesia. *Economic Modelling, 29*, 208-216.

Zahedi, A. (2011). A review of drivers, benefits, and challenges in integrating renewable energy sources into electricity grid. *Renewable and Sustainable Energy Reviews, 15*, 4775-4779.

In: Renewable Energy
Editor: Andreas Poullikkas
ISBN: 978-1-62618-231-8

Chapter 4

HYDROGEN BASED ENERGY STORAGE

Karsten Müller and Wolfgang Arlt
Friedrich-Alexander-University Erlangen-Nürnberg, Germany

ABSTRACT

The transformation of the energy system towards a power generation based on renewable energies necessitates the implementation of storage technologies. A number of short term storage options exist and are already in use. However, large scale, long term storage can only be performed on hydrogen based chemical storage systems. Nevertheless, the low storage densities of hydrogen constitute a severe constraint of these concepts.

To overcome this limitation hydrogen can be liquified, compressed or chemically transformed. A major focus of recent research activities has been the chemical transformation. A conversion into a carbon based fuel by Fischer-Tropsch or related processes is one possibility for this. The existing infrastructure and the high storage densities for this kind of fuels make this option very interesting even though the overall efficiencies are quite poor.

A crucial requirement for the production of these fuels is the availability of a concentrated carbon oxide stream. A second option is the reversible storage of hydrogen on an organic carrier compound. Therefore an unsaturated compound is hydrogenated and later on dehydrogenated again, when the energy is needed. Certain constraints should be considered during the selection of the carrier material. For example, liquid compounds are easier to handle than solid or gaseous substances. Since the melting points of the hydrogenated and dehydrogenated forms often significantly vary, a close look on the phase change behavior is necessary. A big advantage of organic hydrogen carriers compared to other forms of chemical conversions of hydrogen is the fact that the hydrogen is released again as hydrogen. Therefore, the efficient power generation systems like fuel cells that are available for hydrogen can be used. The overall efficiency thus can be significantly increased compared to the conversion into fuels such as Fischer-Tropsch-gasoline.

INTRODUCTION

One, if not the biggest, disadvantage of renewable energies is their fluctuation in power generation. Most renewable energies are not able to allow for constant operation, not to speak of an on-demand production. Furthermore, even predictability of production is limited in terms of accuracy and time, since weather forecast still can only give reliable prognosis for a few days. To deal with this problem the storage of surplus energy is unavoidable. A huge number of different storage technologies exist or are currently under development for the storage of electrical energy. Depending on the application different technologies are meaningful (compare Figure 1).

So called Supercaps, i.e. high power capacitors, are an option for short-time storage, to compensate fluctuations in the electric power grid on the scale of seconds to a few minutes [1]. Works concerning mobile applications are going on, but mainly as an additional storage device in fuel cell driven cars [2]. For fly wheels the situation is similar. The technology is suited for short periods of stationary storage, but difficult for mobile applications. For the storage of energy recovered during braking of cars fly wheels can be an option, but as the main storage technology they seem to be inappropriate. However, there have been prototypes and even a small series of 20 gyrobuses since the 1970s [3, 4]. Nevertheless, this technology never reached significant distribution and is not likely to experience a breakthrough in the near future.The conservation of the angular moment which prohibits rotation of the axle as well as safety issues in case of a breaking axle or bearings set limits to this technology. Pumped hydro is one of the most efficient and cheapest storage technologies, but is totally inappropriate for mobile applications [5]. The low storage density and the fact that potential construction sites are scarce in many countries are additional drawbacks to this technology.

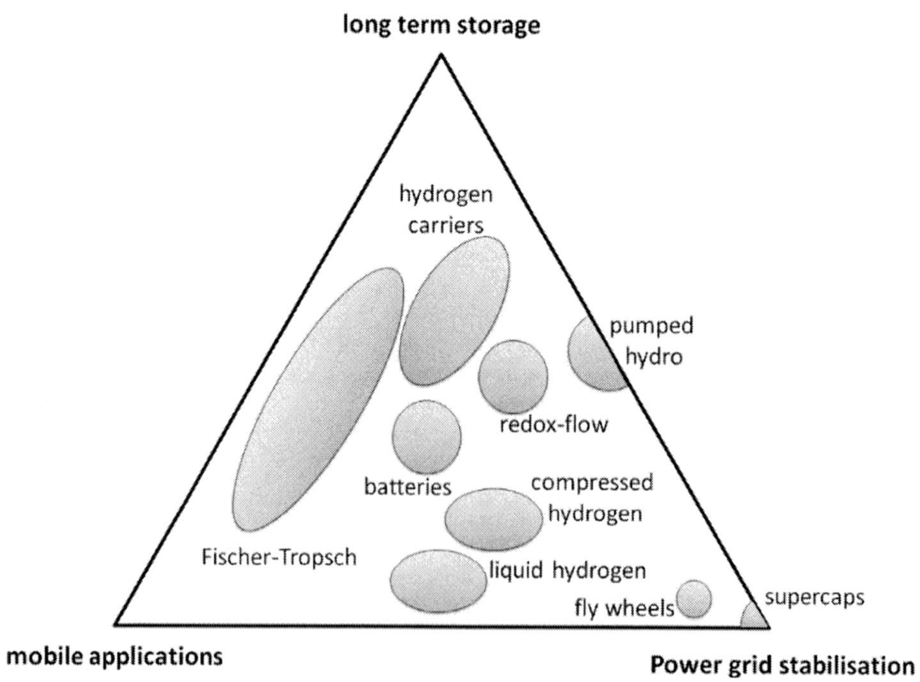

Figure 1. Fields of application of different energy storage technologies.

The use of rechargeable batteries seems to be an obvious possibility for energy storage. Accumulators are charged directly with electric energy and release the energy again directly in form of electricity. Still rechargeable batteries are not a viable technology for all storage purposes, due to a couple of reasons: 1. gravimetric and volumetric storage density is low, 2. materials, such as lithium and cobalt, with limited availability are needed, 3. self-discharge still is a problem in many cases and 4.accumulators often suffer from a memory effect and related problems [6] Flow batteries (redox flow) are an opportunity to overcome some of these drawbacks, but the systems are rather complicated and storage densities are even lower than for most conventional batteries. Further, harmful electrolytes would restrict this technology to a professional environment. Chemical energy storage is a way to overcome these issues. Especially hydrogen is a compound with a high combustion energy that has drawn much attention and carries great potential for future energy storage systems.

STORAGE OF ELEMENTAL HYDROGEN

In hydrogen-based chemical energy storage electric energy is used to produce hydrogen by electrolysis. The storage of elemental hydrogen has the advantage that only a small amount of energy is needed for the chemical part of the storage process. The overall efficiency for hydrogen based energy storage without use of further measures for the storage is determined by the efficiency of the electrolysis and the efficiency of the subsequent power generation. The efficiencies are usually calculated relative to the lower heating value (LHV). The efficiency of electrolysis $\eta_{electrolysis}$ is assumed to be

$$\eta_{electrolysis} = \frac{LHV_{H_2}}{E_{el}} = 0.7. \qquad (1)$$

If an efficiency of $\eta_{Fuel\ Cell} = 55\ \%$ is assumed for the reconversion in a fuel cell the overall efficiency η_{total} can be calculated as

$$\eta_{total} = \eta_{electrolysis} \cdot \eta_{Fuel\ Cell} = 38.5\ \%. \qquad (2)$$

This defines a best-case scenario for hydrogen based energy storage. Storage normally requires further energy input. A compression of hydrogen from 20 bar to 350 bar in two stages with intercooling for example has an efficiency of about 95.8 % relative to the LHV of the hydrogen. In this case, the overall efficiency would be reduced to about

$$\eta_{total} = \eta_{electrolysis} \cdot \eta_{compression} \cdot \eta_{Fuel\ Cell} = 36.9\ \%. \qquad (3)$$

Besides electrolysis, hydrogen could also be produced from e.g. biomass [7]. We estimated efficiency of such a hydrogen production process by a process simulation using the ASPEN Plus software (version 7.1). Considering energy demands for purification about 77 % of the LHV of the saccharide could be regained in form of hydrogen. If hydrogen from such a source (instead of electrolysis) would be used the overall efficiency would be increased to 42.4 % for the best-case scenario (assumption: no energy demand for the storage of hydrogen;

compare equation 2).For comparison: If hydrogen is produced conventionally from methane by steam reforming, an efficiency of up to 54.7 % could be reached (electricity produced by fuel cell relative to the LHV of the applied methane). This estimation considers the energy demand of the steam reforming process, but neglects the energy demand of previous purification of the methane which can be significant especially in case of biogas. However, comparability of electrolysis with hydrogen from biomass or natural gas is limited since totally different forms of original energy are converted. Gravimetric storage density of pure hydrogen would be excellent, but the volume of hydrogen at ambient conditions is unacceptably large. One option is pressurization to 350 bar or even up to 700 bar, but safety concerns and insufficient volumetric storage densities make other variants more interesting. Liquefaction for example can yield higher storage densities. Thus, liquid hydrogen is already in use in many road-test fleets for hydrogen driven vehicles. However, the volumetric storage density of the whole system (including insulation) is still not satisfying. Problems mainly arise from the fact that the critical temperature of hydrogen is only -239.96 °C. Therefore, to reach liquefaction the temperature inside the tank always has to be kept below this value. The resulting temperature difference to the surrounding makes significant heat transfer into the tank during holding time unavoidable. To compensate the heat transfer hydrogen is evaporated and has to be released to prevent a pressure built up. Energy losses during storage thus are caused either by release of hydrogen or by energy demand for cooling [8]. Hence, long term storage necessitates further conversion of the hydrogen into another chemical form. This chapter puts a focus on chemical hydrogen conversion, because these technologies are the most promising approaches for long term energy storage.

PRINCIPLE CONCEPTS OF HYDROGEN CONVERSION

To reach viable storage densities at ambient conditions it is necessary to convert hydrogen into another form. There are two principle ways of conversion: irreversible conversion to form a fuel that is combusted when energy is needed or reversible conversion using an energy carrying compound where the hydrogen is released in case of energy demand. The use of fuels from irreversible hydrogen conversion causes local CO_2-emissions, but the overall process is carbon neutral, since only CO_2 is released that has been converted before. ETS based reversible hydrogen conversion is totally free of CO_2-emissions during operation, since CO_2-emissions are caused only during the manufacturing of the ETS. The same holds for carbon monoxide and nitrous oxide, since these gases are not formed in fuel cells. The storage density depends on the concrete fuel or carrier used. In case of methane storage density further depends on the temperature and pressure during storage. Storage density in Fischer-Tropsch products (gasoline, diesel) is higher than for partly oxidized hydrocarbons such as methanol. Storage density of organic hydrogen carriers is lower, but still significantly higher than for liquid or pressurized hydrogen. The storage capacities for both approaches are theoretically unlimited. However, certain constraints exist concerning the reactive partner (CO_2 or hydrogen carrier, respectively). The removal of CO_2 from the atmosphere is not reasonable due to the energy demand. The amount of energy that can be stored by irreversible conversion is therefore limited by the availability of aconcentrated CO_2stream. In the current technological system CO_2 is produced continuously. The absolute amount of energy that can be stored is therefore unlimited, while the power input is restricted by the number and size of the usable, stationary CO_2-sources. Power uptake in case of reversible conversion is only limited by the capacity installed but installation has no upper limited. The amount of available carrier material can become a limiting factor and therefore the absolute amount of energy stored is limited by the previously produced amount of carrier material.

IRREVERSIBLE CONVERSION TECHNOLOGIES

Irreversible conversion technologies are based on chemical reactions of hydrogen with a low molar mass compound (Figure 2). Nearly all irreversible storage processes need a concentrated stream of a carbon oxide (except for ammonia based storage, which necessitates concentrated nitrogen). Carbon monoxide would be best, but usually only carbon dioxide is available from biogas or flue gas treatment. The carbon dioxide can be converted to carbon monoxide by a reversed water-gas-shift reaction:

$$H_2 + CO_2 \rightleftharpoons H_2O + CO \qquad (4)$$

As can be seen from equation 4 one mol hydrogen is consumed per mol carbon dioxide in this reaction without any further use of the energy stored in the respective hydrogen. The carbon monoxide can be converted with hydrogen from electrolysis into different fuels such as gasoline, diesel, methanol or methane. These fuels can be used in the existing energy infrastructure and, except for methane, stored at ambient conditions. The recovery of the stored energy in all cases necessitates combustion with release of CO_2 to the atmosphere. For a new storage "cycle" concentrated CO_2 has to be gained again from an appropriate source, since a recycle of the CO_2 is not meaningful.

A number of different compounds for the storage of energy by irreversible conversion of hydrogen have been suggested or even implemented on a pilot-scale so far:

- methane (Sabatier-process) [9]
- the Fischer-Tropsch-process [10-12]

  - gasoline
  - diesel

- methanol [13]
- ammonia [14, 15]
- formic acid [16].

Methane and Fischer-Tropsch products can be used directly within the existing energy infrastructure. Investments are therefore limited to the production sites exclusively, while the existing energy conversion technologies in cars, households and industry can be used. Methanol would necessitate some changes in the existing energy system, but still allows for a subsequent use of most existing energy related equipment.

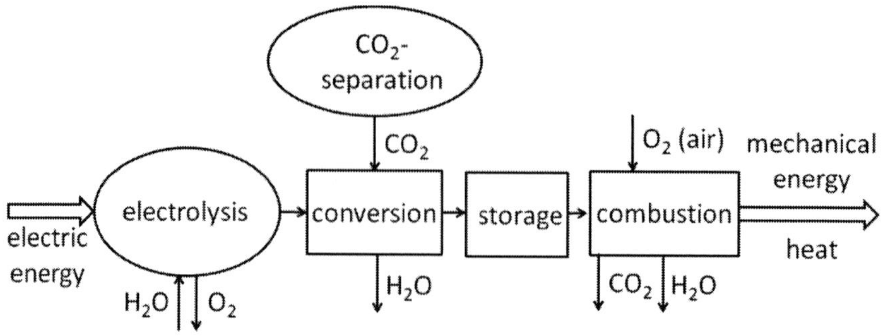

Figure 2. Schematic of the irreversible hydrogen storage concept.

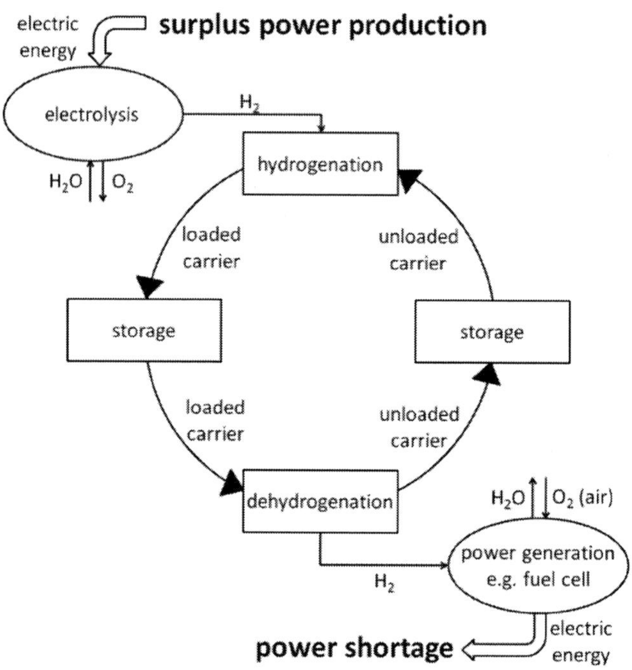

Figure 3. Schematic of the reversible hydrogen storage concept.

A large scale use of ammonia would necessitate significant changes on the existing infrastructure. However, ammonia does not seem to be a viable option for energy storage on a large scale and the concept structurally differs from the other systems as will be discussed later.

METHANE

Methane opens a variety of energetic application options. It can be combusted directly for heating of residential and commercial buildings as well as for industrial heating demand. Besides this it can be converted in a gas motor and thus be used for transport purposes. A large number of methane driven cars is already in use, making the integration into the transport system rather easy. Alternatively electric energy can be regained using gas turbines or special fuel cells. Due to this fact, methane seems to be the most appropriate non-ETS hydrogen conversion technology for the storage of electric energy for later recovery of electric energy.

The production of methane from hydrogen and a carbon oxide is based on the so called Sabatier-reactions:

$$4\ H_2 + CO_2 \leftrightharpoons 2H_2O + CH_4 \tag{5}$$

$$3\ H_2 + CO \leftrightharpoons H_2O + CH_4 \tag{6}$$

The methane production from carbon dioxide (equation 5) can be viewed as a combination of the methane production from carbon monoxide (equation 6) with a preceding reversed water-gas-shift reaction (equation 4). In case of both carbon oxides parts of the hydrogen are already "combusted" during the conversion process (for carbon monoxide one third and for carbon dioxide even half of the hydrogen). As a consequence the so-called Sabatier-reaction is highly exothermal ($\Delta^R h$ = -167.9 kJ/mol for equation 5). This waste heat could be used quite well, since it is produced on a quite high temperature level (about 280 °C). Nevertheless, an efficient utilization is difficult if the Sabatier unit is run on fluctuating power supply. The LHV of the formed methane is 802.5 kJ/mol, while the LHV of the used hydrogen is 4 x 241.9 kJ/mol = 967.6 kJ/mol-CH_4. The LHV of the product is thus 17 % lower than the LHV of the reactants. This loss is still rather small compared to the "loss" of hydrogen during conversion. The relatively high LHV of the product can be attributed to different reasons: 1.) water forms only two third of the combustion products of methane, compared to 100 % in case of hydrogen. Thus, condensation of the flue gas starts at a lower temperature. 2.) The carbon atom is available for the combustion process as well, which was not the case for the starting hydrogen and carbon dioxide.

We have analyzed the overall energy efficiency of methanization based energy storage in previous work [6]. A process simulation has been performed with the ASPEN Plus V7.1 software considering an efficiency relative to the LHV of 70 % for the electrolysis, 42 % for the gas engine, 35 % for the gas turbine and 55 % for a combined cycle power plant. Our simulation showed an efficiency of 72.8 % for the conversion step. The energy demand for the conversion also contains the energy demand for the carbon dioxide supply by an amine treating of biogas. Based on these results the overall efficiency η_{total} can be calculated according to equation 7:

$$\eta_{total} = \eta_{electrolysis} \cdot \eta_{conversion} \cdot \eta_{engine/turbine} \tag{7}$$

Depending on the technology used for reconversion of the stored energy the efficiency was calculated to be about

- 21.4 % for a gas engine,
- 17.9 % for a gas turbine and
- 28.1 % for a combined cycle power plant.

Compared to the efficiencies for elemental hydrogen storage these values are quite poor. Nevertheless, the better storability of methane and utilization in an existing infrastructure makes methanization still an interesting option.

Fischer-Tropsch

Fischer-Tropsch is one of the most appropriate energy storage options for mobile applications. Two aspects of Fischer-Tropsch products make them interesting for this field: 1. the high storage density and 2. the nature of the current transport infrastructure. The volumetric as well as the gravimetric energy density of gasoline and especially diesel is one of the highest at ambient conditions, which is crucial for transport purposes. Furthermore the vast majority of cars and trucks already runs on gasoline and diesel. If these fuels would be produced in a regenerative way, no adaption measures on the consumer end would be necessary.

The Fischer-Tropsch process is closely related to the Sabatier process (the Sabatier reaction can be viewed as a border case of the Fischer-Tropsch reaction). The most important type of reaction is the formation of alkanes:

$$n\,CO + (2n+1)\,H_2 \leftrightharpoons C_nH_{2n+2} + n\,H_2O \quad (8)$$

Besides this reaction a huge number of alternative reactions can occur in a Fischer-Tropsch reactor forming products such as alkenes, alcohols and aromatics. The product distribution of a Fischer-Tropsch process can be described as a function of the chain-growth probability by an Anderson-Schulz-Flory distribution [17]. The value of the chain-growth probability can be controlled by the selection of the catalyst and the process parameters. The product distribution for a chain-growth probability of 0.8 is depicted in Figure 4.

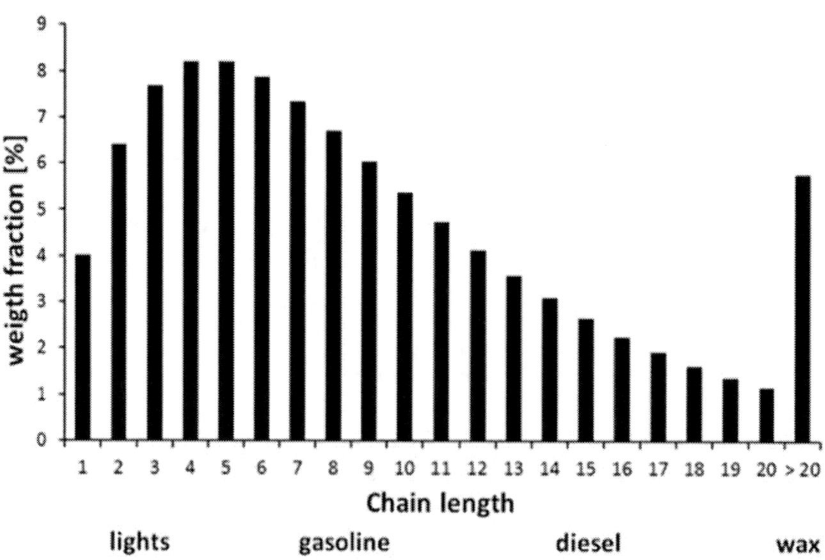

Figure 4. Product distribution of a Fischer-Tropsch reactor for a chain-growth probability of 0.8.

By increasing the chain-growth probability the mode of the distribution can be shifted towards the production of diesel and waxes. The synthesis gas needed for the Fischer-Tropsch reaction can be derived from biomass (for further description of this concept see e.g. [18, 19]). If a Fischer-Tropsch process should be used for the storage of electrical energy hydrogen from electrolysis is mixed with a carbon oxide stream. The carbon monoxide could be formed in-situ (meaning in the same reactor) by a reversed water-gas-shift reaction. The presence of a large carbon dioxide fraction in the synthesis gas however favors the formation of short-chain alkanes instead of the desired long chain products needed for gasoline and diesel [20].

Consequently, a reversed water-gas shift reaction has to be performed previously in a separate reactor. We performed a process simulation for a Fischer-Tropsch process fueled by hydrogen from electrolysis and carbon dioxide from biogas treatment. We assumed an efficiency of 70 % for the electrolysis step as above and an efficiency of 30 % for reconversion in an Otto engine and 40 % for reconversion in a Diesel engine [21]. The chain-growth probability was set to 0.9 and a conversion of 70 % was assumed. Based on this simulation an overall efficiency of 16.7 % was reached. This storage efficiency is even lower than that reached in the respective Sabatier-process. The much more laborious separation process for the complex product mixture and the lower efficiency of the engines compared to a combined cycle power plant have to be held responsible for this fact. The rather low efficiency makes Fischer-Tropsch uninteresting for the storage of electrical energy. Yet, for non-electric use of the stored energy it can be an interesting option, due to the high storage density and the existing infrastructure. In an energy system with a significant excess of fluctuating electricity production by renewable energies electrolysis based Fischer-Tropsch could be an opportunity for mobility. However, it has to be doubted that electrolysis based Fischer-Tropsch will ever be able to cover the whole energy demand of the transport sector. As a supplement to biomass based Fischer-Tropsch it might play a role in a future energy system for transport purposes.

METHANOL

The conversion of hydrogen into methanol is simpler than the Fischer-Tropsch process, since only two products (water and methanol) are formed and product separation is less demanding. On the other hand, the carbon atom in the methanol molecule is partly oxidized, leading to a lower energy density (LHV = 19.9 MJ/kg compared to 43.0 MJ/kg for gasoline) [22]. Compared to the Sabatier process (methanization) methanol has the advantage of being liquid at ambient conditions. Methanol does not play a major role in our current energy infrastructure, but introduction would be possible with only minor adjustments. Fuels for cars can be blended with methanol and methanol fuel cells are an option for the storage of electric energy. Furthermore, methanol produced by renewable energies could be used within the chemical industry. To estimate the efficiency of electric energy storage based on methanol we performed a process simulation (Figure 5) as described above. Therein the CO_2 is compressed in two stages to a pressure of 20 bar, mixed with the hydrogen coming from electrolysis and a recycle stream and converted in a reversed water-gas-shift reactor after preheating.

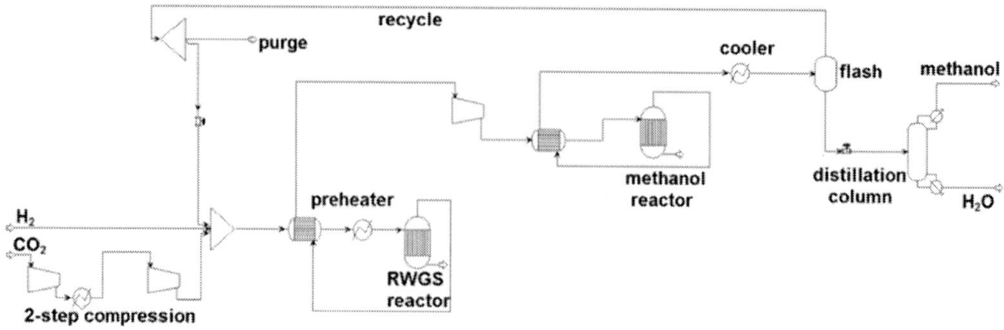

Figure 5. Flowsheet for the ASPEN simulation of the methanol process.

Due to limitations by the reaction equilibrium only partial conversion can be achieved. For the methanol synthesis the mixture is compressed to 35 bar. After the reaction the unreacted gases are removed and recycled, while the liquid phase coming from the flash is separated into pure methanol and the water in a distillation column.Variations of the process are possible, such as the use of a rectification instead of a flash for separating the unreacted gases after the methanol reactor. However this has negligible influence on the overall efficiency. For power regeneration we assumed conversion of the methanol in a direct methanol fuel cell with an efficiency of 37 % [23]. The overall efficiency was estimated to be about 15.4 %. This is one of the lowest values achieved in the group of irreversible conversion technologies. The energy demand of the process itself has only minor influence on the overall efficiency. The small energy content of the produced methanol as well as the not very high efficiency of the power regeneration are the main reasons for the low overall efficiency.

Ammonia

A special case of hydrogen conversion is the use of ammonia as a storage material, since no hydrocarbon fuel is formed. Two kinds of ammonia based energy storage are possible. Lovegrove et al. [24] suggested the use of ammonia in a closed loop for thermal energy storage as a buffer in concentrating solar systems. The solar energy is therein used for the thermal decomposition of ammonia into hydrogen and nitrogen. During cloud transients the exothermal synthesis of ammonia is used to regain the stored heat. Disadvantageous in this system is the fact that elemental hydrogen has to be stored and the storage density is accordingly low.

The second variant is electrolysis of water for ammonia synthesis with subsequent reconversion in a fuel cell. The use of a hydrogen fuel cell after decomposition of the ammonia as well as a direct ammonia fuel cell [25] is possible. Ammonia can be stored at slightly elevated pressures (7-10 bar) and ambient temperature and exhibits a medium storage density (22.2 MJ/kg [15] or 120 kg-H_2/m^3 if the hydrogen is released by catalytic conversion [26]). To estimate the energy efficiency of an energy storage based on the second variant we performed a simulation of the process using ASPEN Plus. The flowsheet for the synthesis step is shown in Figure 6. For our simulation we assumed hydrogen production by electrolysis and nitrogen supply by liquefaction of air.

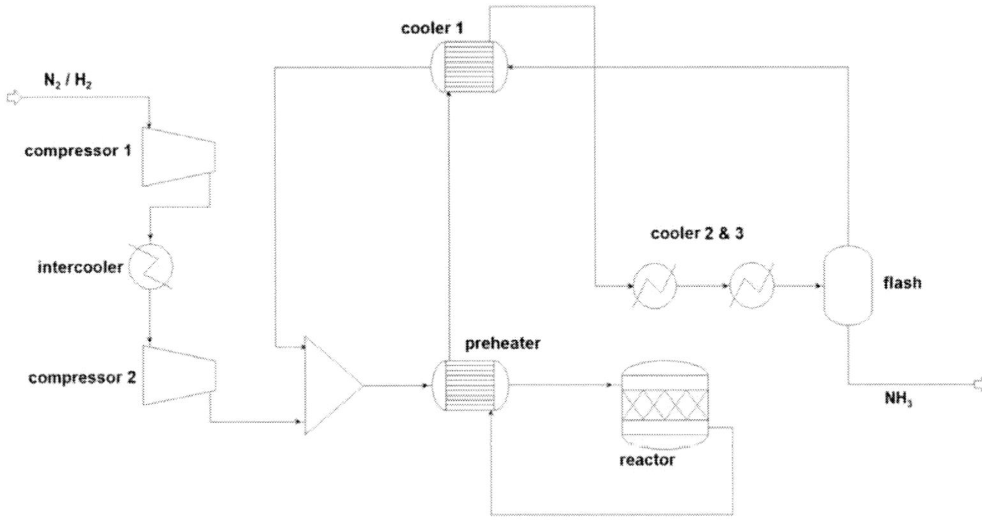

Figure 6. Flowsheet for the ASPEN simulation of the ammonia process.

This gas mixture is compressed in two steps to a reaction pressure of 104 bar.The gas, together with a recycle stream is preheated using the hot stream coming out of the reactor. The product stream is cooled down by heating up cooler streams, heat dissipation to the surrounding and finally by a compression cooling machine to a temperature of -15 °C to separate the ammonia from the non-converted reactants. The ammonia can be stored at about 10 bar and ambient temperature. For power reconversion we assumed a hydrogen fuel cell with an efficiency of 55 % and previous ammonia decomposition at a temperature of 700 °C. Summing up all energy demands of this process an overall efficiency of 30.3 % has been calculated. However, the energy demand of the ammonia decomposition should rather be subtracted from the energy output than be summed up with the energy demand. In this case the energetic efficiency would only be 19.8 %. After the start-up period the heat for the decomposition must not necessarily be applied electrically but could be supplied by combustion of parts of the released hydrogen. For this scenario the overall efficiency was calculated to be about 22.5 %. If waste heat from the fuel cell can be used as well the efficiency could further increase. All these values are within the average to upper range for chemical hydrogen conversion technologies. Some safety issues, such as the toxicity and corrosivity of the hazardous substance ammonia, have to be addressed if ammonia is used for energy storage purposes. In the controlled environment of a factory, where only trained personal is present, the handling of ammonia is not problematic. In the public area, e.g. as a fuel for cars, safeguard measures concerning accidental or intentional release and other events of fault are more demanding.

FORMIC ACID

Formic acid can function as a hydrogen storage material. Elemental hydrogen can be recovered by a catalytic reaction using e.g. iron catalysts [16]:

$$HCOOH \leftrightharpoons CO_2 + H_2 \quad (9)$$

This approach could therefore be viewed as a form of reversible hydrogen conversion, since the hydrogen is regained. However, we do not regard formic acid as an ETS, since a collection and storage of the released carbon dioxide for later formic acid production does not seem reasonable. Furthermore production of formic acid from hydrogen and carbon dioxide is demanding.

Due to the energetically stable nature of the carbon dioxide molecule the thermodynamic driving force of the reaction is very low ($\Delta^R g$ = +35.7 kJ/mol). The reaction equilibrium is therefore far on the side of hydrogen and carbon dioxide even at elevated pressures. To allow for a regeneration of the formic acid from carbon dioxide and hydrogen it is meaningful to remove the free acid from the reaction system. Reaction with an alcohol or an amine is an option for in-situ product removal, but recovery of the formic acid from the salt or ester causes additional effort and energy demand [27]. Production by reaction of carbon dioxide and hydrogen is not mandatory, since formic acid is also a product of biomass processing [16] However, if formic acid is not produced from hydrogen it does not represent a hydrogen storage technology in a narrower sense. It is questionable, if formic acid will be an option for the storage of surplus electricity production. Nevertheless, formic acid represents an interesting carrier material for energy from biological sources.

ETS Based Technologies

Alternatively energy carrying compounds (ETS; from German: "EnergieTragenderStoff") could be used [28] In this case a substance is used that has an energy-rich and an energy-lean form. The energy-lean form of the ETS is transformed into the energy-rich form in times or at places of energy excess. The energy is regained at times or places of demand by converting the energy-rich form back into the energy-lean form. The concept of reversible hydrogen storage based on ETS is depicted in Figure 3. Again, hydrogen is produced by electrolysis from surplus energy.

This hydrogen is stored in a closed loop of a carrier material. Thus, no further input stream into the system is needed. The carrier material, usually an unsaturated organic compound, is hydrogenated, forming the respective saturated compound, which can be stored. If energy is needed the carrier is dehydrogenated and the hydrogen released. The unloaded carrier is recycled and hydrogenated again in succeeding storage cycles. The power generation can be done by a variety of technologies (e.g. gas motors or turbines), but a fuel cell is the most probable option.

Each unsaturated compound can be hydrogenated and – at least theoretically – be dehydrogenated again. Hence, the number of potential hydrogen carriers would be very large. However, some requirements for an appropriate carrier material significantly reduce the number of potential compounds:

- long-term stability
- availability of raw materials
- storage density
- manageability
- dehydrogenation characteristics

Especially the three last requirements disqualify many compounds. To achieve a noteworthy gravimetric storage density the number of double bounds in the unsaturated molecule should be large compared to the molar mass. For the volumetric storage density a condensed form at ambient conditions is further necessary. This non-gaseous criterion is also important for the manageability. In order to be handled easily the carrier should neither by gaseous nor solid but liquid in its hydrogenated and dehydrogenated form (a so called "Liquid Organic Hydrogen Carrier", LOHC).

The most problematic task in the selection of a carrier compound is the dehydrogenation characteristic. Hydrogenation of unsaturated compounds is usually rather simple in terms of catalysis as well as thermodynamics. The huge thermodynamic driving force of the hydrogenation leads to a small driving force for the dehydrogenation. As a consequence the reaction equilibrium strongly favors hydrogenation over dehydrogenation. Dehydrogenation is an endothermal reaction so that increasing temperature can shift the equilibrium towards the dehydrogenated form. Yet, high temperature would have a negative effect on the long-term stability of the carrier and catalyst, cause evaporation during dehydrogenation and lower the overall efficiency. The Free Energy of reaction $\Delta^R g$ as the thermodynamic driving force of the chemical reaction depends on an enthalpic contribution ($\Delta^R h$) and an entropic contribution ($\Delta^R s$).

Reaction equilibrium favors the bigger part of hydrogenation reactions, even though entropy unfavors hydrogenation ($\Delta^R s < 0$; the binding of gaseous hydrogen on a bigger molecule causes the loss of up to five degrees of freedom per hydrogen molecule). The change in entropy during an isothermal hydrogenation of a double bond at standard conditions is around-120 J/mol·K. Therefore Free Energy of reaction $\Delta^R g$ exceeds the enthalpy of reaction $\Delta^R h$ by about 36 kJ/mol at standard conditions. Since hydrogenation reactions usually are highly exothermic they are thermodynamically favorable at moderate temperatures, even though entropy favors dehydrogenation. Increasing temperature can be used to shift the reaction equilibrium towards dehydrogenation. It is therefore desired to find chemical compounds with an enthalpy of reaction for hydrogenation between -40 kJ/mol and about -60 kJ/mol to allow for hydrogenation as well dehydrogenation at reasonable conditions.

Isolated double bounds in a molecule can easily be hydrogenated, but dehydrogenation at 1 bar requires temperatures higher than 800 °C (and even more at elevated pressures). In case of conjugated double bonds the dehydrogenation temperature is slightly decreased, but still far too high for practical use. Only aromatic compounds can be dehydrogenated at reasonable temperatures. Benzene or toluene would accordingly be self-evident hydrogen carrier materials and have already been suggested some decades ago [29, 30]. Both substances exhibit good physical properties, are quite cheap and nearly unrestrictedly available. However, dehydrogenation is still quite demanding. Heterocyclic compounds have therefore drawn much attention in recent years as hydrogen carrier materials. While e.g. oxygen containing compounds show no advantage in their dehydrogenation characteristics, mainly nitrogen containing aromatic heterocyclic compounds seem to be appropriate hydrogen carrier materials [31]. Additional substitution with additional heteroatoms, such as boron, might further enhance dehydrogenation, but cause other issues [32]. Due to its interesting physical and chemical properties N-ethyl carbazole (NEC) has drawn much attention in recent years [33-35].

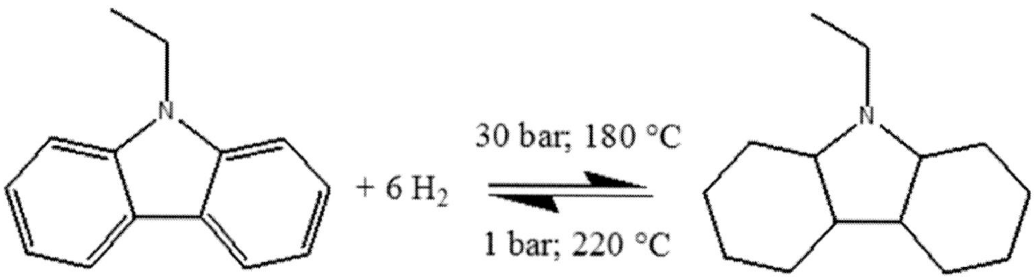

Figure 7. Reaction scheme for hydrogen storage using N-ethylcarbazole.

Up to six hydrogen molecules (= 12 H-atomes) can be stored on a single NEC molecule (Figure 7). Dehydrogenation can be carried out at rather low temperatures (only about 220 °C) and ambient pressure. Thermodynamically hydrogenation would be carried out best at low temperatures, but the reaction kinetics demand a certain minimum temperature. To achieve total conversion despite the unfavorable thermodynamics at high temperature, the pressure has to be increased for the reaction. To estimate the overall efficiency of LOHC based hydrogen storage we again performed simulations using the ASPEN Plus software. In reversible storage systems two steps have to be considered for the process simulation. In the first step hydrogen is produced by electrolysis at elevated pressures with an efficiency of 70 % relative to the LHV. Due to the strong exothermic nature of the hydrogenation reaction no energy demand is caused by preheating of the reactants (except for a start-up period).

Dehydrogenation is highly endothermic and thus a large amount of heat has to be supplied to the second process step. We considered partial preheating of the hydrogenated LOHC by cooling down the hot stream coming out of the reactor, but still energy demand is dominated by heating in the dehydrogenation step. Minor energy demand is caused by the need to remove traces of the LOHC and, where necessary, solvent from the gas stream. We performed simulations for different LOHCs, such as toluene, N-ethyl carbazole and azaborines, and different hydrogen purification technologies, such as partial condensation, pressure swing adsorption and membrane processes. The overall efficiency for power generation in a fuel cell with an efficiency of 55 % ranged between 29 and 31 %. If no heat integration was considered or other power regeneration technologies were assumed the efficiency decreased and reached results between 22 and 29 %. Energy storage systems based on reversible conversion of hydrogen are – unlike irreversible conversion systems – applicable for integration into residential and commercial buildings. This offers the opportunity to utilize the waste heat of the hydrogenation or other utilities for heating the building (or cooling during summer with an absorption chilling machine). The electrical storage efficiency is more or less unaffected by the utilization of the waste heat, while the total energy efficiency can be significantly increased.

CONCLUSION

Long-term energy storage on a large scale can only be carried out by producing hydrogen via electrolysis. The low volumetric storage density of hydrogen makes additional measures necessary. Compression or cooling are the most straight forward solutions, but are not really appropriate for long-term storage. To overcome the issues related to hydrogen loss during the storage period a chemical conversion is meaningful. Irreversible as well as reversible conversion are possible and exhibit their specific advantages for different applications. Some of the irreversible conversion technologies (Fischer-Tropsch, Sabatier process) yield products that can be used directly within the existing energy infrastructure and the storage densities of Fischer-Tropsch products are excellent. On the other hand irreversible conversion depends on large, concentrated carbon oxide streams, is inappropriate for small sites and shows low overall efficiencies. Better efficiencies can be achieved using reversible hydrogen conversion. Organic hydrogen carriers can also be hydrogenate at rather small sites and stored under ambient conditions. The current focus for LOHC applications is stationary (e.g. integrated into an office building) and mobile use in larger units (e.g. ships). The employment in smaller units, such as cars, is currently under development.

ACKNOWLEDGMENT

The authors wish to thank Jin Geng, Florian Fabisch and PedramAbbassi for the support during process simulation and Prof. Peter Wasserscheid for valuable discussions.

REFERENCES

[1] Abbey, C.; Joos, G. *IEEE Transactions on Industry Applications* 2007, 43, 769-776.
[2] Schiffer, J.; Bohlen, O.; De Doncker, R. W.; Sauer, D. U.; Kyun Young, A.; Optimized energy management for fuelcell-supercap hybrid electric vehicles; *Vehicle Power and Propulsion*, 2005 IEEE Conference, 2005
[3] Hagin, F.; Merker, P. First Results From The Man-Gyrobus; *Bundesministerium fuer Forschung und Technologie*: Bonn, 1980.
[4] Zander, E. Experiments On 20 Hybrid Electro-Buses For Public Transport Services; *Bundesministerium fuer Forschung und Technologie*: Bonn, 1980.
[5] Deane, J. P.; Ó Gallachóir, B. P.; McKeogh, E. *J. Renewable and Sustainable Energy Reviews* 2010, 14, 1293-1302.
[6] Müller, B.; Müller, K.; Teichmann, D.; Arlt, W. *Chemie Ingenieur Technik* 2011, 83, 2002-2013.
[7] Lobanova, O.; Müller, K.; Mokrushina, L.; Arlt, W. Chemical Engineering and Technology 2012, 35, 735-742.
[8] Wolf, *J. Mrs Bulletin* 2002, 27, 684-687.
[9] Sterner, M. Bioenergy and renewable power methane in integrated 100% *renewable energy systems*; Ph.D.-thesis; University of Kassel; 2009.
[10] Fischer, F.; Tropsch, H. *Berichte der deutschen chemischen Gesellschaft* (A and B Series) 1923, 56, 2428-2443.
[11] Takeshita, T.; Yamaji, K. *Energy Policy* 2008, 36, 2773-2784.
[12] Jess, A.; Kaiser, P.; Kern, C.; Unde, R. B.; von Olshausen, C. *Chemie Ingenieur Technik* 2011, 83, 1777-1791.
[13] Olah, G. A. *Angewandte Chemie International Edition* 2005, 44, 2636-2639.
[14] Lovegrove, K.; Luzzi, A.; Soldiani, I.; Kreetz, H. *Solar Energy* 2004, 76, 331-337.

[15] Dunn, R.; Lovegrove, K. Ammonia – The Other Hydrogen; Sydney, 2008.
[16] Boddien, A.; Mellmann, D.; Gärtner, F.; Jackstell, R.; Junge, H.; Dyson, P. J.; Laurenczy, G.; Ludwig, R.; Beller, M. *Science* 2011, 333, 1733-1736.
[17] Van Der Laan, G. P.; Beenackers, A. A. C. M. *Catalysis Reviews* 1999, 41, 255-318.
[18] Tijmensen, M. J. A.; Faaij, A. P. C.; Hamelinck, C. N.; van Hardeveld, M. R. M. *Biomass and Bioenergy* 2002, 23, 129-152.
[19] van Steen, E.; Claeys, M. *Chemical Engineering and Technology* 2008, 31, 655-666.
[20] Visconti, C. G.; Lietti, L.; Tronconi, E.; Forzatti, P.; Zennaro, R.; Finocchio, E. *Applied Catalysis A*: General 2009, 355, 61-68.
[21] Pischinger, R.; Klell, M.; Sams, T. *Thermodynamik der Verbrennungskraftmaschine*; Springer: 2009.
[22] Ogden, J. M.; Steinbugler, M. M.; Kreutz, T. G. *Journal of Power Sources* 1999, 79, 143-168.
[23] Ren, X.; Zelenay, P.; Thomas, S.; Davey, J.; Gottesfeld, S. *Journal of Power Sources* 2000, 86, 111-116.
[24] Lovegrove, K.; Luzzi, A.; Kreetz, H. Solar Energy 1999, 67, 309-316
[25] Vajo, J. *J. Ammonia Fuel Cell*; Patent: US 7157166B2.
[26] Assenbaum, D. W.; Beil, J.; Schmitt, D.; Taccardi, N.; Wasserscheid, P. *Chemie Ingenieur Technik* 2010, 82, 1329-1329.
[27] Schaub, T.; Paciello, R. A. *Angewandte Chemie* 2011, 123, 7416-7420
[28] Müller, B.; Arlt, W.; Wasserscheid, P. *Energy and Environmental Science* 2011, 4, 4322-4331.
[29] Cacciola, G.; Giordano, N.; Restuccia, G. *International Journal of Hydrogen Energy* 1984, 9, 411-419.
[30] Grünenfelder, N. F.; Schucan, T. H. *International Journal of Hydrogen Energy* 1989, 14, 579-586.
[31] Crabtree, R. H. Energy and Environmental Science 2008, 1, 134-138.
[32] Müller, K.; Stark, K.; Müller, B.; Arlt, W. *Energy and Fuels* 2012, 26, 3691-3696.
[33] Pez, G. P.; Scott, A. R.; Cooper, A. C.; Cheng, H.; Wilhelm, F. C.; Abdourazak, A. H. *Hydrogen Storage by Reversible Hydrogenation of Pi-Conjugated Substrates*; Patent: US 7,351,359 B1.
[34] Arlt, W.; Wasserscheid, P. *Chemie Ingenieur Technik* 2012, 84, 1240-1240.
[35] Teichmann, D.; Arlt, W.; Wasserscheid, P.; Freymann, R. *Energy and Environmental Science* 2011, 4, 2767-2773.

In: Renewable Energy
Editor: Andreas Poullikkas

ISBN: 978-1-62618-231-8

Chapter 5

GREEN NANOTECHNOLOGY FOR RENEWABLE ENERGY

Kelvii Wei Guo*
Department of Mechanical and Biomedical Engineering,
City University of Hong Kong, P.R.C.

ABSTRACT

The amount of the world's primary energy supply provided by renewable energy technologies is required urgently because the current fossil fuel usage is unsustainable and associated with greenhouse gas production. Moreover, it is well known that the greatest challenge of nanotechnology today is the understanding the long-term impact on the renewable energy and the related environment of structures that cannot be seen even with the most sophisticated optical microscopes and structures that can interact with the basic components of life.

Therefore, the relevant technologies such as hydrogen fuel, solar cell, biotechnology based on nanotechnology suitable to the friendly environment are reviewed aim to provide the vital information about the growing field of green nanotechnology for renewable energy to minimize the potential environmental risks. It is pointed out that the significantly feasible world's eco-energy for the foreseeable future should not only be realized, but also methods for using the current energy and their by-products more efficiently should be found correspondingly, alongside technologies that will ensure minimal environmental impact.

Keywords: Green nanotechnology, Renewable energy, Fuel cell, Solar cell, Biotechnology, Biofuel

* 83 Tat Chee Avenue, Kowloon Tong, Kowloon, Hong Kong; Email: kelviiguo@yahoo.com.

INTRODUCTION

Today's energy supply is largely responsible for the anthropogenic greenhouse effect, acid rain and other negative impacts on health and the environment. The current trend is clearly not sustainable, especially given the enormous demand for energy predicted for the future. Therefore, an adequate and secure supply of energy is essential to the daily lives, which needs to be achieved with minimum adverse environmental effects. However, energy demand worldwide continues to increase, as the world's population grows and the developing world economies expand.

According to the present policies, even if the potential for increasing low carbon sources of energy is realized, it is clear that remaining reserves of fossil fuels will play a significant part in meeting the world's energy needs for the foreseeable future, and ways of reducing emissions must be found [1]. A growing number of scientists and engineers are exploring and tweaking material properties at atomic scale to create designer materials, which might ultimately increase the efficiency of current energy sources or make new energy sources practical on a commercial scale. At the nanoscale, some materials have different physical properties (strength, conductivity, reflectivity, chemical reactivity, *etc.*) than they do in bulk. Nanoparticles are particles that have at least one dimension in the range of 1~100 nm. For reference, the diameter of the human hair is approximately 80,000 nm. At the nanoscale, fundamental mechanical, electrical, optical, and other properties can significantly differ from their bulk material counterparts. Many nanoscale materials can spontaneously selfassemble into ordered structures. Nanostructured materials also have enormous surface areas per unit weight or volume, so that vastly more surface area is available for interactions with other materials around them. That is useful because many important chemical and electrical reactions occur only at surfaces and are sensitive to the shape and texture of a surface as well as its chemical composition [2-8].

Nowadays, nanotechnologists are searching for new catalysts to improve the energy efficiency of cracking fossil fuels. The world's supplies of light, sweet crude oil-the most desirable type free of sulfur-are getting low; remaining stores of heavier crude oil have sulfur, so new catalysts are needed to remove the sulfur as well as to create cleaner-burning fuels. Additional new catalysts are needed for converting coal to diesel and other liquid transportation fuels. Also, most existing catalysts-such as the one in every automobile's catalytic converter-rely on precious metals, notably platinum, because the transition metals have an optimal configuration of electrons available in their outer orbitals for desired chemical reactions [9-13].

Precious metals, however, are both expensive and scarce, a growing concern as industrializing countries demand more automobiles. So the search will be focused on for cheaper substitutes. Moreover, advances in nanomaterials are expected to be especially valuable in power system applications, from superstrong metal alloys for rotating machinery to tough composites that resist corrosion attack [14-17], less-brittle ceramics for power line insulators [18-21], and slick coatings that will reduce biofouling in cooling water intakes [22-25]. Nanoscale research on lattice structures and electron pairing are expected to improve the understanding of superconductivity, leading to new alloy formulations, novel fabrication techniques, and hopefully a material that will superconduct at room temperature *etc.* [26-28]. Considering the remediation of environmental problems, in the development of "green" processing technologies, research is looking into nanotechnology related to hydrogen, and renewable resources such as solar, wind and ocean energy could be possible solutions to diminishing our dependence on depleting petroleum supplies [29-30].

HYDROGEN FUEL/FUEL CELL

Fuel cells operate almost silently, generating low-voltage direct current from the catalytically aided electrochemical reaction between a fuel (hydrogen or a hydrogen-rich carrier such as natural gas) and an oxidizer (oxygen, either taken from the ambient air or carried in separate storage vessels), and its related auxiliary equipment includes: pumps to transport the fuel through the system and to remove the water produced in the reaction, a blower or compressor to cool the stack and to convey oxygen to the fuel cell, and perhaps a humidifier (required by some types, such as PEM cells). Practical fuel cell systems are very complex to design and build-especially small, rugged ones for cars, trucks, and buses, which must stand up to bumps and to temperature variations. This is one reason why it took so long to put the first prototypes on the road. But the basic principle of how a fuel cell works is fairly straightforward.

In a hydrogen-oxygen fuel cell with an acid electrolyte, molecular gaseous hydrogen is fed in on the anode side. There electrons are stripped out of the H_2 molecules, leaving positively charged hydrogen ions. These migrate through the electrolyte layer to the positively charged cathode, where they combine with oxygen to form water. Meanwhile, the liberated electrons (electric current) flow from the anode through a wire or some other metallic material back to the cathode, performing work-for example, operating an electric motor, lighting up a light bulb, or powering a cell phone. Materials other than hydrogen can serve as ionic energy carriers too-for example, carbon (as a carbonate ion) and oxides of various elements. "In theory, any substance capable of chemical oxidation that can be supplied continuously (as a fluid) can be burned as the fuel at the anode of a fuel cell," In normal combustion all the energy generated is released as heat. In a fuel cell, however, part of the free energy of this electrochemical reaction is released directly as electricity; only the remainder is released as heat-obviously a much more efficient process for powering electric equipment and electric cars.

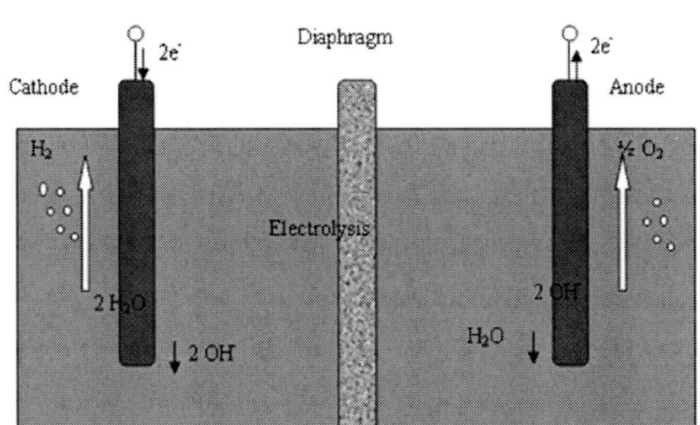

Figure 1. Principle of alkaline fuel cells (AFCs).

An important aspect, one that has assumed much greater importance in recent decades because of the concerns over global warming, is that the preponderant nitrogen component in the atmosphere-about 80 percent of the air is nitrogen-plays no part in cold combustion. Only hydrogen and oxygen react with each other in a fuel cell. By definition, this process precludes the production of nitrogen oxides, a key component in air pollution. In contrast, any open-flame process (such as the combustion of gasoline in conventional car, bus, and truck engines and jet turbines) produces these harmful emissions. Fuel cells, with some variations, come in five different types. The different types of fuel cells are normally named according to the electrolyte they use for their operation. A wider range classification is sometimes made according to the temperature at which the different fuel cells operate.

Alkaline Fuel Cells (AFCS)

Alkaline fuel cells operate at 60-90 °C with efficiencies of 50-60% in converting chemical fuel to electricity. The electrolyte is 35-50% KOH with cheaper catalyst. Electrolysis processes can produce hydrogen by using electricity as the driving force. In alkaline electrolysis, water is split into hydrogen and oxygen using two electrodes in a dilute alkaline electrolyte as shown in Figure 1.

As seen in Table 1, alkaline fuel cells have an electrolyte consisting of an aqueous solution of either sodium or potassium hydroxide. This results in the following cell reactions:

Cathode:

$$2H_2O + 2e^- \rightarrow H_2 + 2OH^-$$

Anode: $2OH^- \rightarrow \frac{1}{2} O_2 + H_2O + 2e^-$

Because of the aqueous electrolyte, temperature of operation at atmospheric pressure is limited to below 100 ^{0}C. Due to the low operation temperature, it allows the use of low-cost materials, such as Teflon-bonded carbon or nickel mesh as electrodes, and polymers in other parts constituting the fuel cell stack. Since the oxygen reduction in an alkaline electrolyte is rather rapid, non-noble metal catalyst can be used to promote the cathode reaction.

The anode reaction is often catalysed by carbon-bound nickel. Consequently, AFCs' main disadvantage is that ambient CO_2 can degrade their performance [31-33] due to the electrolyte's affinity to react with carbon dioxide according to $OH^- + CO_2 \rightarrow HCO_3^-$.

Proton Exchange Membrane Fuel Cells (PEMFCs)

Also known as polymer electrolyte fuel cell (PEFC), or ion exchange membrane (IEM). This results in the following electrode reactions:

Anode: $H_2 \rightarrow 2H^+ + 2e^-$

Cathode: $\frac{1}{2} O_2 + 2H^+ + 2e^- \rightarrow H_2O$

PEMFCs deliver high power density and offer the advantages of low weight and volume, compared to other fuel cells which operate between 50 °C and 80 °C with efficiencies range from 50% to 60%. With a solid polymer as an electrolyte and porous carbon electrodes containing a platinum catalyst, PEMFCs need only hydrogen, oxygen from the air, and water to operate without corrosive fluids. It can be seen that PEMFCs require a noble-metal catalyst (typically platinum) to separate the hydrogen's electrons and protons, and the platinum catalyst is also extremely sensitive to carbon monoxide (CO) poisoning making it necessary to employ an additional reactor to reduce CO in the fuel gas if the hydrogen is derived from an alcohol or hydrocarbon fuel or explore the platinum/ruthenium catalysts more resistant to CO [34-37].

Phosphoric Acid Fuel Cells (PAFCs)

PAFCs use liquid phosphoric acid as an electrolyte with about 55% efficiency with range of 160-220 °C. PAFCs are also typically large and heavy, and less powerful than other fuel cells given the same weight and volume.

The electrodes consist of Teflon-bonded carbon, and noble metal catalyst assist in promoting the electrode reactions:

Anode: $H_2 \rightarrow 2H^+ + 2e^-$

Cathode: $\frac{1}{2} O_2 + 2H^+ + 2e^- \rightarrow H_2O$

Similarly to PEM, the use of noble-metal catalysts makes the PAFC sensitive to CO-poisoning. In addition, PAFCs are expensive because of the expensive platinum as catalyst [38-40].

Molten Carbonate Fuel Cells (MCFCs)

Molten carbonate fuel cells run at 620-660 °C with an electrolyte composed of a molten carbonate salt mixture suspended in a porous, chemically inert ceramic lithium aluminum oxide ($LiAlO_2$) matrix gained the efficiency of 60-65% are considered promising for future power-generation needs.

The reactions are listed as follows:

Anode: $H_2 + CO_3^{2-} \rightarrow H_2O + CO_2 + 2e^-$

$CO + CO_3^{2-} \rightarrow 2CO_2 + 2e^-$

Cathode: $\frac{1}{2} O_2 + CO_2 + 2e^- \rightarrow CO_3^{2-}$

At the same time, due to non-precious metals taken as catalyst at the anode and cathode, the cost reduction is improved significantly. Furthermore, unlike alkaline, phosphoric acid, and polymer electrolyte membrane fuel cells, MCFCs do not require an external reformer to convert more energy dense fuels to hydrogen. Due to the high operating temperatures, internal reforming converted to hydrogen within the fuel cell itself also reduces cost. The primary disadvantage of current MCFCs technology is durability. The high operating temperatures and the corrosive electrolyte used accelerate component breakdown and corrosion with shorter cell life. Therefore, corrosion-resistant materials for components should be explored to increase cell life as well as fuel cell designs [41-45].

Solid Oxide Fuel Cells (SOFCs)

Solid oxide fuel cells operate at the highest temperatures yet: 800-1000 °C with efficiency of 55-65%.

The resulting electrode reactions are:

Anode: $H_2 + O^{2-} \rightarrow H_2O + 2e^-$

$CO + O^{2-} \rightarrow CO_2 + 2e^-$

Cathode: $\frac{1}{2} O_2 + 2e^- \rightarrow O^{2-}$

The electrolyte typically consists of an alloy of rare earth materials, such as yttrium-stabilized zircon dioxide, a ceramic material. Also, SOFCs are the most sulfur-resistant fuel cell type and inert to CO allowed SOFCs to use gases made from coal directly. High-temperature operation results in slow startup and requires significant thermal shielding to retain heat and protect personnel with stringent durability requirements on materials.

Table 1. Types of fuel cells and its characteristics

Fuel cell type	Electrolyte	Charge Carrier	Operating temperature	Fuel	Electric efficiency
Alkaline fuel cells (AFCs)	KOH	OH^-	60-90 °C	Pure H_2	50-60 %
Proton exchange membrane fuel cells (PEMFCs)	Solid polymer	H^+	50-80 °C	Pure H_2 (tolerates CO_2)	50-60 %
Phosphoric acid fuel cells (PAFCs)	Phosphoric acid	H^+	160-220 °C	Pure H_2 (tolerates CO_2, approx.. 1% CO)	~55 %
Molten carbonate fuel cells (MCFCs)	Lithium and potassium carbonate	CO_3^{2-}	620-660 °C	H_2, CO, CH_4, other hydrocarbons (tolerates CO_2)	60-65 %
Solid oxide fuel cells (SOFCs)	Solid oxide electrolyte (yttria, zirconia)	O^{2-}	800-1000 °C	H_2, CO, CH_4, other hydrocarbons (tolerates CO_2)	55-65 %

The development of low-cost materials with high durability at cell operating temperatures is the key technical challenge facing this technology [46-51]. The relationship between the above-mentioned fuel cells and its efficiency with the related operating temperature is shown in Table 1. It is well known that hydrogen is the ideal feedstock for fuel cells. Due to the chemical and physical interactions are mainly governed by surfaces, the nanostructured materials can have substantially distinct properties different from those of the conventional materials. While progress in the development of catalysts for internally converting methane to hydrogen has reduced the temperatures for some fuel cells to the 300~400°C range [52-59].

Advanced nanostructural catalysts may allow hydrocarbon reforming at very low temperatures (below 100°C). Successful development of such catalysts would help substantially in making fuel cells that operate at ambient temperatures. Up to now, the major technical challenges still face a hydrogen economy: storing hydrogen, purifying hydrogen, devising membranes for fuel cells, and producing enough hydrogen on a commercial scale from water instead of from natural gas to support a hydrogen economy genuinely independent of fossil fuels.

For example, for hydrogen storage: one surprising fact about hydrogen is that, unlike most gases, it is most densely stored. Chemically bound in a solid hydride, hydrogen atoms can be densely packed at room temperature. The challenge is designing a hydride that will release the hydrogen quickly when a driver accelerates. One structure under investigation is core-shell nanoparticles: spherical nanoparticles coated with a catalyst. When a hydrogen molecule (H_2) lands on the coating, the catalyst splits the molecule into two hydrogen atoms, which diffuse into the core. Because the nanoscale particles are so small and have high surface area, the atomic hydrogen remains near the surface. The process can be thermally reversed when the hydrogen is needed. For fastest release, other investigators are looking at compounds that "physi-sorb" hydrogen: instead of the hydrogen being chemically bound into a compound, hydrogen atoms or molecules are attracted by electrostatic or Van der Waals forces just to sit on the surface or inside molecular cages until released [53, 55, 57-58, 60-77]. In a word, fuel cells, reacted as hydrogen with oxygen to produce electricity, with the sole exhaust product being nonpolluting water are compact and very efficient in the future energy.

For fuel cell focused on the miniaturization, two main research axes are possible for the global conception of the small fuel cells: either the reduction in the size of existing fuel cells, or the search for new materials and structures usable for their fabrication. Although the first solution seems to be the easiest, it involves the miniaturization of all the auxiliary components and the difficulty to adapt "macrofabrication" techniques to miniature sizes. So research rather tends to apply itself to develop size-adapted methods with suitable materials. One of the technological ways to miniaturize fuel cells is to have recourse to standard microfabrication techniques mainly used in microelectronics and more especially the fabrication of micro- and nano-electro-mechanical systems (MEMS/NEMS). Actually more and more papers show the interest in developing MEMS-based fuel cells, either directly with silicon substrates, or adapting the methods to other substrates such as metals or polymers [78]. These techniques enable notably mass fabrication at low cost (very large number of devices on a very small area) and then could lead to the reduction of the global cost of the miniature fuel cells. Zhang *et al.* proposed a six-cell PEMFC planar stack combined with a small H_2 storage canister [79]. Each cell was made of a home-made MEA (Nafion® 112 membrane with Pt loading on each side) sandwiched between silicon-based flow field plates. These flow fields were fabricated using a classical wet etching process (KOH solution) and forming pin-type patterns with two square through holes for the hydrogen plates and through hole patterns for the oxygen plates. The plates were made electrically conductive by sputtering Ti/Pt/Au metal layers on their surfaces. Six cells were then serially interconnected in the same plane with a fuel distributor fabricated on another piece of wet etched silicon anodically bonded to ultrasonically drilled Pyrex glass. Furthermore, portable electronics also have their own requirements regarding to the power to supply, the packaging, the working temperature and the fuels to use. Thus the manufacturing of miniature fuel cells will be determined by the final applications. As a fuel cell cannot run at higher temperatures than 80 or even 100°C for the user's safety, the choice in the type of fuel cell to use in portable devices is limited to low temperature fuel cells such as PEMFC (for Proton Exchange Membrane Fuel Cell or sometimes Polymer Electrolyte Membrane Fuel Cell) and DMFC (for Direct Methanol Fuel Cell).

Miniature SOFCs (for Solid-Oxid Fuel Cells) functioning at temperatures as low as 500°C (low temperature for SOFCs usually functioning at 800°C or more) have also been reported [80-81]. They are based on thin films technology and use techniques similar as photolithography, etching, physical vapour deposition (PVD). However decreasing the operating temperature involves the decrease of performances of the fuel cell. Besides these temperatures remain inadequate to portable devices because they imply important heat transfer and thus would require an additional cooling system which would increase the overall volume of the fuel cell. More important issue is that all these procedures are involved in chemical reaction or higher cost for equipment. According to the eco-friendly requirements, its related so-called "green" micro- and nano- technology applied to fuel cell fabrication should be given in industrial environment.

Solar Energy/Solar Cell

For solar energy, engineers have been striving to generate power from solar thermal energy for a century. Now, the technology is finally about to come of age. The proportion of solar radiation that reaches the Earth's surface each year is more than 10,000 times the current annual global energy consumption. Meantime, more solar energy strikes the earth on a single day than the world uses annually-which is why the sun is so appealing as an ultimate energy source. So in the photovoltaic community, converting solar photons into electrons cheaply enough to compete with fossil fuels remains a research challenge. With traditional single-crystal silicon solar cells, the best reported maximum efficiency from visible light is 24.7 percent [82-83].

Nanotechnology may help to both raise the efficiency and lower the costs of tapping solar energy. Large single crystals of silicon are costly to grow. Cheaper solar cells may be built from amorphous and microcrystalline silicon; although such technologies are less efficient than single-crystalline silicon (best reported is 12-to-14 percent), the manufacturing processes are about half the cost [82, 84-89]. Non-silicon thin-film materials such as cadmium telluride or CIGS (copper indium gallium selenide) are also cheaper because they can be deposited on steel, glass, or other inexpensive rigid substrates, and have attained an efficiency of 19.1 percent. Researchers are also exploring making organic (plastic) thin-film solar cells that are optically transparent (so they can cover windows) and flexible (so they can wrap around curved shapes), to be able to collect solar energy from the entire surface of an office building rather than just from its roof [90-105]. One crucial technical challenge with flexible solar cells, however, is creating an impervious vapor barrier to prevent water molecules from diffusing inside and lowering the efficiency. Another approach is boosting efficiency at all wavelengths. The sun has a huge amount of energy in the infrared that is not captured with existing single-junction solar cells. The heart of any solar cell is the *p-n* junction, a narrow semiconductor layer where solar photons give rise to electron-hole pairs that flow as current; on either side of the junction are different semiconductor materials, one doped to be *n*-type to collect the electrons that flow in one direction and the other doped to be *p*-type to collect the holes that flow in the opposite direction.

Depending on the choice of semiconductors, the *p-n* junction can absorb solar photons at a specific wavelength. Multijunction thin-film solar cells stack several *p-n* junctions in thin layers so that a solar cell can efficiently produce electricity from a spectrum of wavelengths, including the infrared [106-109]. Potentially even more efficient may be quantum dot solar cells, which absorb solar infrared and create electronhole pairs from thermal gradients. In theory, nanoscale quantum dots of various sizes and chemical compositions could be optimized to harvest solar radiation of all wavelengths from infrared to ultraviolet [110, 111]. Applications in advanced power-generation technologies are also foreseen. The catalysts mentioned above should have direct use in improving the performance of fuel cells, which are just now becoming commercially viable. Hybrid photovoltaic solar cells based on conducting polymers and semiconductor nanorods also hold special promise; by combining the excellent electronic properties of inorganic semiconductors with the process flexibility of organic polymers, researchers are homing in on PV devices with good efficiencies that are easier and much less expensive to manufacture than conventional solar cells. Early prototypes have demonstrated power conversion efficiencies approaching 8% in the laboratory, a level expected to be improved significantly through engineering refinement and new material formulations [112-121].

BIOTECHNOLOGY

The material of plants and animals, including their wastes and residues, is called *biomass*. It is organic, carbon-based, material that reacts with oxygen in combustion and natural metabolic processes to release heat. Such heat, especially if at temperatures >400 °C, may be used to generate work and electricity. The initial material may be transformed by chemical and biological processes to produce biofuels, i.e. biomass processed into a more convenient form, particularly liquid fuels for transport. Examples of biofuels include methane gas, liquid ethanol, methyl esters, oils and solid charcoal.

The success of biomass systems is regulated by principles:

1) Every biomass activity produces a wide range of products and services. For instance where sugar is made from cane, many commercial products can be obtained from the otherwise waste molasses and fiber. If the fiber is burnt, then any excess process heat can be used to generate electricity. Washings and ash can be returned to the soil as fertilizer.

2) Some high-value fuel products may require more low-value energy to manufacture than they produce, e.g. ethanol from starch crops, hydrogen. Despite the energy ratio being >1, such an energy deficiency need not be an economic handicap provided that process energy can be available cheaply by consuming otherwise waste material, e .g. straw, crop fiber, forest trimmings.
3) The full economic benefit of agro-industries is likely to be widespread and yet difficult to assess. One of many possible benefits is an increase in local 'cash flow' by trade and employment.
4) Biofuel-production is only likely to be economic if the production process uses materials already concentrated, probably as a by-product and so available at low cost or as extra income for the treatment and removal of waste. Thus there has to be a supply of biomass already passing near the proposed place of production, just as hydro-power depends on a natural flow of water already concentrated by a catchment. Examples are the wastes from animal enclosures, offcuts and trimmings from sawmills, municipal sewage, husks and shells from coconuts and straw from cereal grains. It is extremely important to identify and quantify these flows of biomass in a national or local economy before specifying likely biomass developments. If no such concentrated biomass already exists as a previously established system, then the cost of biomass collection is usually too great and too complex for economic development. Some short-rotation crops may be grown primarily for energy production as part of intensive agriculture; however, within the widespread practice of agricultural subsidies it is difficult to evaluate fundamental cost-effectiveness.
5) The main dangers of extensive biomass fuel use are deforestation, soil erosion and the displacement of food crops by fuel crops.
6) Biofuels are organic materials, so there is always the alternative of using these materials as chemical feedstock or structural materials. For instance, palm oil is an important component of soaps; many plastic and pharmaceutical goods are made from natural products; and much building board is made from plant fibers constructed as composite materials.
7) Poorly controlled biomass processing or combustion can certainly produce unwanted pollution, especially from relatively low temperature combustion, wet fuels and lack of oxygen supply to the combustion regions. Modern biomass processes require considerable care and expertise.
8) The use of sustainable biofuels in place of fossil fuels abates the emission of fossil-CO_2 and so reduces the forcing of climate change. Recognition of this is a key aspect of climate change policies.

Moreover, in an attempt to alleviate fossil fuel usage and CO_2 emissions, fuels, heat or electricity must be produced from biological sources in a way that is economic (and therefore efficient at a local scale), energetically (and greenhouse gas) efficient, environmentally friendly and not competitive with food production. In the past decade, many scientists apply biological organisms, systems, and processes related to DNA in industrial concerns, forming a new crosscutting technology platform known as biotechnology [122-124].

Biofuels

Finding sustainable feedstock alternatives to petroleum fuel is a key RandD goal, with biologically produced gasoline substitutes topping the list. Results already show environmental benefits from using plant-derived alcohol in engines. Unfortunately, biofuels are commercially produced from sugar, starch and oil-seed based feedstocks currently. For example, bioalcohol is produced from corn starch, soybean, palm fruits, and rape and canola seeds are the common feedstocks for biodiesel production. The further expansion of biofuel production will trigger debate on food/feed versus fuel. Thus, for sustainable biofuel production, non-food feedstock should be used. In the present methods, starch, which has various advantages over other fuels in terms of cost and convenience in handling and manufacture, are also investigated and used as a fuel or a fuel component for a combustor such as a boiler, kiln, dryer or furnace [125-129]. Therefore, for the use of biomass producing fuels, more efficient biomass conversion techniques would help make biofuels more cost-competitive. Land availability and crop selection are major issues in biomass fuel usage. Biomass alternatives can be expected to grow to a significantly larger scale for providing fuel. Land availability may not be a major problem, but land use issues need to be coordinated. The long-term production of biofuels in substantial quantities will require a number of changes. Grain surpluses will not provide sufficient feedstocks for the fuel quantities needed. Producers need to switch to short-rotation woody plants and herbaceous grasses, which can sustain biofuel production in long-term, substantial quantities. Moreover, the increased use of municipal solid waste (MSW) could not only reduce feedstock costs and reserve agricultural land for food production, but also reduce the environmental insult of incineration or landfill expansion. To date, the more widespread use of alcohol could have some safety benefits since alcohol is water soluble, biodegradable, and evaporates easily. However, a major drawback of alcohol compared to methanol is its price which can be almost twice as much as methanol. Although most alcohol is now produced from corn, research has been done on producing this type of alcohol fuel from cellulosic biomass products including energy crops, forest and agricultural residues, and MSW, which would provide much cheaper feedstocks. Conversion of cellulosic materials to alcohol consists of two main steps: hydrolysis and fermentation. Hydrolysis can be accomplished either with acids or with cellulase enzymes. Although the two process routes are currently about equal in cost, acid hydrolysis-a mature technology-is expected to yield only small cost improvements. The cost of the enzymes themselves dominates the enzymatic processes. Moreover, cellulosic alcohol as the alternative to agriculturally-derived alcohol, there are still some limitations associated with their use. A key issue in the case of cellulosic alcohol is the difficult and expensive derivation of fermentable sugars from lignocellulosic biomass. Beyond this, there are the limitations inherent in the fermentation process in terms of rate, efficiency, and the cost of isolating pure alcohol from a dilute aqueous solution. Additionally, alcohol is volatile, toxic, hydrophilic, potentially corrosive to engine components, and of relatively low energy content compared to gasoline or diesel fuel. While it would be expensive to harvest enough natural enzymes to convert biomass into alcohol on a large scale, developing an inexpensive artificial enzyme or other chemical change catalyst through biomimetic research may provide a solution [130-138]. Although biomass has a large amount of advantages as substitutes for fossil fuels as well as electricity and heat, a large increase in biomass energy production has the potential to cause serious environmental problems. Land use issues and concerns about pollution are major concerns. Areas with fragile ecosystems and rare species would need to be preserved. Agricultural lands would also compete with food production. The loss of soil fertility from overuse is a concern. Biomass production would need to be varied and sustainable while preserving local ecosystems. Pollution problems could result from the expanded use of fertilizers and bioengineered organisms on energy farms. The introduction of hazardous chemicals from MSW into the agricultural system could result in increased air and water pollution.

Biomimetic Technology

Biomimetic materials research starts with the study of structure–function relationships in biological materials. Based on the strategies found in nature, bio-inspired materials will be developed. It is very useful in the design of composite materials, or material structures, which typically offering superior properties and functionality, such as superior properties, hierarchical structures, superior assembly mechanisms, superior process control, multi-functionality and adaptability, and holding tremendous potential for improving the capabilities of a wide range of industrial components and systems [139-142].

Enzyme catalyst

Enzyme catalysts illustrate the use of biomimetic chemistry to go beyond the range of biochemical catalysis. One of the distinct practical advantages is converting biomass to alcohol via biomimetic catalysis. Chemical reactions in natural systems are catalyzed by enzymes, which provide unparalleled efficiencies and specifications under near ambient conditions. Artificially produced enzymes may eventually match these high efficiencies and also accomplish reactions far more rapidly and under more aggressive process conditions. Because of the importance of catalytic processes in the chemical and manufacturing industries, research into the development of enzyme-based catalysts has potential for tremendous impact across a wide range of applications [143-152].

Mimicking Photosynthesis

Today, patterning or structuring of individual layers is performed at the process step of the specific layer, e.g., when forming a solar cell. A solar or photovoltaic cell is a device that converts photons from the sun (solar light) into electricity using electrons. In general, a solar cell that includes the capacity to capture both solar and nonsolar sources of light (such as photons from incandescent bulbs) is termed a photovoltaic cell. Fundamentally, the device needs to fulfill only two functions: photogeneration of charge carriers (electrons and holes) in a light-absorbing material, and separation of the charge carriers to a conductive contact that will transmit the electricity. While plants then use the charges to electrochemically synthesize hydrocarbons for food, photovoltaic (PV) cells consolidate individual charges in a circuit to produce DC current. Biomimetic approaches to light gathering and charge separation have shown attractive promise for advancing the efficiency and economy of PV cells [153-162].

Photosynthesis research has led to the device has been incorporated into prototype solar cells assembled from a trimeric ruthenium complex adsorbed on films of nanometer-sized particles of titanium dioxide [163-167]. It is very attractive with higher conversion efficiencies and new cell combinations are being studied for further improvement, research progress in this area is the exciting hotspot and highlighted by dedicated researchers [168-175].

Water Decomposition

Biomimetic photocatalytic-decomposition of water, is interested in from the view point of photo energy transformation. Further, a photocatalyst which shows activity to the photo decomposing reaction of water can be recognized as an excellent photo functional material which provides function such as photo absorption, electrolytic separation or oxidation-reduction reaction at the surface. Effective processes for splitting water into molecular hydrogen and oxygen are of special interest for hydrogen use as a non-carbonaceous fuel, either in direct combustion or as the primary consumable in fuel cells. It has been known for many years that zinc sulfide particles in aqueous suspension catalyze the photodecomposition of water. It should be noted, however, that the systems based on metal hydrides, and particularly complex metal hydrides, are rather costly and require preliminary preparation of reacting solutions and the use of expensive noble metal based catalysts. Metal-based systems also suffer from a number of shortcomings. Metals or alloys, such as, Li, Na, K, Ca, Al-Hg, Al-Li, Al-Na, Si-Al-Ca-Na-Fe-Cu, that directly react with water at ambient temperature are either expensive, or hazardous, or present great safety concerns, including, violent uncontrolled reaction with water. On the other hand, such inexpensive and readily available materials like aluminum (Al) and its alloys and iron (Fe) and its alloys do not react with water at ambient temperature. Al can release hydrogen from aqueous solutions only in the presence of substances, such as, alkali hydroxides: NaOH, KOH, that remove a protective oxide layer from Al surface and, as a result, are consumed in the process, since they are transformed into respective aluminates. More recently, biochemical approach is becoming more and more attractive. More efficient, simple and inexpensive chemical compositions that exceed the performance characteristics, such as, specific energy, power density, of the state-of-the-art hydrogen-generating systems to safely produce hydrogen using water have been studied. However, yields are still low [176-181]. In addition, research on biohydrogen with certain groups of microorganisms that possess functional nitrogenase and/or bidirectional hydrogenases show that with oxygenic photosynthetic microbes, photobiological hydrogen can be produced with high rates. Hydrogen production in this experiment is mediated by an efficient nitrogenase system, which can be manipulated to convert solar energy into hydrogen at rates that are several fold higher, compared with any previously described wild-type hydrogen-producing photosynthetic microbe [182]. Furthermore, if a technique can indeed be developed for splitting water at industrially significant rates, novel process cycles may improve overall economics and provide additional benefits. For example, desalination of seawater-currently accomplished by high-temperature evaporation and condensation or by high-pressure reverse osmosis-could be integrated into the photo-induced decomposition process. The hydrogen split out of seawater could be combusted in a boiler or fed into a fuel cell to produce electricity. The water vapor resulting from combustion (or electrochemical reaction) could then be condensed and purified by conventional water-treatment technology to provide potable water. Such a cycle would, of course need to be economically competitive with the conventional desalination approaches [183-185].

Low Temperature Biofuel Cell

It is well known that the fuel cells now under development must operate at medium to high temperatures-some as high as 900 °C. An ambient-temperature fuel cell would represent an important advance of technology. Biomimetic research into ion transport in plants at the macromolecular level may make such a device possible [186]. The electricity generation function of current fuel cells depends on catalytically stripping a hydrogen ion-a proton-from a hydrogen atom and inducing it to travel from the cell's anode, through an electrolyte, to its oxygen-bathed cathode. Electricity is generated when the electron remaining from each proton-stripped atom travels from the anode through an external circuit to reunite at the cathode. The process creates water and heat in addition to electricity [122-123, 151-152, 187-188]. The ability of some biological systems to transport ions across cellular and intracellular membranes provides a model for a lower-temperature, proton-pumping process [189]. However, the process of proton pumping is not yet fully understood on a molecular level, and a more complete understanding of the photochemistry is required. Still, if biomimetic proton pumping could be applied to developing a low-temperature fuel cell, it would carry with it real advantages in cost, weight, cycle life, and construction materials.

Wind and Ocean Energy

Wind energy is an indirect form of solar energy in contrast to the direct solar energy. Solar irradiation causes temperature differences on Earth and these are the origin of winds. The history of wind power goes back many centuries. Wind power was used for irrigation systems 3000 years ago. It is well known that wind power is the conversion of wind energy into a useful form of energy, such as using wind turbines to make electricity, wind mills for mechanical power, wind pumps for pumping water or drainage, or sails to propel ships. The revival of wind power began with the oil crises of the 1970s. In contrast to the mechanical wind power systems of past centuries, modern wind converters almost exclusively generate electricity. State of the art wind generators have reached a high technical standard and now have powers exceeding 4 MW. The high growth rate of the wind power industry indicates that wind power will reach a significant share of the electricity supply within the next two decades. Furthermore, wind power is one of the most important technologies for stopping global warming. No doubt, wind generators change the landscape, but if the global warming does not get under control, coastal areas that would be protected by the reduction in global warming resulting from wind generator installation will most likely not exist far into the future. A major design criterion for wind power is the need to protect the machine against damage in very strong winds, even though such gale-force winds are relatively infrequent. Wind forces tend to increase as the square of the wind speed. Since the 1-in-50-year gale speed will be five to ten times the average wind speed, considerable overdesign has to be incorporated for structural strength.

Also wind speed fluctuates, so considerable fatigue damage can occur, especially related to the blades and drive train, from the frequent stress cycles of gravity loading and from fluctuations and turbulence in the wind. As machines are built to ever increasing size, the torque on the main shaft becomes a limiting factor. As materials science, it will clearly play an important role in developing coatings, lubricants and lightweight durable composite materials that are necessary for constructing turbine blades and towers that can withstand the stresses-especially those that offshore installations are subjected to (corrosion, wind speeds *etc*.). There is scope to develop embedded sensors/sensing materials which can monitor stability and damage, thus allowing instant safeguarding. Corrosion will also be a problem for offshore turbines exposed to salt, air pollution and UV radiation. The continued development of advanced long lasting protective coatings is required to reduce maintenance costs and prolong the operating life of wind energy devices [190-201].

Waves are caused by the wind blowing over the surface of the ocean. In many areas of the world, the wind blows with enough consistency and force to provide continuous waves. It is important to appreciate the many difficulties facing wave power developments summarized as follows:

1) Wave patterns are irregular in amplitude, phase and direction. It is difficult to design devices to extract power efficiently over the wide range of variables.
2) There is always some probability of extreme gales or hurricanes producing waves of freak intensity. The structure of the power devices must be able to withstand this. Commonly the 50 year peak wave is 10 times the height of the average wave. Thus the structures have to withstand ~ 100 times the power intensity to which they are normally matched. Allowing for this is expensive and will probably reduce normal efficiency of power extraction.
3) Peak power is generally available in deep water waves from open-sea swells produced from long fetches of prevailing wind. The difficulties of constructing power devices for these types of wave regimes, of maintaining and fixing or mooring them in position, and of transmitting power to land, are fearsome. Therefore more protected and accessible areas near to shore are most commonly used.
4) Wave periods are commonly ~ 5–10 s (frequency ~ 0.1Hz). It is extremely difficult to couple this irregular slow motion to electrical generators requiring ~ 500 times greater frequency.
5) So many types of device may be suggested for wave power extraction that the task of selecting a particular method is made complicated and somewhat arbitrary.
6) The large power requirement of industrial areas makes it tempting to seek for equivalent wave energy supplies. Consequently plans may be scaled up so only large schemes are contemplated in the most demanding wave regimes. Smaller sites of far less power potential, but more reasonable economics and security, may be ignored.
7) The development and application of wave power has occurred with spasmodic and changing government interest, largely without the benefit of market incentives. Wave power needs the same learning curve of steadily enlarging application from small beginnings that has occurred with wind power.

The distinctive advantages of wave power are the large energy fluxes available and the predictability of wave conditions over periods of days.

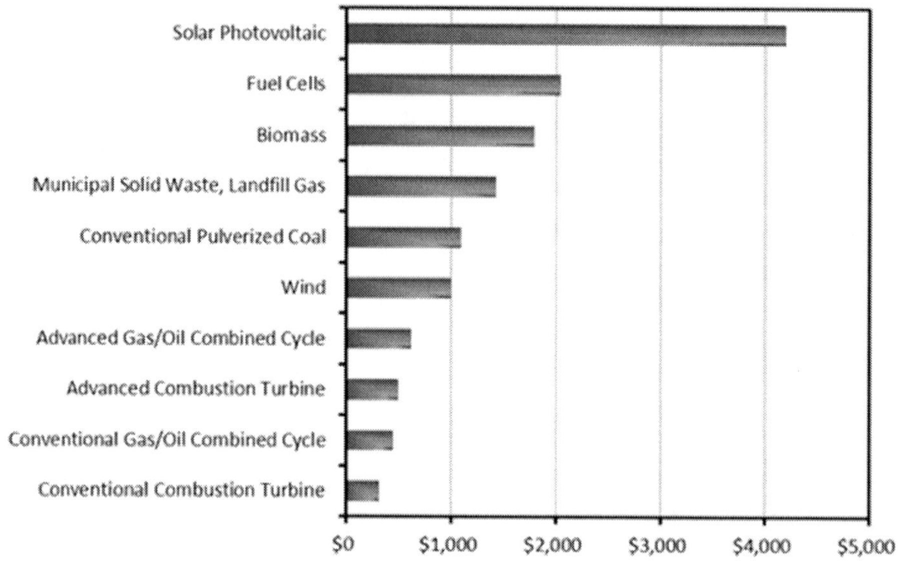

Figure 2. Capital cost per kW of capacity [206].

It is well known that there is tremendous energy in the ocean waves. Wave power **varies considerably** in different parts of the world, and wave energy cannot be harnessed effectively everywhere. To exploit fully the possibilities of electricity production from wave and tidal sources, further advances are required in the most developed ocean technologies including tidal barrages, tidal current turbines and wave turbines. This must be coupled with breakthroughs in less developed technologies that rely on salinity gradient energy, also called blue energy, which work either on the principle of osmosis (the movement of water from a low salt concentration to a high salt concentration) or electrodialysis (the movement of salt from a highly concentrated solution to a low concentrated solution). This requires further developments that reduce the cost or improve the efficiency of membranes to significantly improve the economics of this process. Like wind energy, corrosion is also a problem for wave and tidal energy and will require the continued development of advanced long lasting protective coatings, to reduce maintenance costs and prolong operating life [201-205]. The capital cost of the above reviewed various generating options is shown in Figure 2 [206].

NUCLEAR ENERGY

Since the use of fossil fuels has to be reduced significantly within the coming decades, zero-carbon energy sources are required. One option is nuclear power included nuclear fusion and nuclear fission. For nuclear fission, since nature does not provide uranium in the form that is needed for technical utilization, it must be extracted from uranium ore. Rock with a uranium oxide content of more than 0.1 per cent is a workable uranium ore. Uranium mining produces huge amounts of waste that contains, in addition to some non-toxic components, a lot of radioactive residues. Uranium oxide from uranium ore contains only 0.7 per cent uranium-235. The largest portion is uranium-238, which is not usable for nuclear fission. Therefore, processing plants must enrich the uranium, i.e. the uranium-235 concentration must be increased to 2–3 per cent. If all fossil energy sources used today were replaced by nuclear power, about 10,000 new nuclear power stations would have to be built worldwide. The lifetime of a nuclear reactor is about 30 years, thus all these power stations would have to be renewed after this time. Therefore, a new power plant would have to go on-line every day. In this scenario, politically unstable nations would also acquire, as an unwanted side effect, access to nuclear technology. This would increase the risk of nuclear accidents, sabotage or military use of nuclear energy, resulting in unforeseeable associated costs which make this option increasingly expensive.

As described above, Earth has limited uranium reserves. If the majority of fossil energy sources were to be replaced by nuclear power, the uranium reserves would be depleted in a short time, depending on the nuclear technology employed. Therefore, nuclear fission can only be an alternative to fossil fuels in the medium term. Nuclear fission does not emit any carbon dioxide directly, but the building of the power plant, uranium mining, transport and disposal result in the emission of significant amounts of carbon dioxide. These indirect carbon dioxide emissions are much lower than those associated with the operation of a coal-fired power plant but higher than the indirect carbon dioxide emission of, for example, wind turbines. Transport and storage of radioactive materials bear further risks: uranium and fuel rods must be transported to different processing plants and power stations, and radioactive waste must be transported for further treatment and to intermediate and final storage sites. Toxic and highly radioactive waste such as spent fuel rods is already produced during normal operation of a nuclear power station. Besides other radioactive substances, they also contain about 1 per cent plutonium. One microgram of plutonium, i.e. one millionth part of a gram, is considered to be lethal when breathed in; it will cause death by lung cancer. Hence, one gram of plutonium could theoretically wipe out a whole city. There is no absolute safety guarantee with such nuclear material; the possibility of transport accidents and the emission of radioactive material are very real. Final storage of radioactive waste is also very problematic, because this waste will retain its lethal properties over thousands of years.

For nuclear fusion, many scientists and politicians are working on to develop this technology. To instigate the relevant reaction the particles must be heated to temperatures over one million degrees Celsius. Since no known material can survive these high temperatures, technologies such as locking the reaction materials into intense magnetic fields are examined. Deuterium and tritium for nuclear fusion exist in abundance on Earth, so this technology will not be limited by the availability of the raw materials. However, the question whether this technology will ever work cannot be answered today. Some critics say that the only thing that never changes for nuclear fusion is the remaining time needed for the development of a working reactor, which has been estimated to be 50 years for the past 50 years. Even if this technology will work eventually, there are good reasons against spending more money on this technology. For the development of nuclear fusion, large amounts of money have been spent that could have been used to develop other sustainable technologies. The last and most important reason is the long time that is still needed to bring this technology to maturity. To stop global warming today, working alternatives to the present energy supply are needed urgently. Therefore, to wait for a nuclear fusion reactor that might or might not work at some time in the distant future is not the answer.

Ultimately, there is no unique correct answer. However, renewables are a key part of the solution. Picture the energy supply system as a pyramid whose base today consists of coal and nuclear power. The middle is natural gas, and the top is renewable energy and efficiency.

This pyramid should be inverted so that the base is far more efficient energy consumption. Renewable energy sources and some combined heat and power make up the middle. This part is linked with real time pricing, demanded response, and smart charging and discharging of electrified cars to help balance supply with demand. At the top of the pyramid, the remaining fossil fuels and nuclear power will gradually be phased out. But the world did not wait; it needed more and more horsepower and increasingly drew its power from oil and other fossil fuels. Nuclear energy, as the critical medium term solution to the energy challenges, is a low carbon source of electricity and makes an important contribution to the diversity of the energy supplies. Without the existing nuclear power stations, the carbon emissions would higher and higher. Nuclear energy generation is therefore an important component of the energy mix [207-209]. The technical challenge is for the safe and efficient harnessing of nuclear energy, exploring both fission and fusion technology.

Current nuclear energy capability uses nuclear fission [209-212]. To improve further the efficient and safe utilization of nuclear fission it is necessary to:

1) Full understanding of the physico-chemical effects of radiation on material fatigue, stresses and corrosion in nuclear power stations;
2) Improving methods for spent fuel processing, including developing advanced separation technologies to allow unprecedented control of chemical selectivity in complex environments;
3) Studying the nuclear and chemical properties of the actinide and lanthanide elements; and the radiation effects on polymers, rubbers and ion exchange materials.
4) Storing, disposing and reusing the nuclear waste efficiently;

In addition, the new generation of advanced reactors will have the potential to deliver step-change benefits. Opportunities exist in the design and demonstration of these new reactors. In contrast, nuclear fusion is still at the development stage [213, 214].

Despite some progress, the challenge in fusion research is to generate more energy from fusion than is put in [215-227]. Apart from the continued research into plasma physics, magnetics, instrumentation and so on, the development of high performance structural materials by new techniques capable of withstanding extreme operating conditions is required. For nuclear technology, it is always very nerve-racking. More serious considerations on the nuclear energy exploitation should be taken because of the nuclear crisis caused the contamination after the earthquake and tsunami attacked Japan which crippled the Fukushima Dai-ichi nuclear complex exceeded government safety limits.

CURRENT AND FUTURE DEVELOPMENTS

Oil and Coal as the conventional energy resource will have a long-term role in providing energy diversity and security, providing ways should be found to reduce CO_2 emissions in the longer term controlling SO_2, NOx and carbon in ash. As society moves from an economy based on fossil fuels to a more sustainable energy mix, scientists and engineers will be required not only to develop sustainable energy solutions but also to find more efficient ways of producing, refining and using fossil fuels during the transition. At the same time, increasing the traditional power efficiency and recycling should be sought. Renewable resources such as solar, wind and ocean power exploited by nanotechniques will be promisingly reliable solutions to diminishing the dependence on depleting petroleum supplies. Moreover, regardless of the specific nanotechnology or the specific energy application, finding either new sources of power or new efficiencies will require new breakthrough technologies. In the foreseeable future, research on the remediation of environmental problems and its relevant green technologies will be realized to make the new clean eco energy sources practical on a commercial scale.

REFERENCES

[1] Whitesides GM. The Once and Future Nanomachine. *Scientific American* 2001; 9: 79.

[2] Ashby MF, Ferreira PJ, Schodek DL. Nanomaterials and nanotechnologies: An overview. Nanomaterials: *Nanotechnologies and Design* 2009; 1-16.

[3] Kostoff RN, Koytcheff RG, Lau GY. Global nanotechnology research literature overview. *Technological Forecasting and Social Change* 2007; 74: 1733-47.

[4] Pitkethly MJ. Nanomaterials-the driving force. *Mater Today* 2004; 7: 20-9.

[5] Ashby MF, Ferreira PJ, Schodek DL. Nanomaterials synthesis and characterization. Nanomaterials: *Nanotechnologies and Design* 2009; 257-90.

[6] Ashby MF, Ferreira PJ, Schodek DL. Nanomaterials and nanotechnologies in health and the environment. Nanomaterials: *Nanotechnologies and Design* 2009; 467-500.

[7] Van Hove MA. Atomic-scale structure: From surface to nanomaterials. *Surf. Sci.* 2009; 603: 1301-5.

[8] Zäch M, Hägglund C, Chakarov D, Kascmo B. Nanoscicncc and nanotcchnology for advanced energy systems. *Curr. Opin. Solid. State Mater Sci.* 2006; 10: 132-43.

[9] Palacios MA, G´omez MM, Moldovan M, *et al.* Platinum-group elements: quantification in collected exhaust fumes and studies of catalyst surfaces. *The Science of the Total Environment* 2000; 257: 1-15.

[10] König HP, Hertel RF, Koch W, Rosner G. Determination of platinum emiisions from a three-way catalyst-equipped gasoline engine. *Atmos Environ.* A: General Topics 1992; 26: 741-5.

[11] Reddy BR, Raju B, Lee JY, Park HK. Process for the separation and recovery of palladium and platinum from spend automobile catalyst leach liquor using LIX 84I and Alamine 336. *J. Hazard Mater.* 2010; 180: 253-8.

[12] Lashdaf M, Lahtinen J, Lindblad M, Venäläinen T, Krause AOI. Platinum catalysts on alumina and silica prepared by gas- and liquid- phase deposition in cinnamaldehyde hydrogenation. *Appl. Catal A: General* 2004; 276: 129-37.

[13] Moriarty P, Honnery D. The prospects for global green car mobility. *Journal of Cleaner Production* 2008; 16: 1717-26.

[14] Inman IA, Datta PS, Du HL, Kübel C, Wood PD. *High temperature tribocorrosion.* Shreir's Corrosion 2010; 1: 331-98.

[15] Lekka M, Koumoulis D, Kouloumbi N, Bonora PL. Mechanical and anticorrosive properties of copper matrix micro- and nano-composite coatings. *Electrochim Acta* 2009; 54: 2540-6.

[16] Bjerklie S. Thinking big with nanotechnology: Nano-coatings expected to revolutionize surface finishing. *Metal Finishing* 2005; 103: 46-7.

[17] Hsu M. Nano-lubrication: concept and design. *Tribology International* 2004; 37: 537-45.

[18] Touzin M, Goeuriot D, Guerret-Piécourt C, Juvé D, Fitting HJ. Alumina based ceramics for high-voltage insulation. *J. Eur. Ceram Soc.* 2010; 30: 805-17.

[19] Chaim R, Reinharz Bar Hama O. Densification of nanocrystalline NiO ceramics by spark plasma sintering. *Mater. Sci. Eng.*: A 2010; 527: 462-8.

[20] Bayliss CR, Hardy BJ. *Transmission and Distribution Electrical Engineering.* 3rd ed. New York: Wiley and Sons 2007.

[21] Plojoux J, Leterrier Y, Månson JAE, Templier F. Mechanical integrity analysis of multilayer insulator coatings on flexible steel substrates. *Thin Solid Films* 2007; 515: 6890-8.

[22] Su XJ, Zhao Q, Wang S, Bendavid A. Modification of diamond-like carbon coatings with fluorine to reduce biofouling adhesion. *Surf. Coat. Technol.* 2010; 204: 2454-8.

[23] Sarkar DK, Saleema N. One-step fabrication process of superhydrophobic green coatings. *Surf. Coat. Technol.* 2010; 204: 2483-6.

[24] Dinu CZ, Dinca VC, Soare S, Moldovan A, Smarandache D, Scarisoareanu N, *et al.* Monitorizing nitinol alloy surface reactions for biofouling studies. *Appl. Sur. Sci.* 2007; 253: 7719-23.

[25] Diego MY, Søren K, Kim DJ. Antifouling technology-past, present and future steps towards efficient and environmentally friendly antifouling coating. *Prog. Org. Coat.* 2004; 50: 75-104.

[26] Ansari IA, Shahabuddin M, Alzayed NS, Vajpayee A, Awana VPS, Kishan H. Enhancement of activation energy in nano diamond doped MgB_2 superconductor. *Physica C: Superconductivity* 2010; 470: 369-72.

[27] Mudgel M, Awana VPS, Bhalla GL, Kishan H. Substantial increment in critical parameters of MgB2 superconductor by boron site nano-carbon substitution. *Physica C: Superconductivity* 2010; 470: S646-7.

[28] Stewart MD, Nguyen HQ, Hollen SM, Yin AJ, Xu JM, Valles JM. Enhanced suppression of superconductivity in amorphous films with nano-scale patterning. *Physica C: Superconductivity* 2009; 469: 774-7.

[29] Khan FI, Hawboldt K, Iqbal MT. Life cycle analysis of wind-fuel cell integrated system. *Renewable Energy* 2005; 30: 157-77.

[30] National Nanotechnology Initiative: Leading to the Next Industrial Revolution. *Committee on Technology, National Science, and Technology Council.* Washington, DC 2000.

[31] Mulder G. Fuel cells-alkaline fuel cells. *Encyclopedia Electrochem Power Source* 2009; 321-8.

[32] Bidault F, Brett DJL, Middleton PH, Abson N, Brandon NP. An improved cathode for alkaline fuel cells. *Int. J. Hydrogen. Energy* 2010; 35: 1783-8.

[33] Varcoe JR, Kizewski JP, Halepoto DM *et al.* Fuel cells-alkaline fuel cells | Anion-exchange membranes. *Encyclopedia Electrochem Power Source* 2009; 329-43.

[34] Peighambardoust SJ, Rowshanzamir S, Amjadi M. Review of the proton exchange membranes for fuel cell applications. *Int. J. Hydrogen Energy* 2010; 35: 9349-84.

[35] Shao YY, Yin GP, Wang ZB, Gao YZ. Proton exchange membrane fuel cell from low temperature to high temperature: Material challenges. *J. Power Sources* 2007; 167: 235-42.

[36] Wee JH. Applications of proton exchange membrane fuel cell systems. *Renewable and Sustainable Energy Review* 2007; 11: 1720-38.

[37] Li QF, Jensen JO, Savinell RF, Bjerrum NJ. High temperature proton exchange membranes based on polybenzimidazoles for fuel cells. *Progress in Polym. Sci.* 2009; 34: 449-477.

[38] Appleby AJ. Fuel cells-phosphoric acid fuel cells. *Encyclopedia Electrochem Power Source* 2009; 533-47.

[39] Sammes N, Bove R, Stahl K. Phosphoric acid fuel cells: fundamentals and applications. *Curr. Opin. Solid. State Mater. Sci.* 2004; 8: 372-8.

[40] Aindow TT, Haug AT, Jayne D. Platinum catalyst degradation in phosphoric acid fuel cells for stationary applications. *J. Power Sources* 2011; 196: 4506-14.

[41] Dicks AL. Fuel cells-Molten carbonate fuel cells. *Encyclopedia Electrochem Power Source* 2009; 446-53.

[42] Randström S, Lagergren C, Capobianco P. Corrosion of anode current collectors in molten carbonate fuel cells. *J. Power Sources* 2006; 160: 782-8.

[43] Paoletti C, Carewska M, Lo Presti R, *et al.* Performance analysis of new cathode materials for molten carbonate fuel cells. *J. Power Sources* 2009; 193: 292-7.

[44] Kim JE, Patil KY, Han JH, *et al.* Using aluminum and Li_2CO_3 particles to reinforce the α-$LiAlO_2$ matrix for molten carbonate fuel cells. *Int. J. Hydrogen Energy* 2009; 34: 9227-32.

[45] Zhang HC, Lin GX, Chen JC. Performance analysis and multi-objective optimization of a new molten carbonate fuel cell system. *Int. J. Hydrogen. Energy* 2011; 36: 4015-21.

[46] Minh NQ. Solid oxide fuel cell technology-features and applications. *Solid State Ionics* 2004; 174: 271-7.

[47] Zhou XD, Singhal SC. Fuel cells-solid oxide fuel cells. *Encyclopedia Electrochem Power Source* 2009; 1-16.

[48] Sun CW, Stimming U. Recent anode advances in sold oxide fuel cells. *J. Power Sources* 2007; 171: 247-60.

[49] Yokokawa H, Tu HY, Iwanschitz B, Mai A. Fundamental mechanisms limiting solid oxide fuel cell durability. *J. Power Sources* 2008; 182: 400-12.

[50] Gorte RJ, Vohs JM. Nanostructured anodes for solid oxide fuel cells. *Curr. Opin. Colloid and Interface Sci.* 2009; 14: 236-44.

[51] Ho TX, Kosinski P, Hoffmann AC, Vik A. Effects of heat sources on the performance of a planar solid oxide fuel cell. *Int. J. Hydrogen. Energy* 2010; 35: 4276-84.

[52] Ahmed K. *Fuel cell system*. US7452619 2008.

[53] Bi YP, Lu GX. Nano-Cu catalyze hydrogen production from formaldehyde solution at room temperature. *Int. J. Hydrogen. Energy* 2008; 33: 2225-32.

[54] Umit BD. How green are the chemicals used as liquid fuels in direct liquid-feed fuel cells? *Environment International* 2009; 35: 626-31.

[55] Kadirgan F, Kannan AM, Atilan T, Beyhan S, Ozenler SS, Suzer S, Yörür A. Carbon supported nano-sized Pt-Pd and Pt-Co electrocatalysts for proton exchange membrane fuel cells. *Int. J. Hydrogen. Energy* 2009; 34: 9450-60.

[56] Mitlitsky F, Sridhar KR, Gottman M, Venkataraman S. *Method of optimizing operating efficiency of fuel cells*. US7575822 2009.

[57] Comisso N, Berlouis LEA, Morrow J, Pagura C. Changes in hydrogen storage properties of carbon nano-horns submitted to thermal oxidation. *Int. J. Hydrogen. Energy* 2010; 35: 9070-81.

[58] Ozalp N, Epstein M, Kogan A. Cleaner pathways of hydrogen, carbon nano-materials and metals production via solar thermal processing. *Journal of Cleaner Production* 2010; 18: 900-7.

[59] Morgenstern DA. *Use of metal supported copper catalysts for reforming alcohols.* US7682724 2010.

[60] Paunović P, Radev I, Dimitrov AT, *et al.* New nano-structured and interactive supported composite electrocatalyst for hydrogen evolution with partially replaced platinum loading. *Int. J. Hydrogen. Energy* 2009; 34: 2866-73.

[61] Kadirgan F, Beyhan S, Atilan T. Preparation and characterization of nano-sized Pt-Pd/C catalyst and comparsion of their electron-activity toward methanol and ethanol oxidation. Int. J. Hydrogen. Energy 2009; 34: 4312-20.

[62] Choudhury P, Srinivasan SS, Bhethanabotla VR, Goswami Y, McGrath K, Stefanakos EK. Nano-Ni doped Li-Mn-B-H system as a new hydrogen storage candidate. *Int. J. Hydrogen. Energy* 2009; 34: 6325-34.

[63] Diwan M, Hanna D, Varma A. Method to release hydrogen from ammonia borane for portable fuel cell applications. *Int. J. Hydrogen. Energy* 2010; 35: 577-84.

[64] Chien CC, Jeng KT. Nobel metal fuel cell catalysts with nano-network structures. *Mater. Chem. Phys.* 2007; 103: 400-6.

[65] Zamfirescu C, Dincer I. Ammonia as a green fuel and hydrogen source for vehicular applications. *Fuel Process Technol.* 2009; 90: 729-37.

[66] Jain IP, Lal C, Jain A. Hydrogen storage in Mg: A most promising material. *Int. J. Hydrogen. Energy* 2010; 35: 5133-44.

[67] Basri S, Kamarudin SK, Daud WRW, Yaakub Z. Nanocatalyst for direct methanol fuel cell (DMFC). *Int. J. Hydrogen. Energy* 2010; 35: 7957-70.

[68] Roev V, Suh SH, Kwon KJ, Kim HK. *Supported catalyst, method for preparing the same, cathode electrode comprising the same, and fuel cell comprising the cathode electrode*. US7875569 2011.

[69] An SG, Song MK, Park YM, Kim YM, Han SI, Kweon HJ, Lee SH. *Polymer membrane for fuel cell, method of preparing same, and membrane-electrode assembly for fuel cell comprising same.* US7871736 2011.

[70] Fukuta K, Isomura T, Kishino M. *Separation membrane for fuel battery and process for producing the same.* US7868051 2011.

[71] Ito N. Electrolyte layer for fuel cell, fuel cell, and method of manufacturing electrolyte layer for fuel cell. US7867668 2011.

[72] Atanasoski R, Schmoeckel AK, O'Neill DG, Vernstrom GD, O'Brien DP. *Fuel cell catalyst*. US7867648 2011.

[73] Kwak C, Alexandrovichserov A, Kang SK. *Carrier for fuel cell, and catalyst, membrane-electrode assembly, and fuel cell system including the same.* US7858265 2010.

[74] Wheat WS, Mirkovic VR, Nguyen KH, Krause CL, Stevens JF, Casey DG. *Autothermal reforming in a fuel processor utilizing non-pyrophoric shift catalyst.* US7857875 2010.
[75] Choi YS, Joo SH. *Nanocomposite, nanocomposite electrolyte membrane including the same and fuel cell including the nanocomposite electrolyte membrane.* US7842430 2010.
[76] Chang H, Pak CH, Wang JN. *Carbon nanosphere with at least one opening, method for preparing the same, carbon nanosphere-impregnated catalyst using the carbon nanosphere, and fuel cell using the catalyst.* US7837968 2010.
[77] Idem RO, Kumar P, Sun YP. *Catalysts for hydrogen production.* US7824656 2010.
[78] Marsacq D. Les micropiles a` combustible, une nouvelle ge´ne´ration de microge´ne´rateurs e´lectrochimiques. *Clefs CEA* 2004; 50-1.
[79] Zhang XG, Zheng D, Wang T, Chen C, Cao JY, Yan J, Wang WM, Liu JY, Liu HH, Tian J, Li XX, Yang H, Xia BJ. A preliminary study of a miniature planar 6-cell PEMFC stack combined with a small hydrogen storage canister. *Journal of Power Sources* 2007; 166: 441–4.
[80] Menzler NH, Herzhof W, Buchkremer HP, Stoever D. *High temperature fuel cell stack and production of same.* EP2133950 (A1) 2009.
[81] Boltze M, Rozumek M, Kaeding S, *et al. Fuel cell system and method for influencing the thermal balance of a fuel cell system.* WO2008006328 (A1) 2008.
[82] Green MA, Zhao J, Wang A, Wenham SR. Progress and outlook for high-efficiency crystalline silicon solar cells. *Sol. Energy Mater. Sol. Cells* 2001; 65: 9-16.
[83] Nijs JF, Szlufcik J, Poortmans J, Sivoththaman S, Mertens RP. Advanced cost-effective crystalline silicon solar cell technologies. *Sol. Energy Mater. Sol. Cells* 2001; 65: 249-59.
[84] Rana VV. *Crystalline silicon solar cells on low purity substrate.* US7858427 2010.
[85] Ko JH, Eo YJ, Kim JA, *et al. Hetero-junction silicon solar cell and fabrication method thereof.* WO2009078672 (A2) 2009.
[86] Meier SL, Rohatgi A. *Solar cell having crystalline silicon p-n homojunction and amorphous silicon heterojunctions for surface passivation.* EP2228834 (A2) 2010.
[87] Vitanov P, Goranova E, Stavrov V, Ivanov P, Singh PK. Fabrication of buried contact silicon solar cells using porous silicon. *Sol. Energy Mater. Sol. Cells* 2009; 93: 297-300.
[88] Becker M, Luetke ND. *Silicon solar cell.* WO2010088898 (A2) 2010.
[89] Basu PK, Dhasmana H, Udayakumar N, Khan F, Thakur DK. Regulated low cost pre-treatment step for surface texturization of large area industrial single silicon solar cell. *Sol. Energy Mater. Sol. Cells* 2010; 94: 1049-54.
[90] Aoki S. *Method for manufacturing chalcopyrite thin-film solar cell.* US7871502 2011.
[91] Ahn SW, Eo YJ, Lee KR, Lee DH, Lee HM. *Thin-film solar cell and method of manufacturing the same.* US7863515 2011.
[92] Yang TN, Lan SM, Chiang CC, Ma WY, Ku CT, Huang YH. *Method for making an anti-reflection film of a solar cell.* US7863078 2011.
[93] Farris III CA, Brown AS. *Application specific solar cell and method for manufacture using thin film photovoltaic materials.* US7855089 2011.
[94] Kamezaki H, Torii M, Harada N, *et al. Organic thin-film solar cell.* JP2010219448 (A) 2010.

[95] Hollars D. *Manufacturing apparatus and method for large-scale production of thin-film solar cells*. US7838763 2010.

[96] Tsai CY. *Thin film solar cell and fabrication method thereof.* EP2251908 (A2) 2010.

[97] Sasaki T, Koi Y, Tawada Y, Yamamoto K. *Substrate for thin-film solar cell, method for producing the same, and thin-film solar cell employing it.* US7781668 2010.

[98] Fujioka Y, Takeda T. *Thin film solar cell and manufacturing method thereof.* US7750233 2010.

[99] Borden P. *Method for making an improved thin film solar cell interconnect using etch and deposition process*. US7718347 2010.

[100] Sankir ND, Dogan B. Growth of CdS thin films and nanowires for flexible photoelectrochemical cells. *J. Mater. Process Technol.* 2011; 211: 382-7.

[101] Meillaud F, Feltrin A, Despeisse M, *et al.* Realization of high efficiency micromorph tandem silicon solar cells on glass and plastic substrates: Issues and potential. *Sol. Energy Mater. Sol. Cells* 2011; 95: 127-30.

[102] Rath JK, Brinza M, Liu Y, Borreman A, Schropp REI. Fabrication of thin film silicon solar cells on plastic substrate by very high frequency PECVD. *Sol. Energy Mater. Sol. Cells* 2010; 94: 1534-41.

[103] Haeldermans I, Truijen I, Vandewal K, Moons W, Van Bael MK, D'Haen Haen J, Manca JV, Mullens J. Water based preparation method for 'green' solid-state polythiophene solar cells. *Thin Solid. Films* 2008; 516: 7245-50.

[104] Cheng TC, Cheng CH, Huang ZZ, Liao GC. Development of an energy-saving module via combination of solar cells and thermoelectric coolers for green building applications. *Energy* 2011; 36: 133-40.

[105] Gordon I, Vallon S, Mayolet A, Beaucarne G, Poortmans J. Thin-film monocrystalline-silicon solar cells made by a seed layer approach on glass-ceramic substrates. *Sol. Energy Mater. Sol. Cells* 2010; 94: 381-5.

[106] Strümpel C, McCann M, Beaucarne G, *et al.* Modifying the solar spectrum to enhance silicon solar cell efficiency-An overview of available materials. *Sol. Energy Mater. Sol. Cells* 2007; 91: 238-49.

[107] Li HB, Franken RH, Rath JK, Schropp REI. Structural defects caused by a rough substrate and their influence on the performance of hydrogenated nano-crystalline silicon n-i-p solar cells. *Sol. Energy Mater. Sol. Cells* 2009; 93: 338-49.

[108] Basol BM. *Technique and apparatus for depositing layers of semiconductors for solar cell and module fabrication.* US7736940 2010.

[109] Sterzel HJ. *Photovoltaic cell comprising a photovoltaically active semiconductor material comprising a particular portion of tellurium ions replaced with halogen and nitrogen ions.* US7847187 2010.

[110] Guchhait A, Rath AK, Pal AJ. To make polymer: Quantum dot hybrid solar cells NIR-active by increasing diameter of PbS nanoparticles. *Sol. Energy Mater. Sol. Cells* 2011; 95: 651-6.

[111] Fafard S. *Solar cell with epitaxially grown quantum dot material*. US7863516 2011.

[112] Brabec C, Schilinsky P, Waldauf C. *Inexpensive organic solar cell and method of producing same.* US7868405 2011.

[113] Nelles G, Yasuda A, Schmidt HW, Thelakkat M, Schmitz C. *Hybrid solar cells with thermal deposited semiconductive oxide layer*. US7759573 2010.

[114] Jeong CW, Jeon MS, Kim JH, *et al. Method for fabricating silicon nano wire, solar cell including silicon nano wire and method for fabricating solar cell.* EP2221877 (A2) 2010.
[115] Samuelson LI, Deppert KW. *Nanostructures formed of branched nanowhiskers and methods of producing the same.* US7875536 2011.
[116] Zook DJ, Marcus MS, Liu Y. *Nano-structured solar cell.* EP2264780 (A2) 2010.
[117] Meiss J, Hummert M, Petrich A, Pfuetzner S, Leo K, Riede M. Tetrabutyl-tetraphenyl-diindenoperylene derivatives as alternative green donor in bulk heterojunction organic solar cells. *Sol. Energy Mater. Sol. Cells* 2011; 95: 630-5.
[118] Xiao T, Cui W, Anderegg J, Shinar J, Shinar R. Simple routes for improving polythiophene: fullerence-based organic solar cells. *Org. Electron.* 2011; 12: 257-62.
[119] Yassin A, Rousseau T, Leriche P, Cravino A, Roncali J. Evaluation of bis-dicyanovinyl short-chain conjugated systems as donor materials for organic solar cells. *Sol. Energy Mater. Sol. Cells* 2011; 95: 462-8
[120] Huang JS, Chou CY, Lin CF. Enhancing performance of organic-inorganic hybrid solar cells using a fullerence interlayer from all-solution processing. *Sol. Energy Mater. Sol. Cells* 2010; 94: 182-6.
[121] Günes S, Sariciftci NS. Hybrid solar cells. *Inorg. Chim. Acta.* 2008; 361: 581-8.
[122] Lee JY, Shin HY, Kang SW, Park C, Kim SW. Improvement of electrical properties via glucose oxidase-immobilization by actively turning over glucose for an enzyme-based biofuel cell modified with DNA-wrapped single walled nanotubes. *Biosens. Bioelectron.* 2011; 26: 2685-8.
[123] Lee JY, Shin HY, Kang SW, Park C, Kim SW. Use of bioelectrode containing DNA-wrapped single-walled carbon nanotubes for enzyme-based biofuel cell. *J. Power Sources* 2010; 195: 750 5.
[124] St-Pierre P, Masri N, Fournier MC, White TC. *Modified cellulases with increased thermostability, thermophilicity, and alkalophilicity.* US7785854 2010.
[125] Ramanavicius A, Kausaite A, Ramanaviciene A. Enzymatic biofuel cell based on anode and cathode powered by ehanol. *Biosens. Bioelectron.* 2008; 24: 761-6.
[126] Munasinghe PC, Khanal SK. Biomass-derived syngas fermentation into biofuels: Opportunities and challenges. *Bioresour. Technol.* 2010; 101: 5013-22
[127] Anders VN, Carsten A, Liu J. *Processes for production of a starch hydrolysate.* EP2032713 (A2) 2009.
[128] Lewis LT. *Starch as a fuel or fuel component.* US7815695 2010.
[129] Copeland B, Coil GS, Latta Jr. RA, Lenaghan TA. *Methods for generating energy using agricultural biofuel.* US7263934 2007.
[130] Hennessey SM, Seapan M, Elander RT, Tucker MP. *Process for concentrated biomass saccharification.* US7807419 2010.
[131] Mascal M. *High-yield conversion of cellulosic biomass into furanic biofuels and value-added products.* US7829732 2010.
[132] Roberts IV WL, Lamb HH, Stikeleather LF, Turner TL. *Process for conversion of biomass to fuel.* EP2097496 (A2) 2009.
[133] Kubo W, Nomoto T. *Enzyme electrode, enzyme electrode producing method, sensor and fuel cell each using enzyme electrode.* US7816025 2010.
[134] Liu C, Alwarappan S, Chen zf, Kong xx, Li CZ. Membraneless enzymatic biofuel cells based on grapheme nanosheets. *Biosens. Bioelectron.* 2010; 25: 1829-33.

[135] Reneker DH, Wu ST, Wang P, *et al. Nanofibers with high enzyme loading for highly sensitive biosensors.* WO2009029180 (A1) 2009.

[136] Minteer SD, Akers NL, Moore CM. *Enzyme immobilization for use in biofuel cells and sensors.* US7638228 2009.

[137] Kim JB, Jia HF, Wang P. Challenges in biocatalysis for enzyme-based biofuel cells. *Biotech. Adv.* 2006; 24: 296-308.

[138] Bullen RA, Arnot TC, Lakeman JB, Walsh FC. Biofuel cells and their development. *Biosens. Bioelectron.* 2006; 21: 2015-45.

[139] Zhao QB, Fan TX, Ding J, *et al.* Super black and ultrathin amorphous carbon film inspired by anti-reflection architecture in butterfly wing. *Carbon* 2011; 49: 877-83.

[140] Jiang ZX, Geng L, Huang YD, *et al.* The model of rough wetting for hydrophobic steel meshes that mimic Asparagus setaceus leaf. *J. Colloid. Interface Sci.* 2011; 354: 866-72.

[141] Chen S, Hirota N, Okuda M, *et al.* Microstructures and rheological properties of tilapia fish-scale collagen hydrogels with aligned fibrils fabricated under magnetic fields. *Acta. Biomaterialia* 2011; 7: 644-52.

[142] Arafat MT, Lam XF, Ekaputra AK, Wong SY, Li X, Gibson I. Biomimetic composite coating on rapid prototyped scaffolds for bone tissue engineering. *Acta. Biomaterialia* 2011; 7: 809-20.

[143] DiCosimo R, Panova A, Arthur SD, Chenault HK. *Process for enzymatically converting glycolonitrile to glycolic acid.* US7875443 2011.

[144] Kubo W, Nomoto T, Yano T. *Enzyme electrode, and sensor and biofuel cell using the same.* US7687186 2010.

[145] Sakai H, Nakagawa T, Kakuta M, Tokita Y. *New fuel cell, and power supply device and electronic device using the fuel cell.* EP2216847 (A1) 2010.

[146] Feron D, Bergel A. *Fuel cell, using oxidoreductase type enzymes in the cathodic compartment and possibly in the anodic compartment.* US7875394 2011.

[147] Caimi PG, Chou YC, Franden MA, *et al. Zymomonas with improved ethanol production in medium containing concentrated sugars and acetate.* US7803623 2010.

[148] Boeckh D, Hauer B, Haring D. *Enzymatic synthesis of sugar acrylates.* US7767425 2010.

[149] Kragl U, Kaftzik N, Fliege S, Wasserscheid P. *Enzyme catalysis in the presence of ionic liquids.* US7754462 2010.

[150] Hughes J, Symes KC. Biocatalytic manufacturing of (meth) acrylylcholine or 2-(N, N-dimethylamino) ethyl (meth) acrylate. US7754455 2010.

[151] Kalyanasundaram K, Graetzel M. Artificial photosynthesis: biomimetic approaches to solar energy conversion and storage. *Curr. Opin. Biotechnol.* 2010; 21: 298-310.

[152] Martins VA, Bonfim C, da Silva WC, Crespilho FN. Iron (III) nanocomposites for enzynme-less biomimetic cathode: A promising material for use in biofuel cells. *Electrochem. Commun.* 2010; 12: 1509-12.

[153] Montemagno CD, Schmidt JJ, Tozzi SP. *Biomimetic membranes.* US7208089 2007.

[154] Catchpole KR, Polman A. *Photovoltaic cell with surface plasmon resonance generating nano-structures.* EP2109147 (A1) 2009.

[155] Lee WH. *Method and structure for thin film tandem photovoltaic cell.* WO2010039727 (A1) 2010.

[156] Olson JM. *Multi-junction photovoltaic cell with nanowires.* WO2010120233 (A2) 2010.
[157] Wieting RD. *Patterning electrode materials free from berm structures for thin film photovoltaic cells.* US7863074 2011.
[158] Lagally M, Liu F. Graphite-based photovoltaic cells. US7858876 2010.
[159] Argo B, Vidu R, Stroeve P, *et al. Nanostructure and photovoltaic cell implementing same.* US7847180 2010.
[160] Sivaram S, Agarwal A, Herner SB, Petti CJ. *Method to form a photovoltaic cell comprising a thin lamina.* US7842585 2010.
[161] Němec H, Galoppini E, Imahori H, Sundstrom V. Solar energy conversion-Natural to artificial. *Comprehensive Nanoscience and Technology* 2011; 2: 325-59.
[162] McConnell I, Li GH, Brudvig GW. Energy conversion in natural and artificial photosynthesis. *Chemistry and Biology* 2010; 17: 434-47.
[163] Nam WH, Lee KS, Kim DH, *et al. Method of manufacture Ni-doped* TiO.sub.2 nanotube-shaped powder and sheet film comprising the same. US7867437 2011.
[164] Cao GZ, Zhou XY, Liu J, *et al. Aggregate particles of titanium dioxide for solar cells.* WO2011005440 (A2) 2011.
[165] Nakayama K, Kubo T, Nishikitani Y, Masuda H. *Nanotube-shaped titania and process for producing the same.* US7687431 2010.
[166] Parsons G. *Nano-structured photovoltaic solar cell and related methods.* US7655860 2010.
[167] Park JT, Roh DY, Patel R, Son KJ, Koh WG, Kim JH. Fabrication of hole-patterned TiO_2 photoelectrodes for solid-state dye-sensitized solar cells. *Electrochim. Acta.* 2010; 56: 68-73.
[168] Park H, Kim WR, Jeong HT, Lee JJ, Kim HG, Choi WY. Fabrication of dye-sensitized solar cells by transplanting highly ordered TiO_2 nanotube arrays. *Sol. Energy Mater. Sol. Cells* 2011; 95: 184-9.
[169] Lin MC, Tzeng YC, Lan SM, *et al.* InN/InP/TiO.sub.2 photosensitized electrode. US7655575 2010.
[170] Jo SM, Kim DY, Jang SY, *et al. Dye-sensitized solar cell with metal oxide layer containing metal oxide nanoparticles produced by electrospinning and method for manufacturing same.* EP2031613 (A2) 2009.
[171] Holliman P. *Low temperature sintering of dye-sensitised solar cells.* WO20101304476 (A1) 2010.
[172] Yamaguchi T, Tobe N, Matsumoto D, Nagai T, Arakawa H. Highly efficient plastic-substrate dye-sensitized solar cells with validated conversion efficiency of 7.6%. *Sol. Energy Mater. Sol. Cells* 2010; 94: 812-6.
[173] Wessels K, Wark M, Oekermann T. Efficiency improvement of dye-sensitized solar cells based on electrodeposited TiO_2 films by low temperature post-treatment. *Electrochim. Acta.* 2010; 55: 6352-7.
[174] Roscheisen MR, Sager BM, Pichler K. *Photovoltaic devices fabricated from nanostructured template.* US7605327 2009.
[175] Kayama S, Tanaka J. *Ultrafine particulate titanium oxide with low chlorine and low rutile content, and production process thereof.* US7591991 2009.
[176] Sepeur S, Frenzer G, Oliveira PW. *Particles or coating for splitting water.* WO2008154894 (A2) 2008.

[177] Lewis AE. Review of metal sulphide precipitation. *Hydrometallurgy* 2010; 104: 222-4.
[178] Parida KM, Reddy KH, Martha S, Das DP, Biswal N. Fabrication of nanocrystalline LaFeO3: An efficient sol-gel auto-combustion assisted visible light responsive photocatalyst for water decomposition. *Int. J. Hydrogen. Energy* 2010; 35: 12161-8.
[179] Smolyakov GA, Osinski MA. *Efficient hydrogen production by photocatalytic water splitting using surface plasmons in hybrid nanparticles.* WO2011011064 (A2) 2011.
[180] Shelnutt JA, Miller JE, Wang ZC, Medforth CJ. Water-splitting using photocatalytic porphyrin-nanotube composite devices. US7338290 2008.
[181] Domen K, Hara M, Takata T, Hitoki G. Photocatalysts for decomposition of water by visible light. US6864211 2005.
[182] Bandyopadhyay A, Stöckel J, Min HT, Sherman LA, Pakrasi HB. High rates of photobiological H2 production by a cyanobacterium under aerobic conditions. *Nature Communications* 2010; **1:** 139.
[183] Fukuzumi S, Kishi T, Kotani H, Lee YM, Nam WW. High efficient photocatalytic oxygenation reactions using water as an oxygen. *Nature Chemistry* 2010; 3: 38-41.
[184] Wilkins FC, Jha AD, Ganzi GC. *Method and apparatus for desalination.* US7744760 2010.
[185] Logan B. *Desalination devices and methods.* WO2010124079 (A2) 2010.
[186] Jones R. What can biology teach us? *Nature Nanotechnology* 2006; 1: 85-6.
[187] Tran B. *Nano-electronic array.* US7864560 2011.
[188] Ballantine AW, Kirchhoff DC, McElroy JF, Gordon MP. *Separator plates, ion pumps, and hydrogen fuel infrastructure systems and methods for generating hydrogen.* US7686937 2010.
[189] Pennisi CP, Greenbaum E, Yoshida K. Analysis of light-induced transmembrane ion gradients and membrane potential in photosystem I proteoliposomes. *Biophys. Chem.* 2010; 146: 13-24.
[190] Uphues U. *Methods and apparatus to facilitate lubrication of components.* US7811057 2010.
[191] Strangman T, Raybould D, Jameel A, Baker W. Damage mechanisms, life prediction, and development of EB-PVD thermal barrier coatings for turbine airfoils. *Surf. Coat. Technol.* 2007; 202: 658-64.
[192] Lee CK. Corrosion and wear-corrosion resistance properties of electroless Ni-P coating on GFRP composite in wind turbine blade. *Surf. Coat. Technol.* 2008; 202: 4868-74.
[193] Ngoc TP, Wouterson EM. *Wind turbine blade incorporating nanoclay.* GB2456484 (A) 2009.
[194] Hancock M. *Wind turbine blade reinforced with different types of carbon fibres.* GB2451192 (A) 2009.
[195] KallesØee E, Nysteen L. *Improved coating composition for wind turbine blades.* WO2010122157 (A1) 2010.
[196] Kennedy TJ, Presz JR, Werle MJ. *Coated shrouded wind turbine.* WO2010141867 (A2) 2010.
[197] Khalfallah MG, Koliub AM. Effect of dust on the performance of wind turbines. *Desalination* 2007; 209: 209-20.
[198] Karmouch R, Ross GG. Superhydrophobic wind turbine blade surfaces obtained by a simple deposition of silica nanoparticles embedded in epoxy. *Appl. Surf. Sci.* 2010; 257: 665-9.

[199] Dalili N, Edrisy A, Carriveau R. A review of surface engineering issues critical to wind turbine performance. *Renewable and Sustainable Energy Reviews* 2009; 13: 428-38.
[200] Subrahmanian KP, Dubouloz F. Adhesives for bonding wind turbine blades. *Reinforced Plastics* 2009; 53: 26-9.
[201] Serrano E, Rus G, García-Martínez J. Nanotechnology for sustainable energy. *Renewable and Sustainable Energy Reviews* 2009; 13: 2373-84.
[202] Steinberg PD, Charlton T, Afsar A. *Antifouling coating*. WO2008009067 (A1) 2008.
[203] Morse AP. *Wave energy harvesting and hydrogen-oxygen generation systems and methods*. US7872363 2011.
[204] Pujol T, Montoro L. High hydraulic performance in horizontal waterwheels. *Renewable Energy* 2010; 35: 2543-51.
[205] Muradov NZ, Veziroğlu TN. "Green" path from fossil-based to hydrogen economy: An overview of carbon-neutral technologies. *Int. J. Hydrogen. Engergy* 2008; 33: 6804-39.
[206] U.S. Energy Information Administration. Assumptions to the Annual Energy Outlook 2002. U.S. *Department of Energy* EIA 0554 2002; 68.
[207] Ilić R, Skvarč J, Golovchenko AN. Nuclear tracks: present and future perspectives. *Radiation Measurements* 2003; 36: 83-8.
[208] Van Goethem G, Hugon M, Bhatnagar V, Manolatos P, Deffrennes M. Euratom innovation in nuclear fission: Community research in reactor systems and fuel cycles. *Nucl. Eng. Des.* 2007; 237: 1486-1502.
[209] Liviu PS. Micro-structured nuclear fuel and novel nuclear reactor concepts for advanced power production. *Progress in Nuclear Energy* 2008; 50: 539-48.
[210] Miao P, Odette GR, Yamamoto T, Alinger M, Klingensmith D. Thermal stability of nano-structured ferritic alloy. *J. Nucl. Mater.* 2008; 377: 59-64.
[211] Mansur LK, Ishino S, Konings RJM, Lemaignan C, Werme L. The half-century of nuclear materials. *J. Nucl. Mater.* 2010; 400: 1-2.
[212] Macaskie LE, Mikheenko IP, Yong P, *et al.* Today's wastes, tomorrow's materials for environmental protection. *Hydrometallurgy* 2010; 104: 483-7.
[213] Toschi R. Nuclear fusion, an energy source. *Fusion Eng. Des.* 1997; 36: 1-8.
[214] Jordanova J, Fischer U, Pereslavtsev P. Nuclear performance analyses of the HCLL breeder blanket for a fusion power reactor. *Fusion Eng. Des.* 2009; 84: 985-8.
[215] Kondo K, Murata I, Klix A, Seidel K, Freiesleben H. Problems of lead nuclear data in fusion blanket design. *Fusion Eng. Des.* 2009; 84: 1076-80.
[216] Petersson P, Kreter A, Possnert G, Rubel M. Nuclear micro-beam analysis of deuterium distribution in carbon fibre composites for controlled fusion devices. Nuclear Instruments and Methods in Physics Research Section B: *Beam Interactions with Materials and Atoms* 2010; 268: 1833-7.
[217] D'haeseleer WD. The importance of fusion development towards a future energy source. *Fusion Eng. Des.* 2003; 66-68: 3-15.
[218] Clarke JF, Haas GM. The U.S. program for fusion nuclear technology development. *Fusion Eng. Des.* 1989; 8: 17-22.
[219] Mori S. Japanese program for fusion nuclear technology development. *Fusion Eng. Des.* 1989; 8: 9-16.
[220] Tanaka S. Overview of research and development activities on fusion nuclear technologies in Japan. *Fusion Eng. Des.* 2006; 81: 13-24.

[221] Andreani A, Diegele E, Gulden W, *et al.* Overview of the European Union fusion nuclear technologies development and essential elements on the way to DEMO. *Fusion Eng. Des.* 2006; 81: 25-32.
[222] Strebkov Y, Belyakov V. Overview of fusion nuclear technology in Russia. *Fusion Eng. Des.* 2002; 61-62: 47-59.
[223] Liu Y. Overview of fusion nuclear technology activities at SWIP. *Fusion Eng. Des.* 2010; 85: 2125-31.
[224] Meier WR, Raffray AR, Kurtz RJ, *et al.* Findings of the US research needs workshop on the topic of fusion power. *Fusion Eng. Des.* 2010; 85: 969-73.
[225] Hunt J, O'Connor G. Hybrid fusion, the third nuclear option. *The New Scientist* 2010; 205: 22-3.
[226] Passoni M, Dellasega D, Grosso G, Conti C, Ubaldi MC, Bottani CE. Nanostructured rhodium films produced by pulsed laser deposition for nuclear fusion applications. *J. Nucl. Mater.* 2010; 404: 1-5.
[227] Gao JF, Suo JP. Proposal of self-healing coatings for nuclear fusion applications. *Surf. Coat. Techol.* 2010; 204: 3876-81.

In: Renewable Energy
Editor: Andreas Poullikkas

ISBN: 978-1-62618-231-8

Chapter 6

STOCHASTIC SOLUTIONS FOR THE UNIT COMMITMENT OF POWER SYSTEMS WITH WIND POWER GENERATION

Irina Ciornei [*] ***and Elias Kyriakides***
KIOS Research Center for Intelligent Systems and Networks,
University of Cyprus, Cyprus

ABSTRACT

This chapter focuses on the comparison of two main classes of stochastic unit commitment formulations in the presence of uncertainty due to wind power generation: the scenario based stochastic unit commitment and the robust stochastic unit commitment.

Both methods have an explicit representation of volatility/ stochasticity of wind power production. The advantages and disadvantages of each method will be discussed with an emphasis on their practical implementation for the TSO or the market operator.

The analysis focuses on uncertainty characterization when wind power generation is to be considered and a discussion on the choice of objectives and constraints and their implications on problem dimension and data requirements. An evaluation of the stochastic solutions and their implications on a day-ahead market will be provided.

Keywords: Scenario based stochastic unit commitment, robust stochastic unit commitment, optimization under uncertainty, robust sets, scenario reduction, power systems, wind generation

INTRODUCTION

Wind power generation is by far the fastest growing technology absorbed by the power system sector in the last decade [1].

---

[*] Corresponding Author: Email: ciornei.irina@ucy.ac.cy.

This rate of growth is due to the constant decrease in its levelized cost, which is now comparable with some of the classical thermal power generating technologies such as gas turbines [2]. Further, the commitment of an increasing number of governments around the world towards clean energy solutions contributed significantly to this growth.The practice for the integration of wind energy into power systems varies depending on the market rules and the system conditions (e.g., highly meshed vs. weakly meshed networks, interconnected vs. isolated systems). This chapter will concentrate on the effect that wind generation has on the unit commitment (UC) problem in power systems.

Besides the benefits of wind energy, compared to the conventional soures of energy, wind power generation exhibits certain undesireble technical drawbacks, such as generation variability with high uncertainty over medium to large time span horizons (e.g., more than 36 hours) and lack of load following capability and reserve provision (this is still under discussion for small time periods close to the delivery time).

To better understand the impact that wind power generation can have while operating synchronously with the traditional power network, a brief introduction into the main functions of power system operation should be provided.

Figure 1 presents the basic operation functions of a modern power system, which are usually based on a time separation and distribution of commands (subsystem operation) in a hierarchical manner, as can be seen in Figure 2. This is mainly due to the fact that the energy cannot be stored in large amounts, and therefore different levels of forecasting information are needed to be incorporated as knowns in different operation and control functions of the whole system. The diagram in Figure 1 describes the control and monitoring functions in power system operation in the form of subsystems. Three subsystems are distinguished and they are described below.

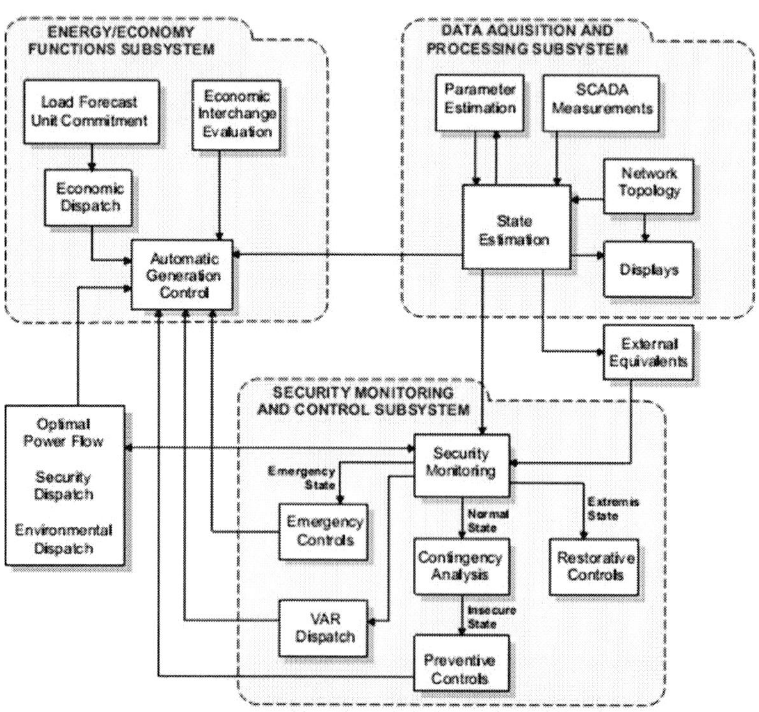

Figure 1. Functional diagram of a modern energy management system [3].

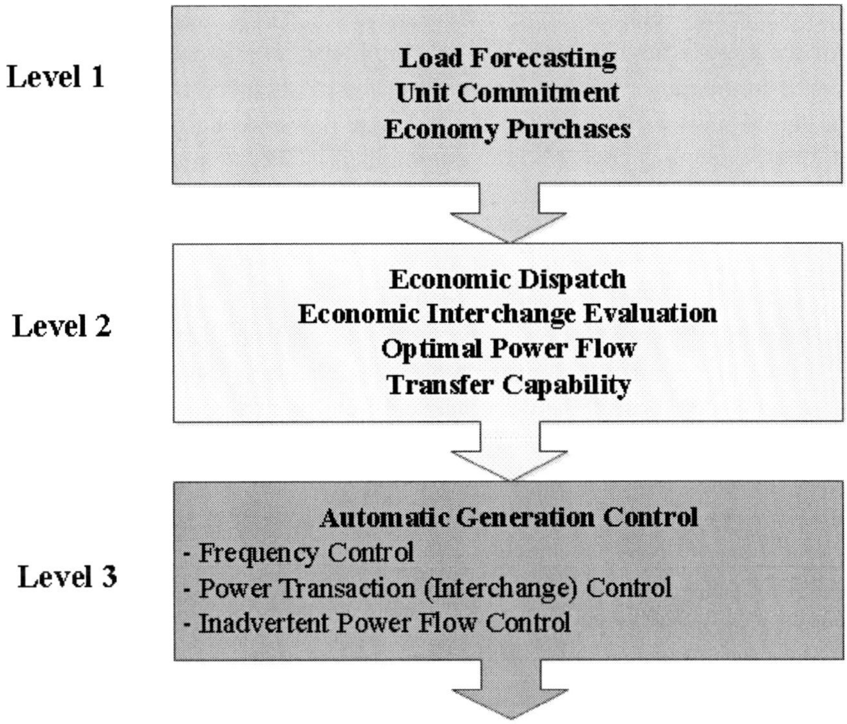

Figure 2. Hierarchical operation of a power system.

Energy/economy subsystem: Its main objective is to economically dispatch the generating units such that to satisfy at all times the forecasted load demand (load forecasting, unit commitment and economic dispatch), while corrections for the deviations from the predicted load value are also handled (AGC); the action of this subsystem spans a time interval of days or weeks.

Data acquisition and processing subsystem: Its main objective is the proper monitoring of the system (e.g., parameter estimation, SCADA measurements, and state estimation); the time action of this subsystem spans intervals of seconds to ten minutes.

Security monitoring and control subsystem: Its main objective is the secure and reliable operation of the system in real time; the action time is from milliseconds to seconds.

All of the above subsystems interact with each other, making the operation of the entire power system a complex task. For example, one erronous undetected signal in one subsystem may triger fault actions/signals in the other interconnected subsystems.

Typically, the power generation problem is based on three different sets of decisions which are dependent on the length of the planning time horizon. The first set consists of the long-term decisions (years) where the decision variables to be determined are the capacity, type, and number of power generators (units) to own. In the medium term (days to months), one needs to decide how to schedule (commit) the existing units for the planning horizon. And finally, in the short term (minutes to hours), the goal is to efficiently determine the amount of power that each committed unit needs to produce in order to meet the real-time electricity demand. The long-term problem is identified as the power expansion problem, the medium-term problem is identified as the unit commitment (UC) problem, and the short-term problem is called the economic dispatch (ED) or generation allocation problem. Note that the mid-term problems may refer also to the maintenance scheduling, when the time horizon is in the range of one year. In this case, the short-term problems refer to both UC and ED, and their time horizon is in the range of minutes to weeks. The ED function determines the economic base points for all units participating in the real time power production as determined from the higher order scheduling function (unit commitment). These output values of the ED are then used by the Automatic Generation Control (AGC) loop to control the units toward these values in real time, thus achieving economic operation. Economic dispatch is typically performed every five to thirty minutes.

One of the main challenging aspects of power systems operation is that electrical energy cannot be economically stored in significant amounts. Thus, electrical power must be consumed at the same moment it is generated. For a reliable supply of power it is therefore essential to maintain the balance between demand (total system load including transmission and distribution line losses) and generation. It is in principle possible to maintain the power balance by adjusting both generation and demand, but historically, mostly the central generation units have been used to follow the demand at all times. The operation of power systems is therefore critically dependent on the capabilities of generators for balancing the load. In reality, one may say that any electric power system is never at its equilibrium. This happens due to the stochastic nature of the demand, due to unplanned equipment failures, or due to stochastic power generation.

Due to the diversity in the characteristics and operation properties of different power generation units in a power system network, scheduling of these generation units is needed based on load forecasts and the economics and technical characteristics of the available generation units. This involves the calculation of the optimal selection of units for power generation (UC) and the optimal loading of the selected generators for subdivisions of time of the initial commitment time span (ED).

Important parameters in unit commitment include start-up and shut-down costs, minimum up and down-times and operating cost. The economic dispatch performs the actual distribution of the total load between committed units, which is optimized for each operating state while taking into account all economic and technical aspects of the units. The cost component which is the most important in the economic dispatch case is the operating cost. The operating cost is dominated by the fuel cost, although labor is also a key component. Thus, the goal of power system economic dispatch is to minimize the fuel cost (or in other words, to maximize system efficiency and minimize system losses that cannot be billed or passed on to customers). From the output of UC and ED an estimation of the associated use of fossil fuels and emission of greenhouse gases can be calculated as well. In summary, in order to reach the goal of economic dispatch, it is important to develop relationships between the cost of power output and the operating costs.

Unit commitment and economic dispatch problems rely on forecasted values for the load demand determined for each dispatch interval. When stochastic power generation such as wind is part of the power system portfolio, the UC and ED problems are highly challenging due to the variability and limited predictability of this generation, which come on top of existing variations and uncertainties of the load. For the operation of power systems with significant amounts of stochastic renewables, the importance of conventional generation will remain or may increase even further in order to guarantee a reliable power supply.

In the last decade several impact studies of wind power integration into the main grid had been performed mainly in Europe and the US. Based on the attitude of the operators and regulators in the EU-27 countries, it seems that there is a pro-renewables stance. A survey report that examined grid integration issues has pointed out that this pro-renewables stance may be due to the so far relatively low penetration of renewables in electricity generation [4]. The absence of curtailment rules and of appropriate compensation schemes in most countries may be an issue in the near future.

Renewable generation impacts electricity markets for both technical and economical reasons. The variability in generation makes renewables more difficult to participate in load following or to readily react to market signals. On the economical part, the operating cost of renewables (especially wind and solar) is significantly lower than that of the conventional sources of energy. This low marginal cost of renewables, creates confusion in the marginal cost-based market model.

Wind generation also impacts the reserve requirements of the power system. However, there is still no clear answer to the amount of reserve that will need to be bought or allocated due to the variability of wind generation. If, for example, the wind power forecasting would be used to determine the reserve requirement to be purchased on the day-ahead market, then most likely this would be a costly solution. In other words, the reserve requirement takes energy out of the energy market merit order (EMMO) and so it will have an impact on the energy price (direct and indirect price signal) as a greater load following component is added to the product. Also, the reserve requirement may be more technically rigorous than required to resolve the ramping concerns due to wind power forecast deviations (higher costs, fewer players due to up/down requirements, units have to be online and synchronized, may reduce the total supply as units must stay within a given range to provide the reserve requirement which may also end up in units operating at partial load that are less efficient. This has a cost in terms of total system efficiency).

According to the "All Island Grid Study" [5] in the presence of increasing wind power capacity installed, the generation from conventional units tends to decrease. In the same report, the power reserve was also considered. The rule used for the purposes of the report states that there should be enough spinning reserves to cover an outage of the largest unit, in combination with a fast decrease in wind generation.

Electricity Market Influence on the UC/ED Structure

Deregulation (other accepted terms are "regulation" or "re-regulation") in the power sector imposed a number of changes not only at the ownership level, but also at the functional/operational level of the power system. Thus, at the economical/trading level, competitive electricity markets facilitate generators and customers (direct power consumers or retailers) to freely contract electricity according to the terms of the contracts (e.g., stipulated price and quantity for periods of time). However, at the physical level, the actual delivery of the electricity is over a constrained transmission network controlled by a system operator.

The deregulation in the power sector also brings a new way of operating the system. Thus, the following new independent bodies were created to ensure independence and fairness in the electricty market:

- *Market Operator* (MO): It is responsible for facilitating a fair competition, transparent and effective access of all market participants to the network and for the scheduling of the generating units (e.g., day-ahead unit commitment). The MO can take different formats such as Market Agents Committee (Latin America), Independent System Operator (USA), electricity pool (Ireland), and Power Exchange (most systems in Europe and Australia).
- *System Operator* (SO, TSO/ISO): It is responsible for managing the network and assuring the security and continuity of power supply.

The Market Operator manages the (commercial) exchange between producers and consumers and supports the UC/ED calculation by matching the market offers and bids until the "gate closure" time at which the SO considers to be so close to real time delivery of power that the scheduling can only be modified/adjusted in a centralized way. "Gate closure" is the time when participants reach their final production schedule, after bilateral trading, spot market (or day-ahead market) and adjustment of the schedules (eventually in the intraday markets).

At the structural and operational levels one can distinguish three major phases: (i) installation of additional generation capacity (e.g., to replace obsolete facilities and to meet demand growth) (ii) scheduling of existing generation capacity (i.e., maintenance management, economic dispatch) and (iii) shorter-term response of generators to meet the actual load (i.e., the use of operating reserves to compensate for unexpected outages or deviations from the predicted generation/demand levels). In all three phases, regulation has to find the right balance among economic efficiency, reliability of supply and environmental and social impact.

The responsibility for decision making for the first two phases may be centralized in some governmental agency, or taken by vertically integrated utilities (subject to supervision), or left to the initiative of private investors. In all cases the System Operator is in charge of the third phase.

In the above context, one may talk about UC/ED at local (generation ownership) level or about the UC/ED at regional or whole market level which is typically performed by independent bodies such as Market Operators (MO) and/or Transmission System Operators (TSO), independent (I)SO, or groups of TSOs/ISOs.

At the generation ownership level one may notice that the UC/ED formulation changes in the sense that the objective of the generating company may be either profit increase or cost minimization under uncertain demand.

At regional/whole market level one may notice that the UC might be performed centrally (following similar structure as in the vertically integrated systems) while certain hierarchy-temporal steps, due to the influence of the market design, need to be followed (e.g., commitment decisions in a day-ahead market performed by the MO and decisions adjustments in the interval between "closure-to-gate" time and actual delivery of electricity, performed by the TSO/ISO).

For a better understanding of the market design influence on the UC/ED structure, an example of the market sequence in the EU electricity market (adapted from [6]) is given in Figure 3.

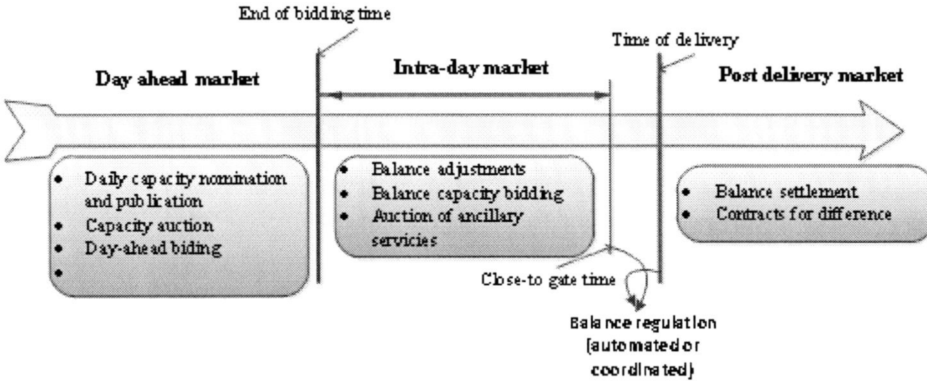

Figure 3. NordPool time division of trading and operation procedures [6].

One of the main differences between the European Power Exchange and the ISO markets is based on the dispatch procedure. European Power Exchange empowers an unconstrained, pure energy market merit order (EMMO) procedure, where the bid and offer curves are balanced without considering the constraints of the system, while ISO markets are based on the Locational Marginal Prices (LMP) concept, where transmission system constraints are allowed. The locational marginal price (LMP) is defined as the cost of purchasing one unit of power at a certain location of the system (bus) at a certain time. It is specific to ISO markets (USA). Technically speaking, LMP is a voluntary (bid-based) security-constrained, economic dispatch that determines energy and transmission congestion prices at specific points based on marginal generation costs (MGC), also known as incremental cost (IC). In other words, LMP is the computational model that determines the optimal generation dispatch, as well as locational energy and transmission congestion prices (security constrained economic dispatch - SCED). The main conceptual differences between the SCED and the LMP market model are: the "offer prices" (generally referred to as "bids") of each generator are substituted for "operating cost" to determine unit dispatch; all customers at a specific point on the grid pay the price of the generator that is dispatched to serve the next MW of load at that point, as affected by local bids and transmission congestion; and the costs for any transmission congestion are directly assigned to customers within the specific location served by the constrained transmission line(s).

Scenario Based Stochastic Unit Commitment

The short term scheduling in power system operation (unit commitment problem) has been an active field of research during the past two decades [7] and four main lines of research have been identified [8]:

(i) development/improvement of optimization algorithms
(ii) inclusion of new modeling features
(iii) consideration of uncertainty
(iv) management of the risk of generation companies

Typically, the unit commitment is performed once for a period of one day, a few days, or one week. The unit commitment decisions are typically under the Market Operator (MO) actions of a day-ahead market clearing. Commitment and generation level decisions are then updated only when necessary. For example, if some substantial and unpredicted changes have occurred (e.g., in the case of wind power generation, deviations from the prediction are most likely to happen), then these decisions are updated under the supervision of the TSO/ISO in the interval between closure-to gate and the actual real time delivery of electricity.

A generic deterministic formulation of the UC problem is given below:

$$\begin{aligned}
&\min_{x,y} c^T x + b^T y && (1) \\
&\textit{subject to:} \quad Fx \le f && (2) \\
&\qquad Hy \le h && (3) \\
&\qquad Ax + By \le g && (4) \\
&\qquad I_u y = \bar{d} && (5) \\
&\qquad x \text{ binary}
\end{aligned} \qquad \text{(P1)}$$

where,

$x$are the first stage decision variables (binary commitment decisions)

y are the second stage (dispatch decisions) (real positive values)

$c^T x$ – includes start-up/shutdown cost, no-load cost for large start-up delay thermal units

$b^T y$ – start-up decision for fast units, dispatch cost, y is the dispatch decision variable and takes real values

(2) - refers to the UC constraints (e.g., minimum up/down time, start-up time)

(3) – dispatch constraints such as balance constraint, reserve ramping constraints

(4) – coupling constraints such as max/min generation capacity

I_u - selects the components from y that correspond to uncertain resources

$\bar{d}$– is the expected value of the uncertain resources (deterministic)

A stochastic programming (SP) problem is a mathematical programming problem, with values of some random/stochastic parameters replaced by distributions. In order to solve such a problem, one needs: a model describing the problem, values of the deterministic (known) parameters, an explicit description of the stochasticity (known distributions, described by densities and/or CDFs, historical data, or some properties of the distributions (e.g., moments). In practice however, the stochastic programming approach can handle only discrete samples of limited size. Therefore, one needs to approximate the distribution. This distribution approximation is often called in the literature a scenario tree, as given in Figure 4. A scenario tree is characterized by scenarios (a path from the root to one leaf – presented in red in Figure 4), stages (a moment in time, when decisions are taken – dashed vertical lines) and periods (a time interval between two stages).

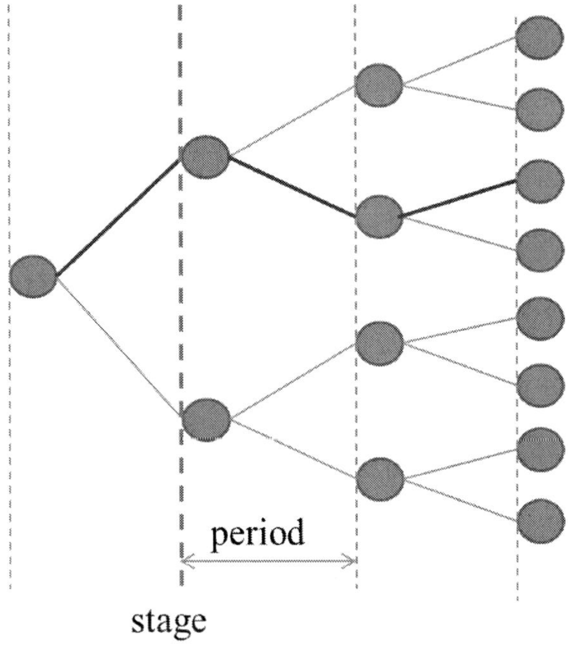

Figure 4. Scenario Tree.

As an example, given the number of uncertain parameters in the system being M (e.g., the number of wind farms), the number of stages being T (e.g., the number of hours in SCUC), and the number of outcomes of each uncertain parameter being N (e.g., one wind farm may have N=2 possible outcomes at one stage), then the number of Scenarios $S = N * (M^T)$ (in the example given in Figure 4 N=1, M=2, T=3 and therefore S=8). The assumption in this calculation is the independence between wind speeds at different wind farms. It can be noticed that the number of scenarios grows exponentially (however it relates to the very conservative assumptions of the number of wind farms and the independence of wind speeds).

A stochastic representation of the UC formulation is given below:

$$
\begin{aligned}
&\min_{x,y} c^T x + b^T y \\
&\textit{subject to:}\quad Fx \leq f \\
&\qquad\qquad\quad Hy \leq h \qquad\qquad\qquad\qquad\qquad\qquad (P2) \\
&\qquad\qquad\quad Ax + By \leq g \\
&\qquad\qquad\quad I_u y = \mathrm{d}(\omega) \qquad\qquad (6) \\
&\qquad\qquad\quad x \text{ binary}
\end{aligned}
$$

where,

d is the total net load (total load demand minus wind generation). It is a random variable denoted by a number of possible realizations $d(\omega)$ (scenarios) of the load demand variable.

The most challenging part in this represenatation is the generation of scenarios, and the choice of reduction techniqes in order to cope with the computational burden and convergence of the optimization solution. The following section of the chapter gives a step by step implementation procedure in this respect.

MODELING WIND POWER UNCERTAINTY

Theoretically, the stochastic processes of wind power can be described by joint probability distribution functions (p.d.f.) which comprise both the marginal and conditional distributions. However, in practice the discrete approximation is to be considered.

Besides the instantaneous probabilistic representation of the wind power production, it should be noted that wind power forecast is generally characterized by the cone of uncertainty (smaller uncertainty with shorter lead time) as presented in Figure 5.

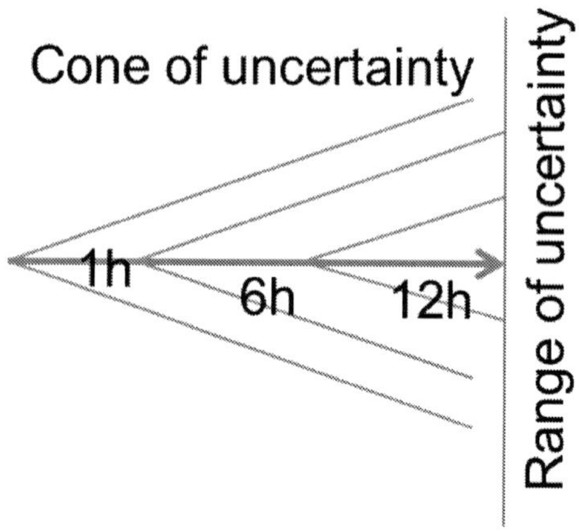

Figure 5. Wind speed uncertainty characteristic.

Techniques to reduce randomness (the range of uncertainty) imply aggregation of wind forecasts over the entire special configuration of the system if the level of correlation among wind plants is low [9]. However, when network constraints are to be considered (location matters) aggregation may not be applied.

The wind speed signal generator used in the current analysis is an adaptation of an autoregressive moving average (ARMA) model proposed in [10] and it can be described as follows:

$$V_{wind}(t) = \overline{V}_{mean} + V_{turbulent}(t)$$

where,

$V_{wind}(t)$ is the wind speed time series;
$\overline{V}_{mean}$ is the mean wind speed at hub height;
$V_{turbulent}(t)$is the instantaneous turbulent part, described by:

$$V_{turbulent}(t) = \delta_t V_t$$

$$V_t = aV_{t-1} - bV_{t-2} + cV_{t-3} + \alpha_t - d\alpha_{t-1} + e\alpha_{t-2}$$

where,

δ_t is the standard deviation
a, *b* and *c* are the autoregressive parameters
e and *d* are the moving average parameters

The Simulink implementation and its outcome for the above ARMA wind speed model is provided in Figure 6 and Figure 7, respectively.

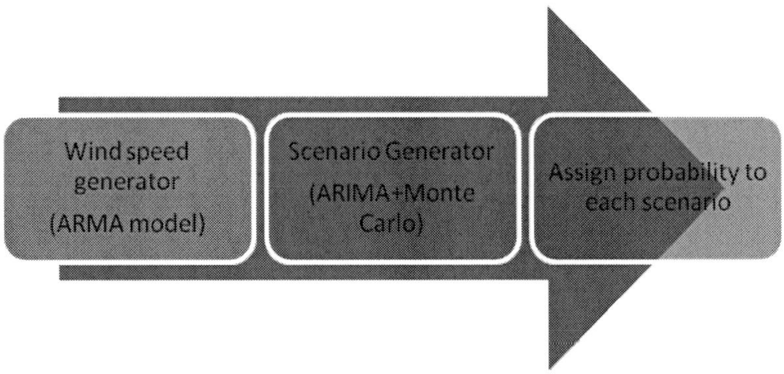

Figure 6. Scenario generation: a flow for practical approach.

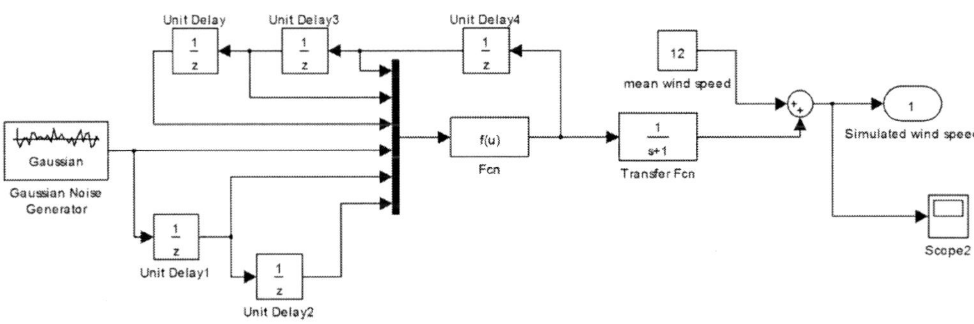

Figure 7. Simulink ARMA model for the wind speed generator.

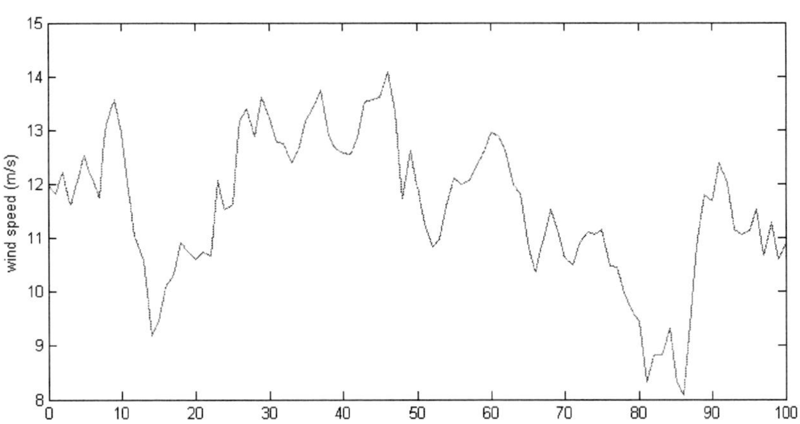

Figure 8. Output of the ARMA wind speed model.

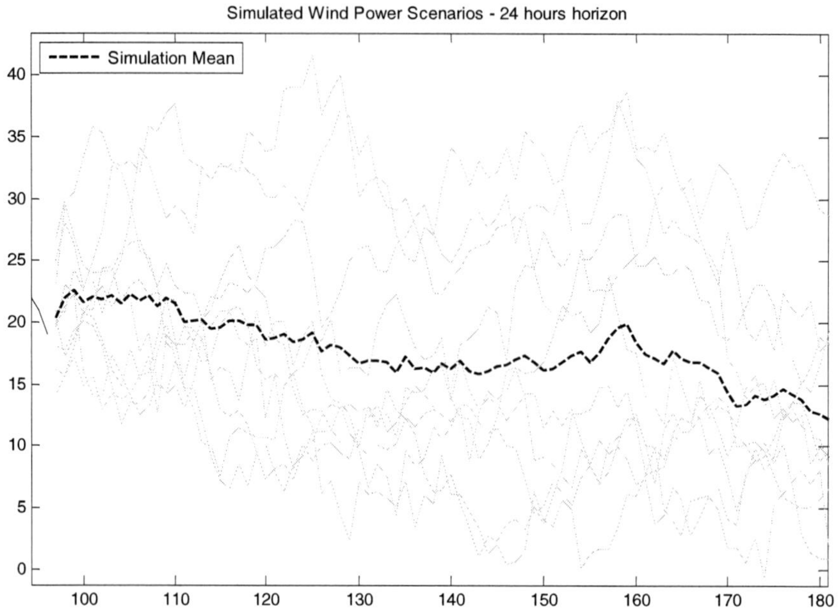

Figure 9. Scenario Generator output.

The second module, ARIMA and Monte Carlo Scenario Generator uses the ARIMA toolbox of Matlab and a Monte-Carlo sampling technique to generate possible wind power outcomes.

The input of this module is the ARMA wind speed generator module and the wind-to-power curve of a generic wind turbine. A sample of 10 generated scenarios for a 24-hour wind power generation is given in Figure 9.

Creation of the Scenario Tree: Scenario Reduction

In order to make stochastic programming useful (applied stochastic programming models), the number of starting simulated scenarios shall be large enough to cover as many possibilities as possible, while the number of scenarios that can be feasibly handled by any commercial or open source solver is much smaller. In the literature, different types of scenario reduction techniques were proposed ranging from stochastic approaches to more statistical rigorous analyses.

The idea of reducing a given set of scenarios and constructing a scenario tree out of the reduced scenario subset is that the reduced scenario tree in a given sense is still a sufficient approximation of the original one. Below, a step by step procedure of a scenario reduction technique adapted from [11, 12] is presented.

The scenario reduction technique to be presented is based on the assumption that all scenarios have a common root in a one-stage tree where branching occurs only after the root node.

A scenario $\omega^{(i)} = \{\eta_0^i, \dots, \eta_{nT}^i\}$ is defined as a sequence of nodes ($\{\eta_k^i\}$) in the scenario tree starting from the root node ($\eta_0 = \eta_0^i$) and ending at one leaf (η_{nT}^i); where, *i*=1..*ns* is the number of scenarios, *k*=1...*nT* is the number of stages in the multi-stage scenario tree. Each node η_k^i is characterized by a given parameter vector $\boldsymbol{x}_k^i = \{x_{k,1}^i, .., x_{k,np_k}^i\} \in \boldsymbol{R}^{np_k}$ with np_k elements. Each scenario has a probability of occurrence which is defined as the product of all probabilities to transition from one stage to another within the scenario. In other words, if $\pi_{j,j+1}^{(i)}$ is the probability to transition from stage *j* to stage *j*+1 within the scenario $\omega^{(i)}$, then

$$\pi^{(i)} = \prod_{j=0}^{nT-1} \pi_{j,j+1}^{(i)}$$

the probability of scenario $\omega^{(i)}$ is denoted as.

The scenario reduction technique is based on the principle that the scenarios which are more close to each other and have the smallest probability to occur are to be deleted until a predefined number of scenarios is reached. The measuring distance between scenarios is defined by the Euclidean norm:

$$d\left(\omega^{(i)}, \omega^{(j)}\right) = \left(\sum_{k=0}^{nT} (\boldsymbol{x}_k^i - \boldsymbol{x}_k^j)^2\right)^{1/2}$$

The steps are as follows:

1) Repeat until the desire number of scenarios is reached:
   a) Delete the scenario $\omega^{(i^*)}$ from the pair of *i*-th and *j*-th scenarios having the minimum distance and the smallest probability
      $\min_{i \neq j} \pi^{(i^*)} d\left(\omega^{(i^*)}, \omega^{(j)}\right), \ \forall i, j = 1..ns,$
   b) Reduce the number of scenarios to *ns*-1
   c) Change the probability of the remaining scenario $\omega^{(i)}$ of the pair $(\omega^{(i^*)}, \omega^{(j)})$ with the found minimum distance to $\pi^{(j)} = \pi^{(i^*)} + \pi^{(j)}$
2) Creation of the multi stage of the scenario tree: backward approach
   a) Initialize the total number of scenarios and the number of stages of the final scenario tree, and calculate the number of branches per node
   b) Inititialize a one-stage tree acording to the predefined number of scenarios (Figure 10). Note that the initial one-stage stays the same for the following time period (the later stages of the multi-stage tree to be created).
   c) Set *n*=*nT* (start backward from the leafs' stage)
   d) *Rep*eat the following until *n*=1;
      i. Determine the series (subsets of scenarios) of admissible nodes

         $\omega_n^{(i^*)} = \{\eta_0, \dots, \eta_n^{(i^*)}\}$, with *i**=1...*ns*, whose inner nodes $\{\eta_i^{(i^*)}, \dots, \eta_{n-1}^{(i^*)}\}$, have to be deleted and afterwards newly set:

      ii. $\min_{i \neq j} \left\{\prod_{k=0}^{n-1} \pi_{k,k+1}^{(i^*)}\right\} d\left(\omega_n^{(i^*)}, \omega_n^{(j)}\right), \forall\, i, j = 1..ns$

iii. Delete the inner nodes $\{\eta_i^{(i^*)}, \dots, \eta_{n-1}^{(i^*)}\}$ of the scenario $\omega^{(i^*)}$ (the nodes lying on the way from the root to the node at the stage *n*-1). Note that the node $\eta_n^{(i^*)}$ of the current stage *n* of the scenario $\omega^{(i^*)}$ remains admissible.

iv. Determine the series of admissible nodes to replace the deleted nodes above according to the minimum distance between $\eta_{n-1}^{(i^*)}$, the predecessor node of deleted inner series of nodes, and all the other scenarios' nodes of the same *n*-1 stage. Therefore, a new way from root to $\eta_n^{(i^*)}$ is defined based on admissible nodes Let's say that the new series of admissible nodes from the root to *n*-1 satge of node $\eta_n^{(i^*)}$ is $\{\eta_0, \eta_1^{(j)} \dots, \eta_{n-1}^{(j)}\}$. This is a common path until stage *n*-1 to both nodes $\eta_n^{(j)}$ and $\eta_1^{(i^*)}$ in stage *n*, therefore a branching occurs between stage *n-1* and *n*.

v. Change probabilities to reach nodes $\eta_n^{(j)}$ and $\eta_n^{(i^*)}$ from their common predecessor $\eta_{n-1}^{(j)}$ to:

$$\pi_{n-1,n}^{(i^*)} = \frac{\pi^{(i^*)}}{\pi_{0,1}^{(i^*)} + \pi_{0,1}^{(j)}}$$

$$\pi_{n-1,n}^{(j)} = \frac{\pi^{(j)}}{\pi_{0,1}^{(i^*)} + \pi_{0,1}^{(j)}}$$

vi. Set *n*=*n*-1. If, for example, scenario S2 in Figure 10 is to have the inner nodes replaced in stage *n*-1 (=2) with the sequence of nodes coming from scenario S9, then, after one iteration, the scenario tree shall look like in Figure 11.

Given that the final scenario tree is formed and there are unique paths (scenarios) from the root node to the leaves, then the stochastic optimization problem (P2) can be rewritten in the equivalent deterministic form (P3) given below:

$$\min_{x,y} \boldsymbol{c}^T x + \textstyle\sum_{i=1}^{ns} \boldsymbol{\pi}^{(i)}\, \boldsymbol{b}^T \boldsymbol{y}(\omega^{(i)}) \qquad \text{(P3)}$$

subject to: constraints (2) to (5) from (P1).

Note that the second term in the objective function of (P3) reflects the expected value of the operation cost under all possible outcomes of the uncertainty in wind power production. Problem (P3) can then be transferred to any mixed integer programming solver (CPLEX, Gurobi, GAMS etc.) in order to obtain the solution. Note that the most critical/challenging part in modeling and solving the stochastic unit commitment is the generation of scenarios and the scenario reduction and creation of the scenario tree. In the next section, another

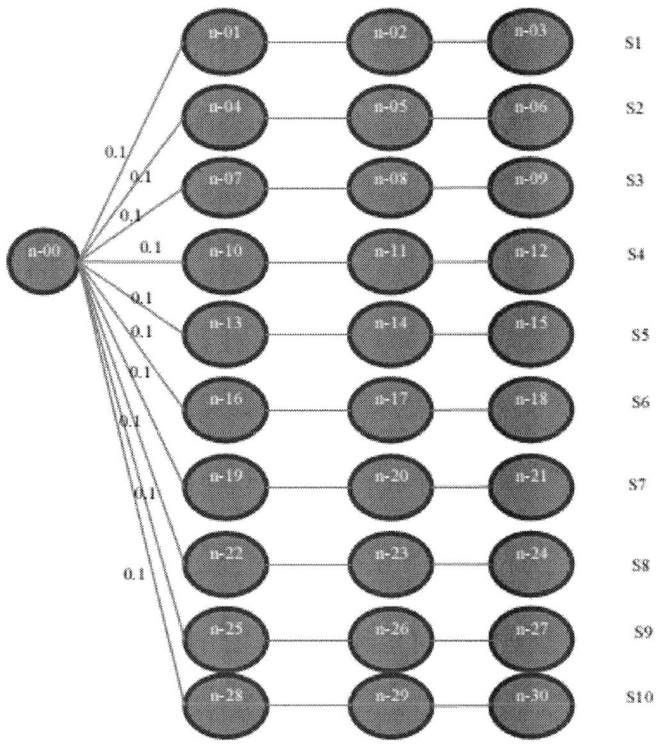

Figure 10. Example of one-stage initial tree of ten scenarios.

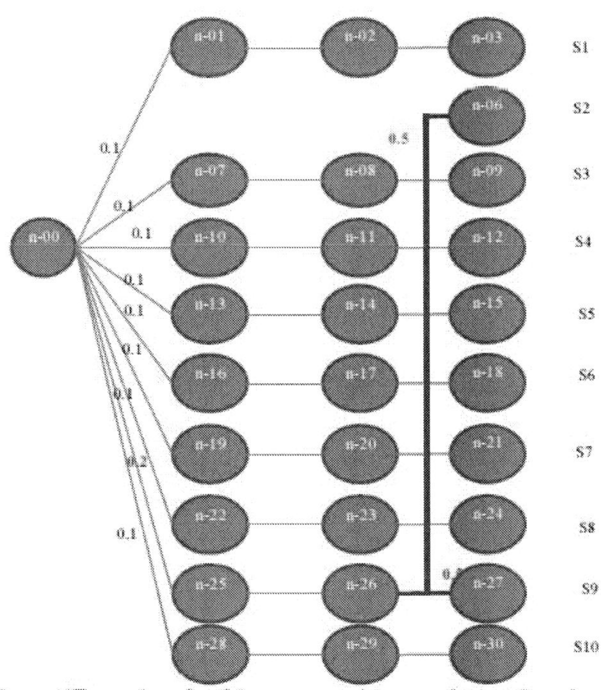

Figure 11. Formation of multi-stage scenario tree after one iteration.

approach, more recently launched [13], which is based on robust stochastic optimization (RO), is presented. This approach is also known as extreme risk adverse optimization. A step by step procedure on how to apply this method is also provided.

STOCHASTIC ROBUST UNIT COMMITMENT

Compared to stochastic optimization (SO) which starts by assuming that the uncertainty has a known probabilistic description, the robust optimization (RO) approach models uncertainty in a deterministic and set-based manner [13].The following questions are to be addressed when dealing with the stochastic robust unit commitment: (a) How to determine the appropriate uncertainty sets? (b) How much flexibility does the designer have in selecting the uncertainty sets? (c) What do these uncertainty sets say about probabilistic feasibility guarantees under various distributions for the uncertain parameters?

In brief, the stochastic problem (P3) is defined in RO terms as follows (P4):

$$\begin{aligned}
&\min_{x,y} c^T \boldsymbol{x} + b^T \boldsymbol{y} && \\
&\textit{subject to:} \quad Fx \leq f && (2) \\
&\qquad Hy \leq h && (3) \qquad\qquad (P4) \\
&\qquad Ax + By \leq g && (4) \\
&\qquad I_u y = D && (7) \\
&\qquad x \text{ binary – UC decision variable} &&
\end{aligned}$$

where,

D is the uncertainty (deterministic) set of nodal net injections (for each time period *t* in the planning horizon *T*. It can be noticed from (P4) that the robust approach impacts the formulation of (5) in (P1) and (P2) and takes the form given in (7, P4).

Let the uncertainty set of nodal net injections (e.g., wind farm power output) in each time period *t* in the planning horizon *T,* be defined by the triple $D^t(\overline{d^t}, \widehat{d^t}, \delta^t)$.

$$D^t\left(\overline{d^t}, \widehat{d^t}, \delta^t\right) = \left\{ \boldsymbol{d^t} \in R^{Nd}: \sum_{i \in Sd} \frac{\left|d^t - \overline{d^t}\right|}{\widehat{d_i}} \leq \delta^t \right\}$$

where,

- $d_i^t \in \left[\overline{d_i^t} - \hat{d}_i^t, \overline{d_i^t} + \hat{d}_i^t\right] \forall i \in Sd$ is the range of uncertainty
- Sd is the set of nodes that have uncertain injections
- Nd is the number of the elements in Sd (nodes with uncertain net injections (e.g., one wind farm)
- $\boldsymbol{d}_t = (d_i^t,\ i \in Sd)$ is the vector of uncertain net injections at time t (e.g., $\boldsymbol{d}^t = d_1^t$)
- $\overline{d_i^t}$ is the nominal net injection of node i at time t

- δ^t is "the uncertainty budget" parameter controlling the uncertainty set and takes values between 0 and Nd.

When, δ^t=0 is equivalent with $D^t = \{\overline{d^t}\}$ is a singleton, similar with the deterministic case. While when $\delta^t = Nd$ is equivalent with the case when D^t is the entire hypercube defined by the intervals for each d_i^t $\forall i \in Sd$.

Practical implementation values for δ^t are recommended to be set to ~$O(\sqrt{Nd}\,)$, according to the Central Limit Theorem.

The problem (P4) can then be rewritten in the equivalent form (P5) below:

$$\begin{aligned}
&\min_{x,y(\cdot)} \boldsymbol{c}^T\boldsymbol{x} + \max_{d\in D^t} \boldsymbol{b}^T\boldsymbol{y}(\boldsymbol{d}) && && \\
&\textit{subject to:}\quad Fx \le f && (2) && \\
&\qquad\qquad\quad Hy\,(d) \le h && (3) && \text{(P5)}\\
&\qquad\qquad\quad Ax + By(d) \le g && (4) && \\
&\qquad\qquad\quad I_u y\,(d) = d,\ \forall d \in D && (5) && \\
&\qquad\qquad\quad x \text{ binary – UC decision variable} && &&
\end{aligned}$$

The objective function in (P5) has two parts: the first part reflecting the unit commitment decisions (the cost of commitment) which are first to be determined for example in a day-ahead market context and the second part which reflects the worst case cost of the dispatch decisions (it can be seen as a second-stage decision). The worst case-scenario is represented as a maximization of outcome over the entire uncertainty set *D*. Note that in the above formulation, the unit commitment decision vector $\boldsymbol{x}$ also takes into account all possible uncertain outcomes represented in the uncertainty set. Therefore, the UC solution remains feasible/ robust for any realization of the uncertain outcome of wind.

As long as the second part of the objective function deals with the dispatch decisions (which state for the operation cost minimization over the dispatch horizon *T*), then (P5) can be rewritten in the form of (P6) below.

$$\begin{aligned}
&\min_{x,y(\cdot)} \boldsymbol{c}^T\boldsymbol{x} + \max_{d\in D^t} \min_{y\in\Omega(x,d)} \boldsymbol{b}^T\boldsymbol{y} \qquad \text{(P6)}\\
&\qquad\quad \text{s.t. } \boldsymbol{Fx} \le \boldsymbol{f}
\end{aligned}$$

where, $\Omega(x,d) = \{\boldsymbol{Hy} \le \boldsymbol{h}; \boldsymbol{Ax} + \boldsymbol{By} \le \boldsymbol{g}; \boldsymbol{I_u y} = \boldsymbol{d}\}$ is the set of feasible dispatch solutions for a fixed commitment decision $\boldsymbol{x}$ and the nodal uncertain net realizations *d* (which is then maximized over the uncertainty set *D*).

The dual max-min term in (P6) is by itself a hard to solve sub-problem. However, in order to solve (P6) a two level algorithm approach shall be used. The outer-level which intends to solve the commitment decision master problem also implies the use of decomposition techniques [14] (e.g., Bender Decomposition cutting plain). The solution of the outer-level problem is based on the information given by the inner-level dual max-min problem of the economic dispatch. The dual max-min problems can be solved using approximation algorithms such as Lagrange relaxation [15]or outer approximation [16] algorithms.

In summary, the two stochastic approaches of the UC/ED problem described above possess a number of implementation challenges, mainly at the representation level of uncertainty and the choice of several parameters. In the case of SSUC, the following implementation challenges are to be mentioned:

- Choice of the best probability associated with each scenario; Due to the substantial share of integer variables in the two-stage model solution might be intractable by the majority of algorithms in stochastic programming, which hinge upon continuity of variables
- Application of decomposition techniques, when and where suitable to avoid intractability
- Scenario reduction techniques are needed to reduce the problem dimensionality
- The whole solution procedure requires a large computational time due to multiple runs for deterministic equivalents of the scenarios

On the other hand, in the case of robust optimization (SRUC), the following implementation challenges are to be emphasized:

- Choice of bounds in each uncertain node injection
- Choice of the uncertainty budget parameter
- Computationally intensive solution for large systems where the number of uncertain nodes are high
- Choice of the approximation algorithm for the dual max-min sub-problem.

Conclusion

This chapter provides a broad overview of the impact of different electricity market designs, and the nature of the power generation mix in formulating and solving the unit commitment and economic dispatch operation problems. Renewable power generation, such as wind and solar, besides their environmental benefits and competitiveness in terms of Levelised Cost of Energy (LCOE) of the technology (mainly wind), pose a number of challenges in terms of the design and the efficient and fair operation of the electricity market, as well as commitment decisions. A comparison in the state of the art of two main stochastic optimization approaches for the UC/ED was provided, giving implementation steps and identifying also the implementation. challenges.

The dispatcher's experience and understanding level of the proposed optimization method, with one or another solution may play an important role in the acceptance/rejection of one or another approach.

REFERENCES

[1] M. Goggin, J. Anthony, and A. Severn.(Date 2008). Wind power clean and reliable. *American Wind Energy Association (AWEA), Fact Sheet,* (issue). www.awea.org/utility <http://www.awea.org/ utility>.

[2] "Open Energy Info," http://en.openei.org/apps/TCDB/.

[3] S. Meliopoulos, "Power system modeling, analysis and control," *Lecture Notes*, Atlanta: Georgia Institute of Technology, 2006.

[4] "Integration of electricity from renewables to the electricity grid and to the electricity market-RES INTEGRATION (National Reports)," 2012; Available: http:// www.eclareon.eu/en/res-integration-national-reports.

[5] P. Meibom, R. Barth, H. Brand, B. Hasche, D. Swider, H. Ravn *et al.*, *All island grid study. Wind variability management study. Workstream 2B*, Department of Trade and Investment, Jan. 2008.

[6] C. Batlle, "Electricity Market Designs," *FSR Summer School,* Florence, Italy, pp. 1-15, June 2012.

[7] N. P. Padhy, "Unit commitment-a bibliographical survey," *IEEE Transactions on Power Systems,* vol. 19, no. 2, pp. 1196-1205, May 2004.

[8] S. Cerisola, Á. Baíllo, J. M. Fernández-López, A. Ramos, and R. Gollmer, "Stochastic Power Generation Unit Commitment in Electricity Markets: A Novel Formulation and a Comparison of Solution Methods," *Operations Research,* vol. 57, no. 1, pp. 32-46, Jan.-Feb. 2009.

[9] L. Söder and H. Holttinen, "On methodology for modelling wind power impact on power systems," *International Journal of Global Energy Issues,* vol. 29, no. 1-2, pp. 181 - 198, Nov. 2008.

[10] E. B. Muhando, T. Senjyu, A. Uehara, T. Funabashi, and K. Chul-Hwan, "LQG Design for Megawatt-Class WECS With DFIG Based on Functional Models' Fidelity Prerequisites," *IEEE Transactions on Energy Conversion,* vol. 24, no. 4, pp. 893-904, Dec. 2009.

[11] R. Barth, L. Söder, C. Weber, H. Brand, and D. J. Swider, *WILMAR Documentation: Methodology of the Scenario Tree Tool*, Institute of Energy Economics and the Rational Use of Energy, University of Stuttgart, Jan. 2006.

[12] J. Dupacova, N. Growe-Kuska, and W. Romisch, "Scenario reduction in stochastic programming: An approach using probability metrics," *Mathematical Programming,* no. A 95, pp. 493–511, Feb. 2003.

[13] D. Bertsimas, E. Litvinov, X. A. Sun, J. Zhao, and T. Zheng, "Adaptive Robust Optimization for the Security Constrained Unit Commitment Problem," *IEEE Transactions on Power Systems,* vol. PP, no. 99, pp. 1-12, 2012.

[14] H. A. Taha, *Operations Research - An Introduction*, 7th ed.: Pearson Education, 2003.

[15] S. Boyd and L. Vandenberghe, *Convex Optimization*: Cambridge University Press, Cambridge UK, 2004.

[16] M. Duran and I. Grossmann, "An outer-approximation algorithm for a class of mixed-integer nonlinear programs," *Mathematical Programming,* vol. 39, no. 3, pp. 307-339, 1986.

In: Renewable Energy
Editor: Andreas Poullikkas
ISBN: 978-1-62618-231-8

Chapter 7

EVALUATION METHODOLOGIES AND INPUT DATA FOR OFFSHORE WIND POWER PLANT ELECTRIC TOPOLOGY ASSESSMENT

***A. Madariaga*[†,*]*, I. Zamora*[†]*, J. L. Martín*[†]**
***I. Martínez de Alegría*[†] *and S. Ceballos*[‡]**
† University of the Basque Country (UPV/EHU)
Engineering Faculty of Bilbao, Spain
‡ Tecnalia Research & Innovation, Derio, Spain

Abstract

Offshore wind power is attracting the interest of experts in a range of disciplines such as electricity, electronics, mechanics, materials, control, project management and financing. The objectives are bold. On the one hand, with the fossil era showing slight signs of decay, renewable sources of electricity are seen as a realistic supply of energy for societies of the XXI century. Marine renewable energies in general, and offshore wind energy in particular, are one of the most promising alternatives in this slow but global shift to renewable sources of electricity. On the other hand, during the last decade onshore wind has proven itself as a powerful economic drive for the countries leading its development. And that drive continues worldwide in areas with vast land and wind resources. In contrast, appropriate onshore sites are dwindling in some countries and governments, companies and academia are focusing on offshore wind.

Keywords: offshore wind, cost-benefit analyses, project planning, research economics

UNESCO fields of Science and Technology: 3306, 3307, 3322

*E-mail address: ander.madariaga@ehu.es

1. Acronyms

[WT] Wind Turbine
[OWT] offshore wind turbine
[WF] Wind Farm
[OWF] offshore wind farm
[OWP] Offshore Wind Power
[OWPP] offshore wind power plant
[EWEA] European wind energy association
[GWEC] Global wind energy council
[RES] Renewable Energy Supplies
[RESE] renewable energy sources of electricity
[O&M] operation and maintenance
[TSO] transmission system operator
[OFTO] offshore transmission system owner
[EC] European Commission
[NREAP] national renewable energy action plan
[TGC] tradeable green certificates
[ES] Electric System
[OS] offshore substation
[CPS] connection point substation
[PMSG] permanent magnet synchronous generator
[SCIG] squirrel cage induction generator
[LV] low voltage
[MV] medium voltage
[HV] high voltage
[AC] alternating current
[DC] direct current
[LVAC] Low Voltage Alternating Current
[LVDC] Low Voltage Direct Current
[MVAC] medium voltage alternating current
[MVDC] medium voltage direct current
[HVAC] high voltage alternating current
[HVDC] high voltage direct current
[LCC] line commutated converter
[VSC] voltage source converter
[OWTES] Offshore Wind Turbine Electric System
[WECS] wind energy conversion system
[CS] collector system
[TS] transmission system
[SW] Software
[HW] Hardware
[SG] synchronous generator
[PS] Power System
[NPC] neutral point clamped

[DCF] discounted cash flow
[NPV] net present value
[CPI] Consumer Price Index
[SG] Synchronous Generator
[1SG] One Stage Gearbox
[3SG] Three Stage Gearbox
[DD] direct drive
[SVC] static VAr compensator
[STATCOM] static synchronous compensator
[PP-MTS] Point to Point MTS
[GR-MTS] General Ring MTS
[ST-MTS] Star MTS
[STR-MTS] Star with a central switching Ring MTS
[OFR-MTS] OSS Ring MTS
[ONR-MTS] CPSS Ring MTS
[EPC] energy production cost
[ETC] energy transmission cost
[IGBT] Insulated Gate Bipolar Transistor
[TCR] thyristor controlled reactor
[p.u.l.] per unit length
[XLPE] cross linked polyethylene
[PE] Polyethilene
[CST] Converter Station
[YZTC] year zero total cost
[ENS] energy not supplied
[RACS] Radial Collector System
[RICS] Single-sided Ring Collector System
[DRICS] Double-sided Ring Collector System
[STCS] Star Collector System
[PF] power factor
[PCC] point of common coupling
[rpm] rounds per minute
[VSI] Voltage Source Inverter
[p.u.] per unit
[GIS] gas insulated switchgear
[MVSC] Multiterminal Voltage Source Converter
[CSI] current source inverter
[NREL] National renewable energy laboratory of the U.S. Department of energy
[COE] cost of energy
[FCR] fixed charge rate
[ICC] initial capital cost
[AEP] annual energy production
[MTS] Multiterminal Transmission System
[NETS] National Electricity Transmission System
[GEP] grid entry point

[IP] interface point
[OTEG] Offshore transmission expert group of the U.K.
[NETS] National Electricity Transmission System
[NETSO] National Electricity Transmission System Operator
[DTI] Department of trade and industry
[Ofgem] Office of gas and electricity markets
[OSP] offshore supply point
[GBSQSS] Great Britain's security and quality of supply standard
[MITS] main interconnected transmission system
[EPS] electric power system
[FR] Failure Rate
[MTBF] mean time between failures
[MTTR] mean time to repair
[CCL] connection capacity lost
[CCP] central collection point
[FACTS] flexible AC transmission system
[MTTF] Mean Time To Failure
[MMC] multi modular converter
[LCOE] levelized cost of energy
[CAPEX capital expenditures
[OPEX operational expenditures
[LAT lowest astronomical tide
[MHWS] mean high water spring
[IEA International Energy Agency

2. Introduction

Offshore wind represents today less than the 2% of global installed wind capacity, and the OWPP commissioned during the last year represented the 2.5% of the annual wind market. Even the most optimistic projections show offshore wind as no more than 10% of global installed capacity by 2020. However, there are different reasons for the increasing interest in the offshore wind market [1]:

- It is a relatively new technology, with significant opportunities for cost reduction, technical innovations and indeed, revolutionary developments which could change the face of renewables in some parts of the world.
- It is particularly suitable for large scale development near big demand centers represented by the major port cities of the world, avoiding the need for long transmission lines, as is so often the case onshore.
- The wind resource offshore is generally much greater, and offshore windfarms make sense in very densely populated coastal regions with high property values, where there are real constraints for onshore development.

The offshore wind capacity installed worldwide today is around 4.6 GW, and a forecast of 9.8 GW by the end of 2014 seems realistic taking into account the projects already under construction [2]. 40 GW in Europe and 30 GW in China could be installed by 2020, although it is not clear to which extent these ambitious targets will be achieved. The countries that are leading the change in scale of the offshore wind business are the United Kingdom, Germany, China, Denmark and Sweden, in terms of installed capacity and in terms of technological leadership. Belgium and the Netherlands have also installed offshore wind farms, whereas other countries such as France, Japan, South Korea, Spain and the US are showing an increasing interest in this market.

Regarding the development of future OWPP, these projects can be described as non-linear processes that last between seven and ten years depending on the scale of the project and on the situation of the supply chain. Project stages are usually overlapped and iterations involving different teams are some times necessary to find good solutions. The list below presents the typical stages in these projects, taken from the comprehensive guide presented in [3].

1. Feasibility
2. Licenses
3. Tender
4. Financing
5. Engineering
6. Manufacturing
7. Installation
8. Commissioning
9. Life-cycle
10. Decommissioning

Several of these phases require cost-benefit analyses to support the decisions to be taken, and the approaches can differ significantly. For example, studies related to the feasibility of the project (stages 1 to 4) need to consider the offshore wind power plant as a whole. It is of mayor importance to get sufficient accuracy in all the aspects considered whereas maintaining the agility that allows to evaluate a sufficient number of alternatives.

On the contrary, the engineering design teams involved after the principal contracts (stages 5 to 8) have many of the variables fixed due to the supply agreements signed among the developers and the leading companies of their supply chain. These design teams have the challenge of providing a technically and economically feasible solution to the system they are designing, bearing in mind the contracts to be signed afterwards with their suppliers. And they should do it without loosing the overall perspective of the project and respecting the conditions fixed in the contracts, because changes introduced after the contracts usually lead to significant increases in the costs.

These cost-benefit studies are interesting also in relation to academic research. The technical or economical benefit derived from an improvement in one of the systems that make up the OWPP should be counterbalanced with the increase in cost necessary to include that design innovation in the project. And this should be done maintaining a global perspective of the project, because partial approaches usually overestimate the benefits derived from the technical improvements, and because time can be lost studying alternatives non-realistic due to determinants related to the project logistics, maintenance or economics. However, the technical complexity of these systems and the correlation between economic, environmental and technical variables makes it a challenge to obtain realistic assessments. Figure 1 attempts to represent the complexity of these analyses.

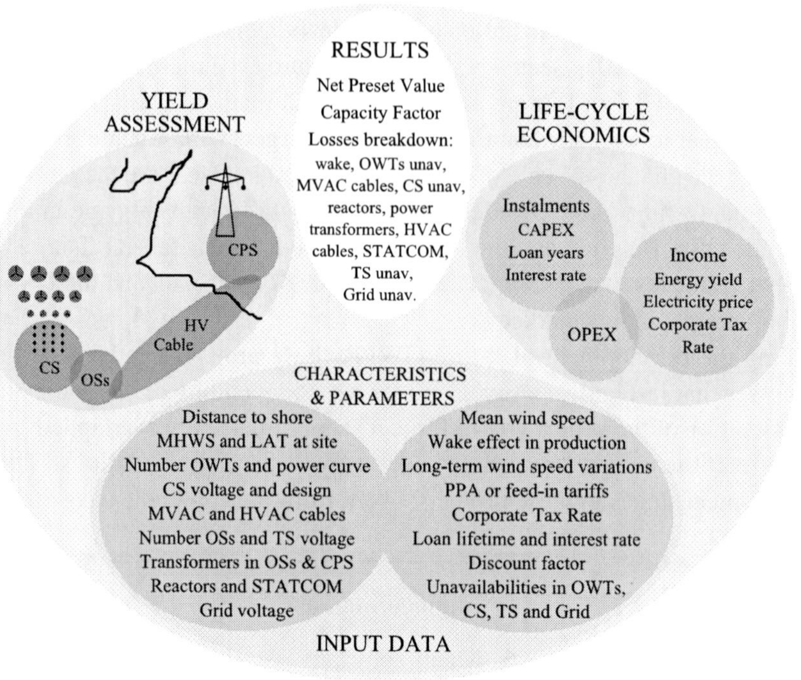

Figure 1. OWPP electric topology evaluation complexity at a glance

Ideally, these analyses of the alternative OWPP electric topologies should be performed using a methodology that considers both technical and economical issues, which has a scope that is as wide as possible, that has a simple formulation and which is open to the consideration of new topologies proposed by the offshore wind research community. During the last decade, a number of studies focused on this issue have been published, and the approaches differ significantly. The *criteria* proposed by the authors to perform the comparisons are divergent, as well as the formulation used for the assessments, the input data considered and the results of each study.

In this context, the main objectives of this chapter are to present the most relevant of these studies, to determine their strengths and weaknesses and to gather reliable input data and results useful for further analyses. With these objectives in mind, the rest of the chapter is organized as follows.

Section 3. focuses on the comparison criteria used in studies that have considered both the technical and economical aspects of these projects. As has already been mentioned, it is crucial to maintain a wide perspective and, therefore, approaches focused only on particular aspects of the projects are not considered in this section. The studies are organized in three categories depending on the information presented:

- Approaches that describe the electric power production assessment in detail but which neglect to consider the time value of money in the economic analysis.
- Approaches that consider the time value of money and other economic aspects but do not comment on the procedure followed to assess the electric energy production.

- Non-analytical stochastic optimization methodologies applied to project planning for the optimization of OWPP collector and transmission systems.

Section 4. deals with the input data demanded by the assessment methodologies considered. As the input data needed in the different approaches is not the same, this section is not focused on a particular methodology. On the contrary, it organizes data taken from a number of sources. There are three subsections dedicated to the following aspects:

- The capital costs or CAPEX.
- The electric energy production assessment, considering issues such as electric power losses and the unavailability of the subsystems that make up the windfarm.
- Economic parameters and OM costs.

Section 5. gathers the results of the key studies cited throughout the chapter, whose conclusions have been corroborated, in most cases, by offshore windfarms commissioned during the last years. These results are presented in two subsections. The first one is focused on the results of studies based on the economic evaluation of the alternatives, and the second one is focused on the results of studies based on the electric power losses in each topology.

Finally, section 6. summarizes the chapter and highlights the relevant information presented. It also points out the strengths and weaknesses of the methodologies analyzed.

3. Comparison Criteria in Different Methodologies

The OWPP electric topology evaluation has been of interest to researchers because of the large scale launch foreseen for this technology. The analysis of the studies published shows an evolution in approaches during the last decade and, therefore, these approaches can be classified into three groups.

The first group [4–12] emphasised the thorough evaluation of the electric power losses in all the systems that make up the OWPP. These loss assessments were set out in the context of methodologies that evaluated the annual energy yield of the plant. Afterwards, basic economic analyses were performed for the offshore wind project under consideration. These approaches performed the analyses on an annual basis, which led to two main drawbacks:

1. They neglect the influence of the time value of money.
2. Variations in key parameters throughout the lifetime, such as the mean wind speed and the electricity price, cannot be considered.

The second group [13–17] presented different approaches for enhancing the economical evaluations. The time value of money and other economic variables were taken into account (discount rate, taxes, OPEX, etc.). However, most of these studies do not detail the methodology followed to assess the energy yield of the OWPP. Some of them follow the methodologies of the previous group, whereas others just mention simulation packages well known in other areas of electric power system research. There are two main drawbacks of the approaches taken in this group:

1. With little detail about their formulation, it is difficult to judge a proposal or to apply it to different case studies. It is a matter of trusting the authors and assuming the implications of the results presented.

2. Although the time value of money is considered, the scenarios derived from the correlated variation of the wind speed and the electricity price are not.

The third group [18–24] is focused on the application of non-analytic stochastic methodologies to OWPP planning, i.e. the techno-economic evaluation of alternative electric topologies. These techniques are usually employed by utilities in electric power system operation, although it is not clear to what extent they are appropriate for planning stages. My opinion about this current trend will be presented in subsection 3.3.

3.1. Approaches Describing the Production Assessment in Detail but Which Do Not Take Into Consideration the Time Value Of Money

The EPC is the criteria used by Lundberg [4–6] to compare alternative OWPP layouts. This EPC is defined as how much it costs to produce and deliver a unit of energy to the onshore grid, for a given set of boundary conditions (transmission length, rated power, average wind speed, etc.).

This EPC is obtained, on an annual basis, by dividing the total investment cost of the windfarm by the total energy delivered (1). The total investment cost is calculated assuming that the whole investment is made in the first year and paid off during the life time of the windfarm in constant quotas. A foreseen profit is also considered in the analysis.

$$EPC = \frac{I}{P_{out,avg} \cdot T} \cdot \frac{r(1+r)^N}{(1+r)^N - 1} \cdot \frac{100}{100-p} = K \cdot \frac{I}{P_{out,avg}} \tag{1}$$

Where:

EPC	Energy production cost (EUR/kWh)
I	Investment paid today (EUR)
$P_{out,avg}$	Average power delivered to the grid (kW)
T	Operational time of the OWPP in one year (hours)
r	Interest rate (%)
N	Life time of the project (years)
p	Annual profit (%)
K	Constant

The energy delivered to the grid is calculated by multiplying the average power delivered to the grid by the average number of operational hours during one year and by the lifetime of the windfarm. The annual average power is calculated deducting the losses from the aerodynamic capture (2). The annual average aerodynamic capture ($P_{in,avg}$) is defined as the average power all turbines in the OWPP convert from the wind to mechanical power on the turbine shafts. This average aerodynamic capture is determined using (3), where the momentary power capture in the rotor shafts (P_{mec}) is calculated and the variations in the wind speed over a year are described by the Weibull probability distribution function ($f(\omega_s)$).

$$P_{out,avg} = P_{in,avg} - P_{losses} \tag{2}$$

$$P_{in,avg} = \int_{cut-in}^{cut-out} P_{mec}(\omega_s)\, f(\omega_s)\, d\omega_s$$

$$P_{mec} = \frac{\pi \rho \omega_s^3 R^2}{2} C_p(\lambda, \beta) \tag{3}$$

$$f(\omega_s) = \frac{k}{c}\left(\frac{\omega_s}{c}\right)^{k-1} e^{-\left(\frac{\omega_s}{c}\right)^k}$$

Where:

		C_p	Aerodynamic efficiency
P_{mec}	Mechanical power (W)	λ	Tip-speed ratio ($\lambda = R\,\omega_t/\omega_s$)
ρ	Air density (kg/m^3)	β	Pitch angle
ω_s	Wind speed (m/s)	$f(\omega_s)$	Weibull probability function
ω_t	Rotor speed (m/s)	k	Shape parameter
R	Rotor radius (m)	c	Scale parameter

The losses do not appear explicitly in (1), although they affect the EPC indirectly by decreasing the average output power. To include the losses more directly in the EPC, [4] suggests dividing this formula into two parts as it is shown in (4). The input power is defined as the power collected in the shafts of all the full variable speed OWT present in the OWPP. Extra power losses for fixed speed OWT are introduced since they do not operate to the optimum λ value for low wind speeds.

$$EPC = EPC_{invest} + EPC_{loss} = K\frac{I}{P_{in,avg}} + K\frac{I}{P_{in,avg}}\frac{P_{loss,avg}}{P_{out,avg}} \tag{4}$$

(4) can be applied separately to each component in the OWPP calculating their individual contributions to the EPC. This is due to the fact that the investment comes from the sum of the costs of all components and the mean power loss is the sum of the average power losses in each component [4]. It is worth stressing that the author shows here a *criterion* to take decisions on individual but highly correlated components of the OWPP, considering technical and economic aspects together.

Ackermann et al. [9–11] present an economic comparison of different TS for offshore windfarms. Investment costs, system losses and reliability issues are considered in the analysis of HVAC, LCC-HVDC and VSC-HVDC technologies. The evaluation criterion is the ETC, which can be seen as an adaptation of the previously commented on EPC to the evaluation of transmission systems. The ETC is an estimation of how much it costs to deliver a unit of energy produced by the windfarm to the onshore grid throughout the lifetime of the OWPP. The procedure is summarized in (5), (6) and (7).

$$ETC = \frac{R}{E_D} \frac{100}{100 - p} \tag{5}$$

$$R = \frac{r(1+r)^N}{(1+r)^N - 1} I \tag{6}$$

$$E_D = P_{out,avg} \left(1 - \frac{L}{100}\right) T \left(1 - \frac{U_n}{100}\right) \tag{7}$$

Where:

R Annual investment instalment (EUR)

E_D Energy delivered to the onshore grid (kWh/year)

L Average power losses in the TS (%)

U_n Unavailability of the TS (%)

In [9] the ETC of each transmission system is presented as a function of the transmission length, the mean wind speed of the site and the rated power of the OWPP. Two major parameters are not taken into account in this study: the cost of the offshore platform and the cost of upgrading the grid onshore, also known as network connection cost. VSC-HVDC offshore platforms would be smaller than the ones used in LCC-HVDC but much larger than the ones used in HVAC systems. As VSC-HVDC allows independent control of active and reactive power the network connection cost could be significantly lower compared to LCC-HVDC and HVAC systems.

3.2. Approaches Describing the Economic Evaluation in Detail But Which Do Not Present A Production Assessment Methodology

Walling and Ruddy [13] propose a technique focused on identifying the electrical system layout delivering the lowest total life cycle cost. I will call their approach the *ABC method*. This method tries to condense the complexities of the windfarm business model into a form that can be conveniently used to optimize substation and collection system design for maximized profitability. This method is based on three economic evaluation factors: the *A factor*, related to fixed losses, the *B factor*, related to variable load losses and the *C factor*, related to the ENS due to constraints imposed by electrical system unavailability. This methodology is based on practises widely established among the utilities in the U.S., broadening the scope of application by the introduction of the third factor related to unavailability.

Concerning the derivation and further use of A and B factors, the method suggested in [13] has the recommendation of the IEEE PES Wind Plant Collector System Design Working Group [15]. However, the third factor has not been included in the recommendation, and its application can still be considered an open issue. A very similar method is presented in [14], but is called *techno-economical optioneering*. Apart from showing the application of the method to different case studies, the interest of the paper is that it shows examples of unavailability assessments for different electrical systems.

As an example of this *ABC method*, below I present a survey of its application to the CS cable optimization. Other case studies, as well as detailed formulation and data for the calculation of B and C factors, can be found in [13, 14].

The *B factor* translates load-dependent lost energy production, measured at rated load, into an initial capital cost equivalent given in USD/kW. The meaning of this B factor is that an incremental capital investment in B to reduce rated load loss in 1 kW is justified at the rate of return used in calculating the B factor.

The derivation of the B factor is based on the equality of the present value of annual costs associated with the capital investment to save 1 kW of load-dependent losses (PV_{cap}) assessed using (8), with the present value of the impact on net after-tax revenue and purchased energy costs (PV_{rev}) assessed using (9).

$$PV_{cap} = B - BT \sum_{n=1}^{life} \left[\left(\frac{p}{f} \right)_n^i D(n) \right] + B \left(\frac{p}{a} \right)_{life}^i (1-T)P \tag{8}$$

Where:

B Initial capital investment

T Income tax rate

$\left(\frac{p}{f} \right)_x^y$ Present value of a future cash flow in year x using the compound interest rate y

$D(n)$ Tax depreciation of capital asset in year n

$\left(\frac{p}{a} \right)_x^y$ Present value of a uniform set of future cash flows, from year 1 until year x, at a compound interest rate y

P Property tax rate

$life$ Economic lifetime of the windfarm

i Desired after-tax return on investment

$$PV_{rev} = \left(\frac{p}{a} \right)_{life}^i 8760\, K_{loss} C_{ew} (1-T) + \left(\frac{p}{a} \right)_{life_{ptc}}^i 8760\, K_{loss} C_{ptc}$$

$$K_{loss} = \frac{1}{8760} \int_0^{8760} \overline{P(t)}^2 \, dt \tag{9}$$

Where:

K_{loss} Loss factor (average losses divided by losses at rated production)

$\overline{P(t)}$ Per-unit power output of the windfarm at hour t

C_{ew} Selling price, per kWh of wind generation

C_{ptc} Production tax credit, per kWh of wind generation

Afterwards, this B factor can be used to determine the total evaluated life-cycle cost as the sum of the installed cable cost plus the loss evaluation, which is the loss at full load multiplied by the B factor. Choosing the cross-sectional area of CS cables by means of this evaluated cost leads to a better economic optimum than a conventional thermal rating based selection (figure 2).

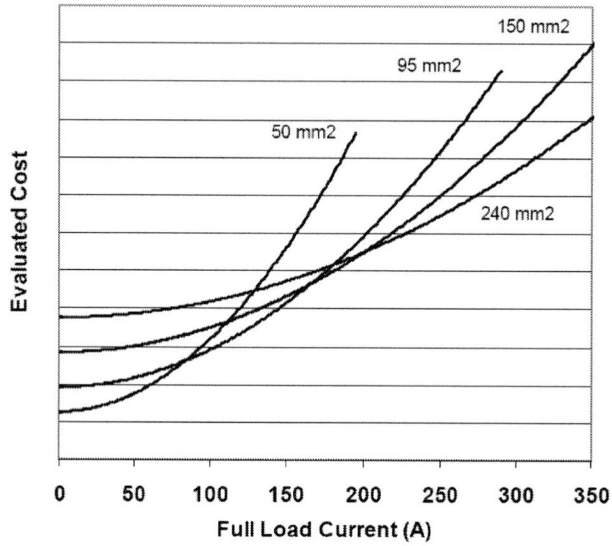

Figure 2. Evaluated life-cycle costs for different submarine cables [13]

A strict interpretation of this formulation would require the recalculation of the B factor for each chosen cable. This is due to the influence of the cross-sectional area of the cable on the losses of the CS and, therefore, on the calculation of the K_{loss} that affects the B factor. However, this interpretation would lead to an iterative algorithm and complex computations the method attempts to avoid. On the contrary, it seems that the authors suggest calculating A, B and C factors for the basic layout, using them as constant coefficients to optimize particular areas of the system, such as the cross-sectional areas in cables, the rating of power transformers and the type/number of these transformers (three-phase transformer *vs.* three-phase bank of single-phase transformers, transformer windings with galvanic isolation *vs.* autotransformers, etc.).

The scaling model presented by the NREL in [25] calculates a LCOE as a way to assess the total system impact of any change in the design of a wind turbine. To do so, the NREL spreadsheet based tool uses simple scaling relationships to assess the cost of wind turbine components and subsystems for different sizes and configurations. Its basic formulation is shown in (10).

$$LCOE = \frac{FCR \times ICC}{AEP} + AOE$$
$$AOE = BLC + \frac{O\&M + LRC}{AEP} \quad (10)$$

Where:

LCOE	Levelized cost of energy (USD/kWh)
FCR	Fixed charge rate (1/year)
ICC	Initial capital cost (USD)
AEP	Net annual energy production (kWh/year)
AOE	Annual operating expenses
BLC	Bottom or seabed lease cost
O&M	Levelized OM cost
LRC	Levelized replacement/overhaul cost

The FCR introduced in the calculation attempts to limit the impact of financial factors, the cost of capital for example, so that the true impact of technical changes can be assessed. This FCR can be understood as the annual amount per dollar of initial capital cost needed to cover the capital cost, a return on debt and equity, and various other fixed charges. It is gauged from a hypothetical project, modelled using a pro-forma cash flow spreadsheet model, and includes construction financing, financing fees, return on debt and equity, depreciation, income tax, and property tax and insurance. The NREL sets this FCR to 0.1158 per year in the results presented in [25].

A different approach is chosen by Breseti et al. [16] to compare HVAC and HVDC transmission systems. The criterion in this case is the YZTC. This YZTC takes into account the costs of investment, the OM costs, also known as OPEX, and the values of the losses and of the ENS. All the investment costs are calculated using the models provided in [4].

The yearly total energy losses are evaluated by means of an iterative power flow procedure, but no details about these simulations are disclosed. The windfarm yearly energy production is calculated considering a Rayleigh distribution simplified to 18 operational points. Yearly variable energy losses are computed at each of the 18 operational points of the production duration curve, while yearly constant energy losses are assumed to have an operational time of 8400 h. The losses in the converter station are not mentioned.

It makes sense to comment on three interesting aspects of this methodology:

- A full set of sensitivity analysis is performed, evaluating the factors in which variation could affect the results of the economic assessment significantly. The following parameters are considered: the transmission distance, the cost of cable installation and protection, the cost of the converter station and the availability of the converter station.

- Details are given about the criteria followed to determine the rating of the compensation equipment required for the HVAC alternatives. For onshore compensation with a fixed shunt reactor, this is rated to the average between the reactive power production of the cable at no load and at full load. For compensation at both ends of the submarine cable, the offshore reactor is rated to absorb half the total reactive power generated by the cable at no load, whereas the onshore reactor is rated to the average of the maximum and minimum power supplied shoreward by the cable when the offshore reactor is in service.

- An analysis of the cost-effectiveness of a redundant connection between OS is presented (cable system *C* in figure 3). It is based on the statement that the interconnection is cost-effective only if the connection expenses are lower than the net present gain from increased energy production, which is evaluated by considering the yearly production curve of the OWPP and the unavailability of the cable connection to the shore.

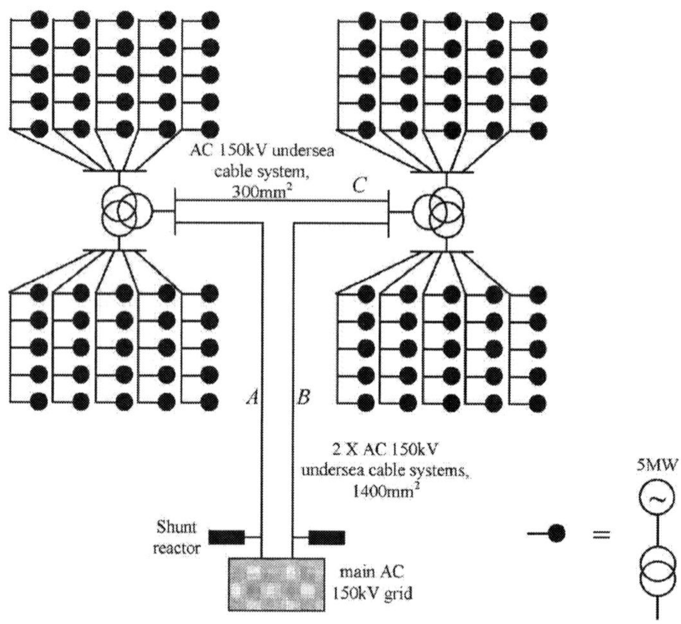

Figure 3. HVAC link with interconnected OSs for a 500 MW OWPP [16]

Van Eeckhout et al. [17] propose the DCF as the criterion for the economic comparison of HVAC and HVDC transmission systems. The inputs for the calculation are the initial investment costs, the annual costs and the foreseen revenue. The annual costs and revenues are then yearly discounted, resulting in the NPV of the monetary difference between the investment in HVDC or HVAC as the transmission technology. Only the operating costs of maintenance and losses are incorporated in the economic comparison. The differences in reliability and in insurance costs are omitted.

Another interesting point about this study is that it attempts to determine the influence of the TS's ownership on the economic evaluation of OWPP. To do so, the authors consider two different scenarios. One in which the OFTO and the OWPP owner are the same company and another in which they are different companies. The results show different break-even distances between technologies in both cases. However, the results of the analysis could be distorted by the fact that different WECS are considered in both scenarios and, therefore, the validity of the results is unclear.

3.3. Reflection on Non-Analytical Stochastic Methodologies Applied to CS and TS Optimization

During recent years, the interest of the research community in the application of non-analytical stochastic optimization techniques to techno-economical studies in the field of electric power systems has clearly increased. And the study of OWPP electric topologies is not outside that trend. However, it is not clear whether those approaches can be appropriately applied today to CS and TS optimization. This subsection gathers different views on two issues related to the application of such techniques on grid integration studies for OWPP: the extension of methodologies used in power system operation to planning studies, and the suitability of these methodologies for incorporating unavailability issues in the techno-economic analyses of OWPPs.

3.3.1. Methodologies Used in the Electric Power System Operation Applied to System Planning

The dimensions and complexity of electric power systems discourage the use of exact optimization techniques in power system operation. The non-analytic stochastic methods are an effective option for finding rapid, robust and good solutions [26]. However, we should bear in mind that when utilities and TSO decide to use these methods, they have at their disposal extensive databases with the key operational data of the grids under study. Therefore, they usually have the possibility of checking the accuracy of the non-analytical stochastic optimization methods with actual data.

However, the evaluation of alternative electric topologies for OWPP, also known as grid integration studies for CS and TS optimization, is a planning task. And it is not clear whether techniques useful in system operation are appropriate for system planning. In [18] for example, it is suggested that stochastic optimization methods lead to more meaningful information for OWPP design than deterministic techniques because they consider the stochastic nature of the problem. A number of papers have been published recently following these ideas [19–24]. On the contrary, other authors suggest that these extremely detailed and elaborated approaches may be suitable when system composite reliability analysis is being carried out, but they are hardly justifiable at the early stages in decision making for selecting electric system layouts [13, 14].

The data needed to check, tune and apply appropriately these non-analytic stochastic optimization methods to OWPP planning is difficult to obtain in the offshore wind industry due to the incipient stage of the technology (if available at all at the moment). Therefore, the applicability to a real case study of the results obtained from those non-analytic stochastic optimization analyses is not clear at all. In some of the papers cited above, the analysis seems to be focused on the mathematical optimization procedure itself more than on the engineering problem they are intended to solve.

We should also bear in mind that most of the key evaluations of OWPP electric topologies published during the last decade follow deterministic methods for the layout evaluation and for the unavailability assessment [4, 9, 10, 16]. And they seem to have driven decisions in the offshore wind industry during the first decade of this budding business. If that is going to change, it will be necessary to rely on studies checked with actual data, which seems to be difficult in the next few years.

3.3.2. Non-Analytical Approaches to Evaluate Stochastic and Highly Correlated Phenomena

Utility substation and infrastructure designs focus on maintaining continuity of service, i.e. reliability, because power flow disruptions as brief as a few seconds can be devastating to manufacturing operations [27,28]. This translates technically into redundant schemes and automatic transfer concepts that recognize faults, clear them and reconfigure the system properly. Therefore, utilities and TSO usually see the inclusion of redundancies in electric power systems as a trade-off between the necessary investment and the gained security of supply.

However, OWPP are an entirely different problem, and their design should be focused on delivering energy efficiently, i.e. availability[1]. A design for availability would not necessarily include redundancy and automatic transfer schemes, because short durations of inoperability do not greatly affect the revenue generation of the OWPP. Simpler infrastructures requiring manual or manually directed reconfigurations can provide nearly equivalent energy production with substantially lower initial and operating costs [13]. However, an outage[2] that constraints or eliminates OWPP output for a long period has a strong economic penalty, because available energy cannot be delivered to the marketplace and will never be recovered. Therefore, availability is one of the major drives in the design of the electric infrastructure of OWPP [31].

Historically, statistical deterministic approaches have been used to determine the availability of utility assets and industrial plants [32]. Some authors propose applying these methods to estimate the unavailability of OWPP electric systems [9, 13, 14, 16]. The assessed unavailability is afterwards used to evaluate the economic value of the ENS due to forced outages throughout the lifetime of the OWPP. Applied only to TS, these methods show little difficulty in their formulation. That may be the reason why most of the studies on unavailability in OWPP following deterministic methodologies have up to now focused on the TS, neglecting the influence of other systems like the OWT or the CS.

However, an availability assessment of an entire OWPP taking into account the OWT, the CS, the TS and weather conditions in conjunction is a real issue. The number of total and partial forced outages to be considered in the study increase considerably, and so does the number of system states to be considered in the statistical analysis. Moreover, the economic effect of each possible fault depends on the weather conditions at the site until the system is restored. The randomness of fault locations and weather conditions in OWPP availability assessment are highly correlated.

Other authors [18–24] try to tackle this problem with non-analytic stochastic methods. These studies also consider the CS, taking into account the already mentioned stochastic and correlated characteristics of forced outages, i.e. the location of the fault and the wind

[1]According to [29], the availability of a certain component or system in the context of offshore wind projects can be calculated using (11).

$$Availability = 1 - Unavailability = 1 - \frac{MTTR}{MTBF} \cdot CCL(\%) \tag{11}$$

Where the MTBF and the MTTR are usually given in years, and the CCL due to planned or unplanned downtimes is given as a percentage of the capacity of the component or system

[2]The state of a component or system when it is not available to properly perform its intended function [30].

conditions during the MTTR associated with that fault. However, as also previously mentioned, the application of these methodologies depend highly on the availability of data, which are difficult to get. A few companies (Dong, Vattenfall, E.On) have current access to actual data from operating OWF and, logically, they do not disclose it. Moreover, the MTBF and MTTR of most of the components installed in new OWPP are unknown, because new technical solutions have been incorporated into the latest projects commissioned. In this context, bearing in mind the objective of this thesis, it does not seem a good idea to add uncertainties within the methodology itself.

4. Input Data in Previous Methodologies

This section focuses on the input data demanded by the evaluation methodologies previously described. Details about the CAPEX, the energy yield assessment and a number of parameters which need to be fixed in these studies (e.g. lifetime, OM costs, energy price, etc.) are commented on in the following three subsections.

4.1. Capital Costs or Capital Expenditures (CAPEX)

The cost models presented in [4] have been used in prestigious OWPP layout evaluations performed during the last decade [9, 16]. These models include formulae to assess the cost of OWT, transformers, cables, converters, switch gear and OS's structure, and are given in 2003 SEK.

1. The cost of the OWT is assessed using (12). The model includes a MV transformer but it does not include the foundation, and it is based on data from turbines in the range from 0.5 to 2.5 MW and with different WECS. The author uses the same cost model for AC and DC wind turbines because they have the same components (figures 4.6 and 4.11 in [4]).

$$C_{wt} = -1.46 + 8.38\,P_n \tag{12}$$

   Where C_{wt} is the cost of the wind turbine (MSEK) and P_n is the rated output active power (MW).

2. The cost of the transformers is given using (13). The model is based on data from transformers with rated power in the range from 6 to 150 MVA, high voltage in the range from 47 to 140 kV and low voltage in the range from 10 to 77 kV.

$$C_{tr} = -1.2080\,10^6 + 2143\,S_n^{0.4473} \tag{13}$$

   Where C_{tr} is the cost of the transformer (SEK) and S_n is the rated output power (VA).

3. The AC cable costs can be evaluated (14).

$$\begin{aligned} C_{ac} &= A_p + B_p\, e^{\left(C_p S_n / 10^8\right)} \\ S_n &= \sqrt{3}\,U_n I_n \end{aligned} \tag{14}$$

Table 1. Parameters for the cost models in AC and DC cables [4]

AC cables				DC cables		
Rated voltage (kV)	A_p $(\cdot 10^6)$	B_p $(\cdot 10^6)$	C_p	Rated voltage (kV)	A_p $(\cdot 10^6)$	B_p
22	0.284	0.583	6.15	5	-0.346	0.4080
33	0.411	0.596	4.10	40	-0.314	0.0618
45	0.516	0.612	3.00	160	-0.100	0.0164
66	0.688	0.625	2.05	230	0.079	0.0120
132	1.971	0.209	1.66	300	0.286	0.00969
220	3.181	0.11	1.16			

Where C_{ac} is the cost of the AC cables (SEK/km), U_n is the line to line rated voltage (V), I_n is the rated current (A) and A_p, B_p and C_p are constants (table 1).

The DC cable costs are assessed using (15).

$$\begin{aligned} C_{dc} &= A_p + B_p P_n \\ P_n &= U_n I_n \end{aligned} \tag{15}$$

Where C_{dc} is the cost of the DC cables (SEK/km), U_n is the pole to pole rated voltage (V), I_n is the rated current (A) and A_p and B_p are constants (table 1).

The installation cost for both AC and DC cables is considered to be the same: 800 SEK/m for land cables and 2400 SEK/m for submarine cables. The cost of the cable compensating inductors is set to 2/3 of the cost of a transformer with the same VA rating.

4. The same 1 SEK/VA cost is considered for all the converters independent of its rated power and of its type (AC/DC or DC/DC).

5. Two types of GIS are considered depending on their location: those sited in the OS and those sited on the OWT' tower bases (16).

$$\begin{aligned} C_{sg-os} &= 320 \cdot 10^3 + 6U_n \\ C_{sg-owt} &= 100.33 \cdot 10^3 + 2.8726U_n \end{aligned} \tag{16}$$

Where C_{sg-os} is the cost of the switch gear installed in the OS (SEK), C_{sg-owt} is the cost of the switch gear installed in the OWT (SEK) and U_n is the line to line rated voltage of the switch gear (V).

6. All foundations for the OWT are assumed to be monopiles and their cost is set at 7.5 MSEK, independent of the water depth and the turbine size. The cost of the platform for the OS is given by (17), considering living quarters for workers, heliport, MV and HV switch gear and transformers or converters.

$$C_{pl} = 20 \cdot 10^6 + 0.7P_n \tag{17}$$

Where C_{pl} is the cost of the platform for the OS (SEK) and P_n is the rated active power of the OWPP (W).

In [9] cost data for HVAC, LCC-HVDC and VSC-HVDC transmission systems are presented. Table 2 summarizes the cost of components used in HVAC as a function of TS's voltage and rated power. Table 3 summarizes the cost of components used in VSC-HVDC transmission systems. Special attention should be paid to the difference between the HVAC and HVDC cable lying costs presented in both tables. For the HVDC-VSC cable system, a cost of 200,000 EUR/km is set with the assumptions that only one cable can be installed during one route of the cable installation vessel and that a bipole needs two HVDC cables. For the HVAC cable system, a cost of 100,000 EUR/km is set. All cost data is given in 2004 EUR, converting those found in other currencies using the official exchange rate of November 2004 and converting those referred to previous years to 2004 using an annual inflation rate of 2 % [10].

Table 2. HVAC transmission systems costs [9]

	Cables			
Voltage rating (kV)		132	220	400
Cost[1] (MEUR/km)		1.50	1.65	1.95
	Switch gear			
Voltage rating (kV)	33	132	220	400
Cost (MEUR)	0.058	0.124	0.183	0.303
	Transformers			
Power rating (MVA)	300	400	500	600
Cost (MEUR)	2.43	3.00	3.54	4.06

[1] Plus 0.1 MEUR/km per each cable laying.

Table 3. HVDC transmission systems costs [9]

	HVDC-VSC station		
Total cost (EUR/VA)	0.11		
	Polymeric extrusion DC cable pair		
Power rating (MW)	220	350	500
Cost[1] (MEUR/km)	0.3037	0.4453	0.6086

[1] Plus 0.2 MEUR/km per cable pair laying.

In [17] an economic analysis for HVAC and VSC-HVDC transmission systems is presented, considering a 300 MW OWPP including 60 turbines of 5 MW. For HVAC technology, the authors chose two parallel copper 3-core cables with a 500 mm^2 cross sectional area. For VSC-HVAC technology, a 376 MVA converter module is chosen with ±150 kV

DC cables. The considered investment cost input data is presented in table 4. Regarding cable installation costs, the author states that the laying process of DC cables is easier than that of AC cables, leading to lower installation costs due to the flexibility of the cables (single core versus 3-core) and the thinner insulation of DC cables (which is in clear contradiction to the statement made in [9]). All the costs are given in 2009 EUR.

Table 4. HVAC and HVDC capital costs [17]

Item	HVAC	VSC HVDC
Substation (MEUR)	10	45
Cable (kEUR/km)	1500	600
Cable installation (kEUR/km)	340	215
Offshore substation rig[1](MEUR)	13	24
Onshore land use (kEUR/km)	50	125

[1] A cost of 1000 EUR/m^3 is assumed for both options.

Two operating costs, maintenance and losses, are considered in the economic comparison. The differences in reliability for both transmission technologies are not taken into account. The HVAC transmission needs shunt inductances at both ends of the line to compensate the capacitance of the submarine cables and an additional dynamic compensator such as a SVC or a STATCOM for power factor control. The compensation investment cost for a 50 km transmission is taken to be 7.5 MEUR, which seems to include only dynamic compensation equipment, neglecting the shunt reactors usually installed at the OS. A summary of the investment costs, the efficiency of the main components and the maintenance costs reported in the study are presented in table 5.

Table 5. CAPEX, efficiency and OPEX in operating OWTs [17]

Item	HVAC + DFIG	HVDC + SCIG
Generator Investment Cost (kEUR)	250	400
Converter Investment Cost (kEUR)	—	60
Generator efficiency (%)	98.4	98.0
Converter efficiency (%)	—	98.33
Turbine maintenance[1](kEUR/year)	120	170

[1] 50 km from shore.

The investment costs included in [16] are for 33 kV AC offshore grid cables and their installation, HV cables for the TS and their installation, transformers, electronic converters, AC switch gears, compensation reactors and the support structure for the OS. Apart from the cost equations taken from [4], other interesting data is provided.

The economic evaluation of OM costs is limited to the transmission systems (submarine cables and substations for HVAC and HVDC technologies). The cost of the offshore cable installation is approximately estimated at 50 kEUR/km, independent of the voltage level. A price of 889 EUR/kW is adopted for constant-speed turbines and for variable-speed turbines. Considering that the paper was accepted for publication in October 2006 and the cited references for those variable-speed turbines are probably DFIG type. The costs are given in 2006 EUR.

A detailed breakdown of the costs associated to OWT in offshore wind projects is presented in [25]. As commented on in the previous subsection, the aim of the model is to help turbine designers to evaluate the influence of a certain design decision on all the costs related to that OWT throughout its whole life-cycle. Therefore, the costs considered are not only those directly related to the components present in the turbines, but those related to installation, financial and other aspects of wind projects. The following list organizes all the cost components described in the model:

- Wind Turbine
  - Rotor (blades, hub, pitch mechanism and bearing, spinner and nose cone)
  - Drive Train (low-speed shaft, bearings, gearbox, mechanical brake and high-speed coupling, generator, variable-speed electronics, yaw drive and bearing)
  - Nacelle (frame, electrical connections, hydraulic and cooling systems, cover)
  - Tower
  - Control, Safety and Monitoring
- Balance of Station
  - Foundation
  - Transportation
  - Port and staging equipment
  - Turbine installation
  - Electrical connection
  - Permits and licenses
  - Engineering and Site assessment
  - Personnel access equipment
  - Scour protection
  - Surety Bond for decommissioning
- Other Costs
  - Warranty Premium
  - Replacement and overhaul
  - OM
  - Bottom Lease

The cost models provided are in most cases functions of one or more of the following parameters: the rated power of the turbine, the mass of the component and the assessed AEP. These cost functions can be very valuable when evaluating the cost difference between different WECS for the same rated power. However, without details of the scope of the proposed *formulae* it is difficult to determine the accuracy of the assessments made when extrapolating to 5 MW or bigger OWT. Table 6 presents an estimation of the project costs for a 3 MW OWT installed in shallow waters near the shore. Data provided are primarily based on a 500 MW windfarm using one hundred and sixty seven OWT with a rotor diameter of 90 m and a hub height of 80 m. This OWPP would be installed in 10 m water depth 9 km from shore. Array spacing would be 7 rotor diameters by 7 rotor diameters. All the costs are presented in 2005 USD.

Table 6. Project cost breakdown for a 3 MW OWT [25]

Concept	Cost (kUSD)	Mass (Mg)
Rotor	477	50.96
Drive Train	1027	43.61
Nacelle	398	44.94
Control, safety and monitoring	60	
Tower	415	200.76
Adaptation to offshore environment	321	
Turbine Capital Cost	**2,698**	**340.27**
Monopile foundation	1,114	
Transportation	281	
Port and staging equipment	74	
Turbine installation	371	
Electrical interface and connections	926	
Permits, engineering and site assessment	119	
Personnel access equipment	64	
Scour protection	204	
Surety bond for decommissioning	176	
Balance of Station Capital Cost	**3,331**	
Offshore Warranty Premium	357	
Initial Capital Cost	**6,386**	**340.27**
Replacement and overhaul (per year)	16	
OM costs (per year)	30	
Sea bed lease (per year)	5	
Annual Operating Expenses	**51**	

The economic assessment carried out by Quiñonez-Varela et al. [33] is summarized in (18). Only capital costs for AC windfarms are considered, but they are broken down in detail. The costs of the sub-sea cable system (C_{CB}), the overall collector system (C_{CS}) and the whole windfarm (C_{WF}) are considered. The following components are considered: wind turbines, 33 kV switchgear (*e.g.* remote operated isolators, generator incomers, hub transformer feeders, hub bus tie and installations), 33 kV cables (cross-sections of 95, 185, 500, 630 and 2000 mm^2 depending on the design) and cable trenching, 33/132 transformer and the full costs of installation of all equipment.

$$\begin{aligned} C_{CB} &= C_{CA}(\theta)\, N_{CA}\, L_{CA} \\ C_{CS} &= (C_{SW}\, N_{SW}) + (C_{TX}\, N_{TX}) + C_{CB} \\ C_{WF} &= (C_{WT}\, N_{WT}) + C_{CS} \end{aligned} \quad (18)$$

Where:

$C_{CA}(\theta)$ Cost of sub-sea cable (i.e. cost per km and installation) as a function of its cross sectional area θ.

N_{CA}, L_{CA} Total number and length of cable sections for each cable type (i.e. of an specific θ).

C_{SW}, N_{SW} Cost of switch gear components, including installation, and the total number of units installed, respectively.

C_{TX}, N_{TX} Cost of transformers, including installation, and the total number of units installed, respectively.

C_{WT}, N_{WT} Cost of OWT, including installation, and the total number of units installed, respectively.

Only the considered CAPEX breakdown (18) and the final figures of the study are presented (table 7), but it is worth considering them because the costs of the components were available from manufacturers and consultants within the DOWNVIND consortium.

Table 7. CAPEX as a function of the CS design [33]

	CAPEX (GBP/kW)		
Collector system design	CS cable	Overall CS	Total OWPP
Radial	30.44	83.36	818.56
Single sided ring	74.16	131.39	866.59
Double sided ring	120.24	175.32	910.52
Star	24.80	81.26	816.46
Hybrid CS	44.20	98.20	

The OWPP considered in the analysis consists of 200 OWT rated at 5 MW and arranged in seven-turbine clusters. The offshore voltage level within the CS is 33 kV and 132 kV for the TS. The step-up transformers at each turbine are 690 V/33 kV rated at 6.25 MVA, and the transmission transformers at the OS are 33/132 kV rated at 200 MVA. The cable runs between wind turbines are 900 m long, and the cable runs between each cluster and the OS are 5 km long. The 33 kV cables in each string have tapered ratings and various standard XLPE cable types are used with cross sectional areas between 95 and 2000 mm^2. The 132 kV cables to shore are three-core XLPE insulated with a cross-sectional area of 800 mm^2 and a length of 25 km. Unfortunately the individual costs of the components are not disclosed.

Finally, other interesting references are commented on briefly in the list below:

- The OTEG includes all the input values considered in their cost-benefit analyses as an appendix of [34]. No new type of data is added to those already commented on throughout this section (failure rates, repair times, no-load and full-load losses, capital costs, OM costs and reference energy prices for the ENS). Although the cost breakdown presented is a little bit ambiguous sometimes this data is valuable because it is securely based on Round 2 projects in the UK.

- Dicorato et al. [35] present a general model to evaluate the total investment for an OWF depending on the chosen layout. Single cost centres for HVAC transmission systems are thoroughly detailed and a good survey of the investment costs taken from different sources is presented. Moreover, that data is compared to up-to-date information obtained by the authors and cost formulae are presented for all the cost centres. However, almost no mention of HVDC systems is made and the scope of each formula is not clarified, which makes their further application difficult.

- The Ministry of Industry of the Spanish Government [36] provides the unit values of reference for the investment and operation and maintenance costs of facilities in the Spanish National Electricity Transmission System, including overhead and buried 132/220/400 kV single and double circuit lines and compensation equipment.

- Stoutenburg and Jacobson [37] gather cost data in VSC-HVDC transmission systems, providing clear references to the sources considered. It is worth commenting on the capital cost values for 376 and 574 MW ±150 kV converter stations, for the offshore platform and for the 150 kV bipole cables. Cable capital cost and installation cost are broken down for each type of cable considered (from 630 to 1400 mm^2).

- Banzo et al. [38] disclose interesting data in the context of a proposed stochastic optimization model for OWF electric layout planning. The main author is works for Iberdrola, one of the utilities leading the development of RESE worldwide. The paper considers actual data in relation to Barrow offshore wind farm commissioned in 2006, presenting investment costs for 33 and 132 kV cables and data of failure and repair rates of cables and transformers. The total cost of Barrow (34.96 million EUR) is broken down into three main items: the investment costs (25.15 million EUR), the costs associated with ENS (5.87 million EUR) and the cost associated with electric power losses (3.94 million EUR).

4.2. Electric Power Production Assessment

The electric energy yield assessment of an entire OWPP usually considers the following three aspects:

1. The assessment of the energy yield of an individual OWT.

2. The evaluation of the electric power losses along the CS and the TS.

3. The unavailability of the systems and the assessment of the ENS.

4.2.1. Turbine Annual Average Production

Given an annual mean wind speed at a site, the procedure to assess the energy production of an individual OWT on an annual basis is widely accepted. When only the mean wind speed at the site is known, the Rayleigh probability density function is used to assess the number of hours a year the wind is expected to blow within a certain range. When historic data from masts is available, it is normal to employ the more accurate Weidbull probability density function for the same purpose. Details have already been provided in subsection

4.1. and the application of this procedure, set out in a methodology to assess the annual energy production of an entire OWPP, is shown in section 4.2.

4.2.2. Electric Power Losses in Collector And Transmission Systems

Brakelmann [7,8] suggests a calculation procedure for the efficiency of AC power transmission from OWPP to the grid, by means of long three-phase MVAC and HVAC submarine cables. The author states that a power of 500 MW can be transmitted over a distance of 100 km by means of a 245 kV HVAC cable, whereby usual limits for the voltage variation ($\leqslant 10$ %) as well as for the phase variation along the line ($\leqslant 30°$) are not exceeded.

In [7], the current distributions in the CS and in the TS are studied. Procedures to assess the momentary losses of the whole system and their mean value during a certain period are presented. In [8], the author analyses the dependency of the power losses in HVAC cables with temperature and proposes a procedure to consider that influence on the analyses. A summary of the methodology proposed in both papers is presented below, organized in three calculation steps:

1. Current distribution along MVAC and HVAC submarine cables.
2. Momentary power losses in the OWPP.
3. Mean value of the power losses during a year.

First step: current distribution along MVAC and HVAC submarine cables.

A radial collector system is organized in electric clusters, along which the power outputs of the turbines are superposed. This means a section-by-section increasing load current in each cluster of the CS. Therefore, the number of turbines feeding into a cluster is limited by the current carrying capacity of the cables and by the output rated power of the turbines. Figure 4(a) shows the current distribution in the MVAC cable stretches conforming a radial cluster, as well as the conductors' temperature in each stretch. The case study considers ten 3 MVA turbines interconnected by means of 20 kV 800 mm^2 three-core XLPE submarine cable.

Regarding the current distribution along the TS, the conductor current can undergo remarkable variations along the route, depending on the momentary load situation and on the reactive power compensation strategy. Usually this current increases strongly towards the end of the line, where the capacitive power is fed in. Figure 4(b) presents current distributions along a 100 km 245 kV 3∗1200 mm^2 XLPE submarine cable. They are represented as a function of the transmitted active power and of the reactive compensation strategy (full lines for compensation at both ends of the cable and dotted lines for onshore compensation).

The losses in the TS should be calculated by considering the current distribution along the line and the influence of the conductor's temperature by means of an appropriate calculation of the line parameters. The voltage ($\underline{U}$) and the current ($\underline{I}$) in the HVAC cables can be determined using (19).

$$\begin{aligned} \underline{U}(z) &= \underline{U}_2 \cosh(\underline{\gamma} z) + \underline{Z}_W \underline{I}_2 \sinh(\underline{\gamma} z) \\ \underline{I}(z) &= \underline{I}_2 \cosh(\underline{\gamma} z) + \frac{\underline{U}_2}{\underline{Z}_W} \sinh(\underline{\gamma} z) \end{aligned} \tag{19}$$

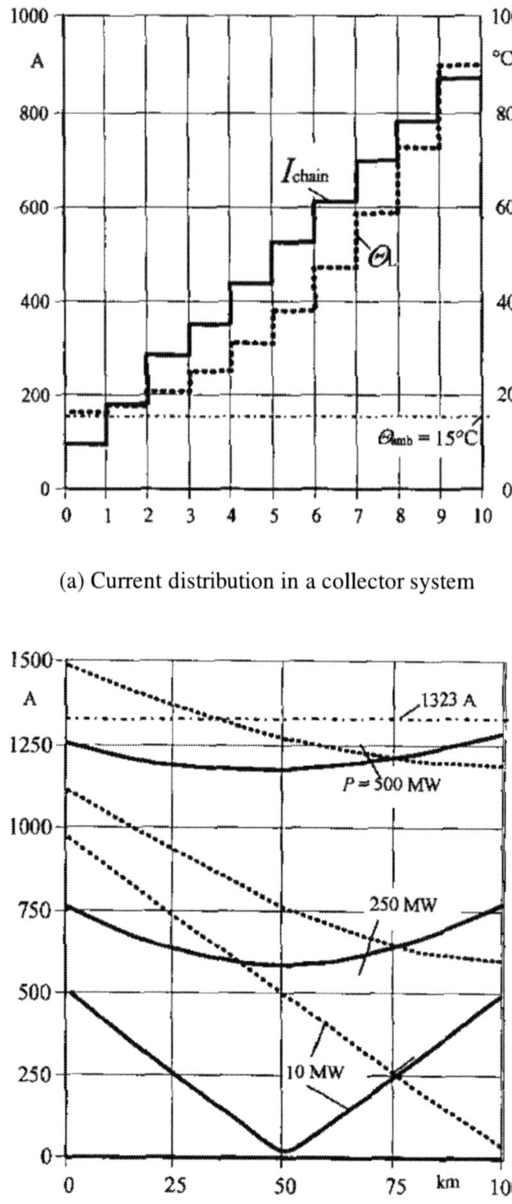

(a) Current distribution in a collector system

(b) Current distribution in a transmission system

Figure 4. Current in MVAC and HVAC three-core submarine cables [7]

Where $\underline{U}_2$ and $\underline{I}_2$ stand for the feed-in voltage and current respectively and z represents the distance from the OS. The surge impedance of the cable ($\underline{Z}_W$) and the transmission coefficient ($\underline{\gamma}$) are calculated using (20), taking into account the parameters of the cable.

$$\underline{Z}_W = \sqrt{\frac{R' + j\omega L'}{G' + j\omega C'}} \qquad \underline{\gamma} = \sqrt{(R' + j\omega L')(G' + j\omega C')} \tag{20}$$

Where L', G' and C' stand for the pul values of the inductance, the leakance and the capacitance of the cable respectively. The equivalent pul resistance (R') represents all current dependent losses of the cable. It is temperature dependant and it can be determined using (21).

$$R' = R'_{L-20}\left[1 + \alpha_T(\theta_L - 20)\right](1 + \lambda_1 + \lambda_2) \tag{21}$$

Where R'_{L-20} represents the DC conductor pul resistance at 20° C, α_T stands for temperature coefficient of the conductor resistivity, θ_L represents the conductor temperature, λ_1 stands for the screen losses coefficient and λ_2 for the armour losses coefficient.

Second step: momentary power losses in the OWPP.

The losses of transformers and compensation coils are assumed to be 1.2 % for nominal load, with 33.3 % voltage-independent losses. The pul losses of the submarine cables (P'_1) can be determined using (22).

$$P'_1 = P'_{max,I}\left(\frac{I}{I_N}\right)^2 \nu_\theta + P'_D \qquad \nu_\theta = \frac{c_\alpha}{c_\alpha \quad \alpha_I \, \Delta 0_{Lmax}\left[1 - \left(\frac{I}{I_N}\right)^2\right]} \qquad c_\alpha = 1 - \alpha_{20}(20 - \theta_U) \tag{22}$$

Where $P'_{max,I}$ represents the nominal current-dependent ohmic pul losses and P'_D represents the current-independent dielectric pul losses. The squared current ratio is used and a correction term for the temperature-dependence (ν_θ) is introduced. θ_U stands for the ambient temperature. It should be noted that these losses correspond to a certain operating range of the OWPP and to a particular point of the submarine cable. The currents for each section of the cluster can be considered constant (figure 4(a)).

For the losses in the HVAC cables, the pul losses (P'_1) can be calculated using (23), considering the current distribution along the line ($I(x)$) and integrating along the line length l_0. The procedure could be simplified by a stepwise summation for discrete intervals of length Δx.

$$P'_1 = \frac{P'_{max,I}}{l_0 I_N^2} \int_{x=0}^{l_0} I^2(x)\,\nu_\theta(x)\,dx + P'_D \tag{23}$$

Third step: mean value of the power losses during a year.

Describing the relative frequency of the wind velocity ($h(\nu)$) by means of a Rayleigh-distribution, the cumulative percentage of the wind velocity ($S_V(\nu)$) can be assessed using (24).

$$S_V(\nu) = \int_0^{\nu} h(\nu')\nu' d\nu' = 1 - e^{-\left(\frac{\pi}{4}\frac{\nu}{\nu^{-2}}\right)}$$
$$h(\nu) = \frac{\pi}{2}\frac{\nu}{\nu^{-2}} e^{-\left(\frac{\pi}{4}\frac{\nu}{\nu^{-2}}\right)} \tag{24}$$

The asymptotic limit of the cumulative percentage of the delivered active power ($S_P(\nu)$), for a given mean wind velocity ($\bar{\nu}$), is the mean value of the electric power output of the windfarm ($\bar{P}$). It can be calculated using (25).

$$\bar{P} = S_P(\infty) = \int_0^{\infty} h(\nu)\, P(\nu)\, d\nu \tag{25}$$

The asymptotic limit of the cumulative percentage of the losses ($S_{P_l}(\nu)$), for a given mean wind velocity ($\bar{\nu}$), is the mean value of the electric power losses in the OWPP ($\bar{P}_l$). It can be calculated using (26).

$$\bar{P}_l = S_{P_l}(\infty) = \int_0^{\infty} h(\nu)\, P_l(\nu)\, d\nu \tag{26}$$

Finally, the mean loss rate ($\bar{\varepsilon}_V$) results from (27), understood as the time-independent mean value of the relative losses.

$$\bar{\varepsilon}_V = \frac{\bar{P}_l}{\bar{P}} \tag{27}$$

Lundberg [4] presents models to evaluate the steady state power losses in most of the components of an OWPP. Loss models for gearboxes, generators, transformers, cables and electronic power converters are provided. Three aspects of the report deserve careful attention: the analytic expressions for the estimation of losses in HV cables, the way the electronic converters losses are characterized when evaluating OWPP topologies and the mechanic power losses in the gearboxes.

1. All AC submarine cables in the study are assumed to be three core cables with XLPE insulation and their parameters are taken from ABB's data sheets. No details about DC submarine cables are given, although the simplicity of the model proposed for losses assessment makes them unnecessary. Approximate models for the calculation of losses in both types of cables are given in (28) with parameters detailed in table 8.

$$P_{loss_{AC}} = P_0\, l + C_0\, l^3 + P_k\, l \left(\frac{S_{in}}{S_n}\right)^2$$
$$P_{loss_{DC}} = 58742\, l \left(\frac{P_{in}}{P_n}\right)^2 \tag{28}$$

Where:

P_{loss}	Losses in three-core XLPE cables (W)
P_0, C_0	No-load parameters (table 8)
P_k	Load parameter (table 8)
S_{in}	Input power of the AC cable (VA)
S_n	Rated power of the AC cable in VA ($3\,V_{rated}\,I_{rated}$)
P_{in}	Input power of the DC cable (W)
P_n	Rated power of the DC cable in W ($V_{rated}\,I_{rated}$)
l	Length of the cable (km)

Table 8. Loss model parameters for AC cables [89]

Insulation voltage (kV)	P_0	C_0	P_k
11	5.01	0.0212	57656
22	13.08	0.0354	57656
33	21.48	0.0421	57656
45	38.40	0.0694	57656
66	70.71	0.1069	57656
132	200.87	0.1726	49470
220	530.30	0.2982	51211

2. Two types of electronic converters are considered in the report: VSC in WECS and TS, and several types of DC/DC converters for WECS with their output in DC. The analysis starts with IGBT and diode on-state voltage drop and switching energy curves. Then, a detailed calculation procedure is followed to determine the conduction and switching losses for each type of bridge. Conduction and switching losses are presented in figure 5 as a % of the rated power for the whole wind speed range (dashed line for the PMSG, solid green line for the IGBT rectifier and solid grey line for the full bridge DC/DC).

   It should be noticed that representing the losses of electronic converters in such a way allows them to be included easily in a calculation procedure for the losses in the whole OWPP. Detailed representation of the losses in a converter demands the consideration of other parameters such as the modulation index and the firing angle in addition to the active and reactive power that is crossing the converter. However, such models would be too complicated for their practical inclusion in a general model of the windfarm. This is a good example of an appropriate compromise between the accuracy of a model and its purpose.

3. The rotational speed of a generator is typically around 1000-1500 rpm and the rotational speed of the rotor for a 2 MW wind turbine is approximately 19 rpm at rated wind speed. As the typical maximum increase of speed per step in a normal gearbox is 6:1, 3 steps are needed to increase the low speed of the rotor to the generator

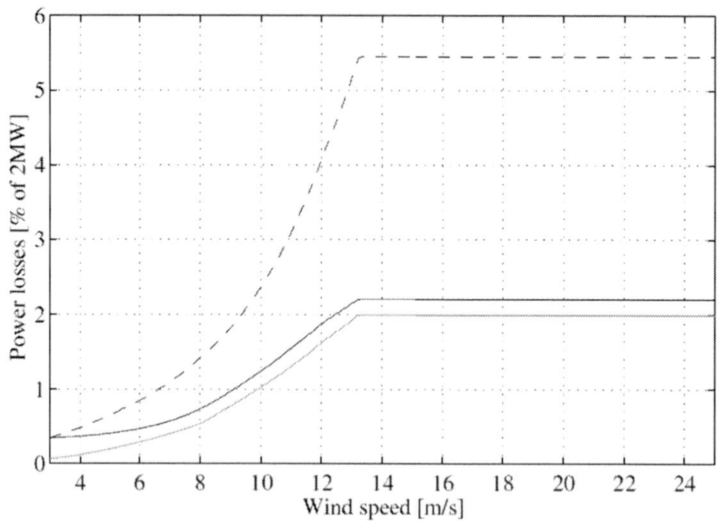

Figure 5. Electric losses breakdown in a 2 MW WECS [4]

speed. Mechanical losses in the gearbox for a wind turbine with rated power between 0.5 and 3.0 MW are given by (29).

$$P_{loss} = 0.02\,P_{in} + 0.005\,P_n\,\frac{\omega_{in}}{\omega_n} \tag{29}$$

Where P_{in} and P_n are the input and rated power respectively (W), and ω_{in} and ω_n are the input and rated rotational speeds respectively (rad/s).

Negra et al. [11] present calculation procedures for the electric power losses in HVAC, LCC-HVDC and VSC-HVDC transmission systems. The three technologies are evaluated considering different power ratings up to 1000 MW and distances to shore up to 200 km. The loss evaluation procedure suggests starting by calculating the output of all the turbines in the OWPP over time using the aggregated model based described in [39]. Then, a different loss assessment procedure is used for each technology.

For HVAC technology, the cable loss calculations follow the procedure suggested in [7,8], whereas the losses in transformers and TCR losses are evaluated as in [6], assuming that the latter have the same no load losses as a transformer with the same VA rating, and half of its load losses. For VSC-HVDC technology, an original procedure is presented to assess the losses of VSC units based on the little information disclosed about ongoing projects (usually the power input and output of the whole link). This procedure can be summarized in the following six steps considering the system presented in figure 6.

1. Obtain loss data available from some VSC-HVDC transmission links as the subtraction of the output power P_{out} from the input power P_{in}.

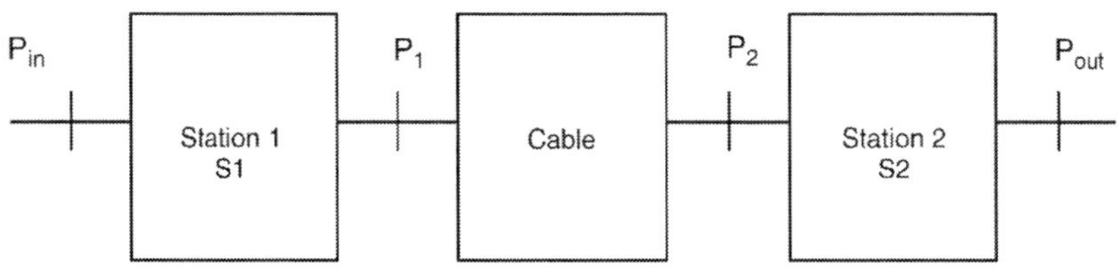

Figure 6. VSC-HVDC transmission scheme for losses assessment [11]

2. Assume that the percent losses (x_s) of both VSC stations are equal and determine it using (30) and (31).

$$\frac{R}{V_C^2}(1-x_s)^3 P_{in}^2 - (1-x_s)^2 P_{in} + P_{out} = 0 \tag{30}$$

$$R = P'_{Lmax}\frac{\nu_\theta}{I_N^2} \tag{31}$$

3. Calculate the power lost in the HVDC cable following and adaptation of Brakelmann's methodology detailed in equations (32) to (35).

$$P_{cab} = P'_{Lmax}\frac{I^2}{I_N^2}\nu_\theta \tag{32}$$

$$P'_{Lmax} = R_U\, c_m\, I_N^2\, l_{cable} \tag{33}$$

$$c_m = 1 + \alpha_{20}\left(\Delta\theta_{Lmax} + \theta_U - 20\right) \tag{34}$$

$$\nu_\theta = \frac{c_\alpha}{c_\alpha + \alpha_{20}\,\Delta\theta_{Lmax}\left[1-\left(\frac{I}{I_N}\right)^2\right]} \tag{35}$$

4. Since the data provide the input and the output power for the whole TS, it is necessary to solve the calculations (30 to 35) in a loop to obtain the value of the current in the cable and calculate the resistance.

5. Losses for VSC stations of different rated power are estimated from the losses of a 350 MW converter station.

6. When more than one VSC converter system is considered, the total power is divided between the different VSC systems based on the lowest overall transmission system losses.

Although it is a very useful proposal to assess key parameters when little data is available, some doubts arise when taking a close look at the procedure. On the one hand, if the procedure assesses the average production and losses in a year, the converter losses should be characterized for all the power range of the link. With a strategy similar to that proposed in [4], at least the no-load and the full-load power losses are necessary, assuming a parabolic variation of these losses. But only x_s is mentioned. On the other hand, after analysing the procedure described by Holttinen and Norgaard in [39], it is not clear if it is appropriate for the purposes it is intended for in this case. The aim of [39] is to establish a procedure to extrapolate the aggregated production of a large number of OWT spread in a large area (probably in several OWPPs), starting from data of a certain offshore site. Therefore, it is an extrapolation procedure focused on the study of the integration of huge amounts of wind energy into the power system. What is more, the D=50 km chosen in [11] for the extrapolation procedure, demands a layout in which the 200 OWT are organized in only five rows, which is not realistic.

It should also be considered that a final adjustment of the multi-turbine power curve made in [39] is intended to ensure that the annual production estimated with the proposed procedure coincides with that obtained adding the individual production of each OWT. This suggests calculating the aggregated power of an OWPP by adding the average productions of all the considered OWT. By doing so, we are assuming the number of hours per year that considered OWT are running within a certain speed range. This assumption seems realistic based on the fact that a 1 GW offshore wind power plant based on 5 MW OWT will cover an area of about 150 km^2, with a maximum distance between OWT of around 20 km.

Quiñonez-Varela et al. [33] compare the steady-state losses of several CS designs for a 1 GW offshore wind power plant. The momentary per-turbine MVA losses in the collector system (S_L), for a given mean wind speed and operating PF, are calculated using (36).

$$S_L = \frac{\sqrt{(P_E - P_F)^2 + (Q_E - Q_F)^2}}{N_{WT}} \tag{36}$$

This procedure weights the summed active and reactive power coming from the turbines in a string (P_E and Q_E) against the simulated power flows of active and reactive power arriving from a string to the offshore main hub (P_F and Q_F). The latter are obtained with PSS/E, although no details about these simulations are given. Figure 7 summarizes the results obtained for different CS designs and PF.

Afterwards, the average energies generated by OWT and lost in the CS are calculated on an annual basis. As the original 20-year historic wind speed record considered by the authors was logged at a heigh of 10 m, the first step is to correct the wind speeds to 90 m, a typical tower-height of the 5 MW OWT. Then, the energy yield of the windfarm (E_T) is calculated using (37) by combining the *binned* annual distribution of wind speeds (0.5 m/s steps) with the turbine power curve.

$$E_T = \sum_{i=1}^{n} [H(U_n)P(U_n)N_{WT}] \tag{37}$$

Where $H(U_n)$ is the number of hours in wind speed bin n, $P(U_n)$ is the turbine power output at the wind speed bin n and N_{WT} is the total number of turbines in the windfarm.

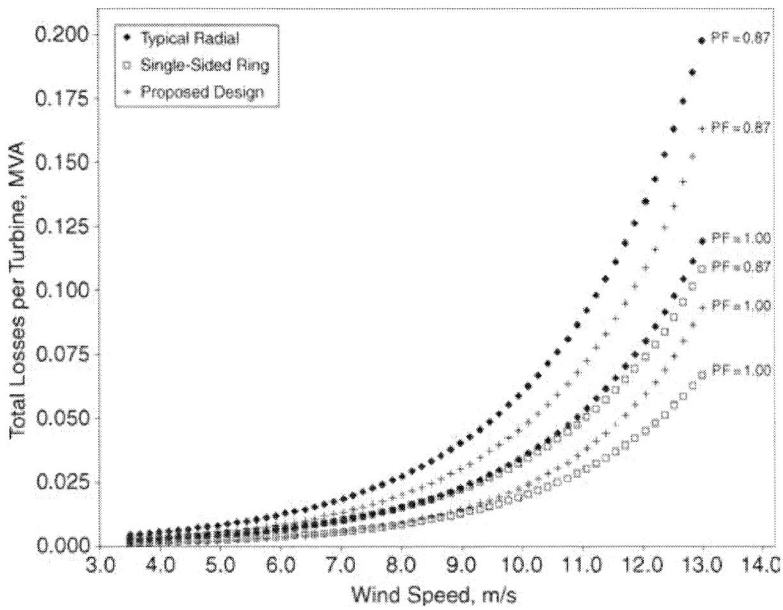

Figure 7. Per-turbine CS losses for different designs [33]

The total MVA losses during normal operation (E_L) are calculated using (38)[3]. The characteristics of power losses are approximated via exponential growth models (A and B parameters) for each CS design and PF.

$$E_L = \sum_{i=1}^{n} \left[H(U_n) A \, e^{(U_n/B)} N_{WT} \right] \tag{38}$$

Finally, Meier et al. [12] present an analysis focused on the calculation of the annual energy production of a 201 MW OWPP based on sixty seven 3 MW OWT. The losses in the CS and the TS are calculated following Brakelmann's procedure [7,8]. The paper gives data about loss characteristics of power transformers (table 9). To evaluate the losses in the TCR used to compensate the reactive power of HVAC cables, the author suggests assuming 5 kW per MVAr of rated reactive power.

4.2.3. Availability in Offshore Wind Power Plants

As mentioned in subsection 3.3., the studies of unavailability in OWPP following deterministic methodologies have up to now focused on the TS, neglecting the influence of other systems like the OWT or the CS. They assess the ENS due to forced and/or unforced outages and afterwards an economic value is considered for that energy. Three of these studies deserve careful attention, and details are provided below.

Ackermann et al. [9] estimate with statistically deterministic methods the unavailability of different OWPP transmission systems. The energy unavailability is defined in this study

[3]This equation is presented here as it makes sense according to the details given in the paper, although as it is presented in [33] lacks the term $H(U_n)$.

Table 9. Losses in power transformers [12]

50 Hz three-phase	P_{nl} (kW)	P_l (kW)	Efficiency (%)	Loss ratio (p.u.)
33/1 kV, 3 MW	6	12	99.40	0.50
132/33 kV, 200 MW	115	650	99.62	0.18
150/33 kV, 200 MW	130	650	99.61	0.20
220/33 kV, 200 MW	190	650	99.58	0.29
500 Hz single-phase	P_{nl}	P_l	Efficiency	Loss ratio
33/1 kV, 3 MW	2.1	2.5	99.85	0.84
33/1 kV, 3 MW	47.5	70	99.94	0.68

as the percentage of energy produced by the windfarm that cannot be transmitted as a result of failures (forced outages) in the TS. It is assumed though that maintenance (scheduled outages) takes place during periods with low wind speeds and thus its contribution to the unavailability of the system is neglected. As an example, the procedure to evaluate the unavailability of the LCC-HVDC bipole depicted in figure 8 is described below.

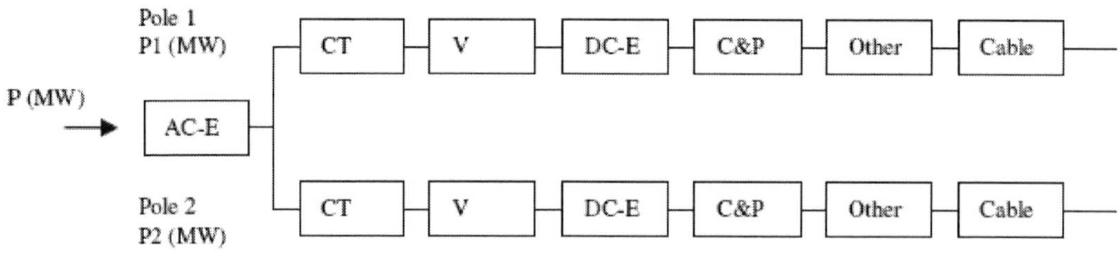

Figure 8. LCC-HVDC bipolar TS scheme for unavailability study [9]

1. Define the components and system considered in the analysis.
    - AC auxiliary equipment (AC-E)
    - Converter transformer (CT)
    - Valves (V)
    - DC equipment (DC-E)
    - Control and protection (C&P)
    - Other (O)
    - Submarine cable (Cable)
2. Obtain data about the probability of the non operation of each individual component (tables 10 and 11).

Table 10. Unavailability data for LCC-HVDC components [9]

Component	Simbol	Probability of not operating
AC auxiliary equipment	$F(O_{AC-E})$	0.001397
Converter Transformer	$F(O_{CT})$	0.009251
Thyristor Valves	$F(O_V)$	0.001175
DC equipment	$F(O_{DC-E})$	0.000647
Protection and Control	$F(O_{C\&P})$	0.000545
Other	$F(O_O)$	0.000124
Submarine cable	$F(O_C)$	0.004462[1]

[1] Per 100 km.

Table 11. Unavailability data for HVAC components [9]

Voltage level	Probability of not operating			
(kV)	Transformers	Breakers	Shunt Reactors	Cables[1]
33	–	0.004608	–	–
132	0.003571	0.001811	0.000701	0.004462
220	0.002460	0.001892	0.000701	0.004462
400	0.003045	0.001426	0.000701	0.004462

[1] Per 100 km.

3. The probability that a pole will not be operating due to a fault on its components is given by (39). It is assumed that states in which failures occur in serial connected components are disjoint and that states in which failures occur in parallel lines are independent from each other. The probability of having a simultaneous fault in both parallel poles is given by (40). Notice that the common AC-E is not included.

$$F(O_{P_i}) = F(O_{CT}) + F(O_V) + F(O_{DC-E}) + \\ + F(O_{C\&P}) + F(O_O) + F(O_C) \quad (39)$$

$$F(O_{P1\&P2}) = F(O_{P1})F(O_{P2}) \quad (40)$$

4. With two possible operating states for each node, the conjunction of the two parallel poles can operate in four different modes: *on - on* in mode 1, *on - off* in mode 2, *off - on* in mode 3 and *off - off* in mode 4. If the TS is operated for a period of time T, the periods in which the system is expected to be operating in each mode are given

by (41).

$$\begin{aligned} T_{mode1} &= [1-F(O_{P1})]\,[1-F(O_{P2})]\,T \\ T_{mode2} &= [1-F(O_{P1})]\,F(O_{P2})\,T \\ T_{mode3} &= F(O_{P1})\,[1-F(O_{P2})]\,T \\ T_{mode4} &= F(O_{P1})\,F(O_{P2})\,T \end{aligned} \tag{41}$$

5. The non-transmitted energy is calculated for the four modes separately. During mode 1 and mode 4 it can be easily deduced that $P_{nontr_{mode1}} = 0$ and $P_{nontr_{mode4}} = P_{avg}$. For mode 2, if it is assumed that the produced power is y, where y is greater than P_1, the non-transmitted power will be $(y-P_1)$. According to Rayleigh distribution there is a specific probability that y will be produced by the windfarm and the average value of the non-transmitted power under those conditions can be calculated using (42). A similar procedure for mode 3 is synthesized in (43).

$$P_{nontr_{mode2}} = \int_{P_1}^{P_1+P_2} (y-P_1)\,F(y)\,dy \tag{42}$$

$$P_{nontr_{mode3}} = \int_{P_2}^{P_1+P_2} (y-P_2)\,F(y)\,dy \tag{43}$$

It should be noticed that the losses in the TS are not considered when calculating the non-transmitted energy in each mode.

6. Finally, the unavailability of the two poles in parallel ($Un_{pole1\&2}$), the unavailability of the common AC auxiliary equipment (Un_{AC-E}) and the unavailability of the entire system (Un_{total}) are assessed using (44).

$$\begin{aligned} Un_{pole1\&2} &= \frac{100}{P_{avg}\,T} \sum_{i=1}^{4} (P_{nontr_{modei}}\,T_{modei})\,T \\ Un_{AC-E} &= 100\,F(O_{AC-E}) \\ Un_{total} &= Un_{pole1\&2} + Un_{AC-E} \end{aligned} \tag{44}$$

Bresesti et al. [16] evaluate the cost of ENS for different TS in a similar way to that described above. The impact of the unavailability of each link under consideration is assessed giving a certain EUR/MWh value to the losses and to the ENS (table 12). The unavailability is calculated through a reliability analysis of the TS considering that the 33 kV system is always available and, therefore, neglecting its failures in the comparison.

Apart from the availability results shown in table 12, the most significant reliability data considered in the analysis is listed below.

Table 12. Unavailability and ENS in different TSs [16]

Case study	Transmission type	Failure mode	Unavailability of hte link	Energy Not Supplied [GWh]	
100 MW	AC	link	0.0061	2.33	
	DC	link	0.0241	9.22	
200 MW	AC	link	0.0061	4.66	
	DC	link	0.0241	18.43	
500 MW	AC 150 kV	cable	0.0082	4.95	8.73
		transformer	0.0040	3.78	
	AC 400 kV	cable	0.0041	7.85	11.64
		transformer	0.0040	3.78	
	DC	CSTs/cable	0.0443	26.63	30.42
		transformer	0.0040	3.78	

- Equal availability figures are considered for both HVAC and HVDC cables. It is assumed that the cable system can experience one failure with a 30 day MTTR during the lifetime of the link.
- 0.00344 year^{-1} is considered to be the failure rate of the distribution transformers, with a 21 day MTTR and independent of its voltage and power rating.
- Due to the lack of VSC-HVDC reliability figures based on operating experience, an overall availability of 99 % for each converter station is assumed, including electronic converters, transformers, reactors, filters, control system and the auxiliary services.

The availability considered for the converter stations has a huge influence on the results of the analysis, but no support data are mentioned in the paper. However, the authors present a set of sensitivity analyses over this availability figure and other uncertain parameters of the assessment.

Finally, Beddard and Barnes [40] present an up to date availability analysis of a VSC-HVDC link based in MMC technology. The methodology is similar to that described above. However, the effort to assess realistic MTTR and MTTF for the components and systems considered in the deterministic statistical analysis performed is especially outstanding. A new concept is included in the MTTR assessment called the *mean time to access*, which is an attempt to clearly define the influence of the stochastic nature of weather conditions in OM strategies and in availability analyses. The different components are also classified according to the transport needed for its maintenance (helicopter, small vessel, large barge with crane, etc.).

4.3. Economic Parameters and O&M Costs

This subsection gathers values scattered in different studies and which have a huge influence on their. Among this values there are:

- The lifetime of the OWPP.
- The prices considered for the purchased and lost energies.
- The OM costs.
- Economic rates such as the interest on commercial loans, the discount factor used to consider the time value of money, the inflation considered to update CAPEX data and the profit expected by the developers.

Regarding the lifetime considered for the OWPP, the values are in the range between 20 and 30 years. A lifetime of 30 years is considered in [9], 25 years in [4] and 20 years in [16, 17, 25]. Bearing in mind the risks still associated with offshore wind projects [41], it does not seem appropriate to consider more than 20 years in the analysis.

In relation to the electricity price, in [16] the losses and the ENS are given an economic value of 97 EUR/MWh during the first five years of operation and a value of 40 EUR/MWh for the following years. The base electricity price considered in [17] is 40 EUR/MWh and sensitivity studies are made in the range between 40 and 120 EUR/MWh.

In reference to the OM costs, in the HVAC and VSC-HVDC transmission system comparison presented in [16], cable maintenance costs are estimated equally for both solutions at 200 kEUR/year and the yearly maintenance costs of the substations are assessed at 0.4 % of the total investments for the correspondent link. In a similar case study reported in [17], the annual maintenance cost of a VSC-HVDC link is taken to be 0.5 % of the capital cost of the components, whereas the lifetime maintenance cost for HVAC equipment is assumed to be 15 % of the investment costs. Also in [17], the maintenance costs of the DFIG turbines are assumed to be 30 % lower than for the SCIG turbines.

In relation to the rates, different values are needed depending on the structure of the economic assessment. For example, in the studies that do not consider the time value of money, the interest rate goes from a 7 % in [4] to a 3 % in [9], with a profit of 10 % and 2 % respectively. In the studies that consider the time value of money the discount factor is in the range between the 7 % fixed in [16] and the 5 % used in [17]. A 2 % inflation is considered in [10, 16].

5. Results of the Key Studies

This section gathers the results presented in the most relevant studies mentioned throughout the chapter. These results are organized in two categories, with one centred on the results of comparisons based on the electric power losses (subsection 5.1.) and the other focused on the results of comparisons based on economic evaluations (subsection 5.2.)

5.1. Comparisons Based on the Electric Power Losses in Each Alternative

The calculation procedures described in [11] for the power losses in different transmission systems, are applied to a number of case studies with the following parameters:

- 500 and 1000 MW offshore wind power plants with 5 MW turbines.
- 9 m/s mean wind speed with Rayleigh probability distribution function.
- D=25 for 500 MW, D=50 for 1000 MW and turbulence intensity I equal to 10 % for the aggregated windfarm model in [39].
- Distances from shore 50, 100, 150 and 200 km.
- The XLPE cables considered for 500 and 1000 MW HVAC alternatives are: 3 or 5 132 kV $3*1000\ mm^2$, 2 or 4 220 kV $3*1000\ mm^2$, 1 or 2 400 kV $3*1200\ mm^2$.
- Converter stations for 500 and 1000 MW VSC-HVDC alternatives (always with 150 kV XLPE cables):
  - One 500 MW or two 250 MW or 600 MW.
  - Two 500 MW or (600+440) MW or (500+600) MW or two 600 MW.

Figure 9 shows that an HVAC system leads to the lowest TS losses for distances up to between 55 and 70 km, depending on the size of the windfarm which influences the configuration of the TS. For longer distances, LCC-HVDC becomes the solution with lowest losses. It should be noted that, as the author chooses the alternative with lower losses for each transmission technology, the chosen HVAC option has a voltage of 400 kV which is not commercially available today for XLPE three-core submarine cables [42,43]. LCC-HVDC is not being currently considered as an option for the transmission systems of OWPP. However, bearing in mind that MMC technology is expected to put VSC converter losses on the same level as that of LCC converters, it is worth considering these results.

The conclusion of the detailed study of the efficiency of an AC windfarm described in [7] are summarized in figure 10. The list below summarizes the characteristics of the case study:

- clusters of ten 3 MVA OWT inter-connected with 20 kV $3*800\ mm^2$ XLPE submarine cable.
- 500 MW OWPP, with a 100 km long TS and 0.8 load factor.
- 245 kV $3*1200\ mm^2$ XLPE submarine cable, laying at 1 m depth in a sea soil with a specific thermal resistance of 0.6 $K \cdot m/W$ and a soil temperature of 15°C.

Overall efficiencies above 92 % can be achieved for nominal load (figure 10(a)). The losses breakdown information for all the operating range of the OWPP is also of interest, although some doubts arise regarding the losses in the internal grid of the windpark. Considering Rayleigh wind probability distribution function, the efficiency of the TS is evaluated by the determination of the time-independent mean value of the loss rate. Figure 10(b)

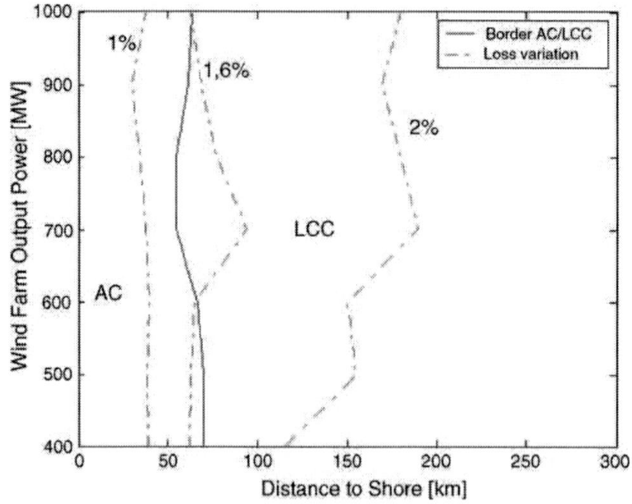

Figure 9. Losses comparison of HVAC and HVDC technologies [11]

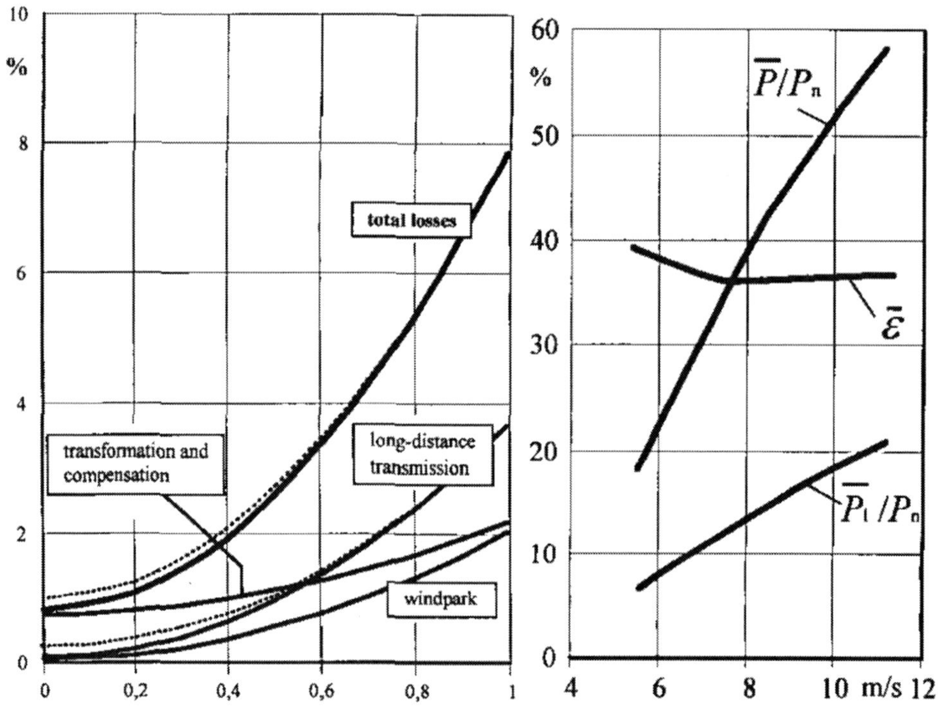

(a) Momentary relative losses (P_1/P_n) like a function of the transmitted power (P/P_n)

(b) Delivered power ($\bar{P}/P_n$), total losses ($\bar{P}_1/P_n$) and total loss rate ($\bar{\varepsilon}_V = \bar{P}_1/\bar{P}$)

Figure 10. Efficiency of a HVAC link in a 500 MW OWPP [7]

shows the strong influence of the mean wind velocity on the energetic gain of the OWPP and on the losses of the TS.

Although no detail is given in [17] about the electric power losses assessment methodology, the results presented seem to be backed up by ABB. This is the reason why figure 11 is included in this section. It summarizes the electric power losses in the transmission link of a 300 MW offshore wind power plant for HVAC and VSC-HVDC technologies. Figure 11(a) presents the momentary losses as a function of the power production, whereas figure 11(b) presents the average losses as a function of the transmission length. According to the latter, the break even distance between HVAC and HVDC technologies is about 80 km, considering the electric power losses of the link with both alternatives.

The total losses for the alternative CS designs presented in [33] are summarized in table 13. Two power factors are considered in the analysis: the worst-case of PF=0.87 lagging and a typical case of PF=0.95 lagging.

Table 13. Annual total losses for different CS designs [33]

	Energy (GVAh)		Percentage of production (%)	
CS design	PF=0.87	PF=0.95	PF=0.87	PF=0.95
Radial	90.7	63.1	2.7	1.9
Single sided ring	49.6	34.1	1.5	1.0
Double sided ring	72.5	51.4	2.6	1.9
Star	86.0	63.9	2.2	1.5
Single return network	71.7	48.1	2.1	1.4

The results of the comparison between different WECS made in [4] are presented in table 14. The first fours rows correspond to AC turbines and the last five rows to DC turbines. As all the turbines have the same rated shaft power and rotor diameter, the differences in the average electric power production depend only on the speed control and on the losses along the drive train. The calculation follows the procedure described in section 3..

Based on these results, the author concludes the following statements:

1. WECS with good performance at cut-in wind speed have high average productions at low average wind speeds, whereas WECS with good performance at rated wind speed have high average productions at high average wind speeds.

2. The fixed speed SCIG and the DFIG are the preferred AC turbines.

   - The DFIG has a good average production for all the average wind speeds.
   - The fixed speed SCIG has better performance at high wind speeds but worse at lower wind speeds.
   - The PMSG with a full-power stator converter has better performance at low mean wind speeds, mainly due to the lower losses of the generator at low speeds when compared with asynchronous machines with a gearbox.

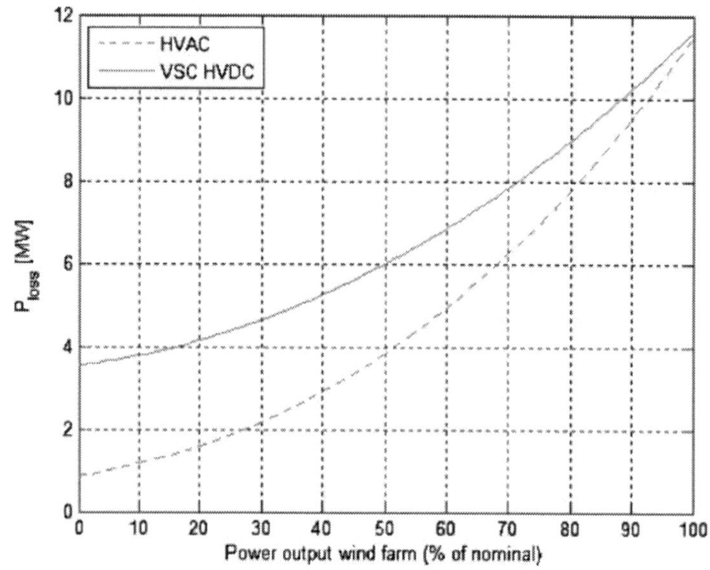

(a) Losses as a function of the generated power for a 50 km transmission

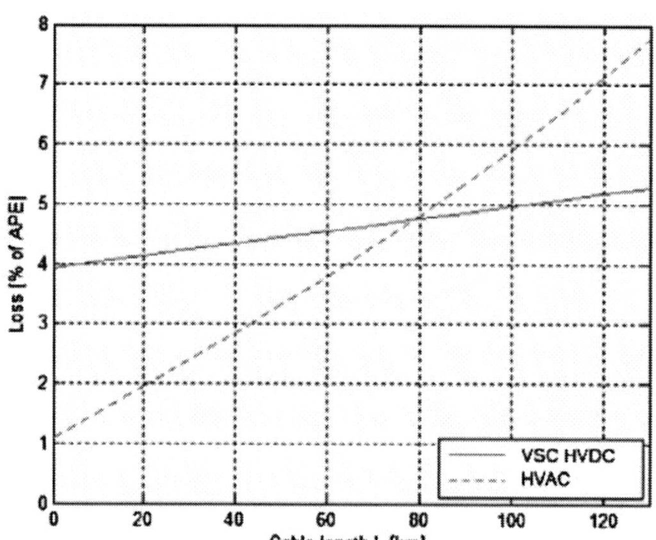

(b) Grade point losses as a function of the cable length

Figure 11. VSC-HVDC and HVAC transmission system losses [17]

3. All DC turbines perform well at low mean wind speeds because of the lower losses of the PMSG. At rated power, their efficiency depends on the WECS topology.

Table 14. Average power production of different WECSs [4]

	Output power (p.u.)		Mean output power (kW)		
WECS	3 m/s	Rated	5 m/s	8 m/s	10 m/s
SCIG[1], fixed speed	0	0.997	271	769	1037
DFIG[2], rotor converter	0.009	0.992	275	771	1038
SCIG, stator converter	0.011	0.964	270	754	1012
PMSG[3], stator converter	0.013	0.967	281	769	1027
PMSG, diode, boost 5 kV	0.015	0.975	288	781	1040
PMSG, diode, boost 40 kV	0.015	0.994	292	793	1058
PMSG, diode, full bridge	0.003	0.971	282	774	1033
PMSG, IGBT, full bridge	0.012	0.968	281	770	1028
PMSG, diode, FBiB	0.014	0.975	286	778	1038

[1] Squirrel Cage Induction Generator.
[2] Doubly Fed Induction Generator.
[3] Permanent Magnet Synchronous Generator.

- The alternative with a 10 kV generator and a boost converter with an output power of 40 kV is preferred. The key point is the lower losses in the diode rectifier and in the boost converter due to the lower currents.
- All the other DC alternatives have 690 V PMSG and the differences in their performance depend on the electronic converter system. The alternative with a diode rectifier and a boost converter offer the best results, whereas the alternative with a IGBT based rectifier and a full bridge offer the poorest performance.

With these results in mind, the author discarded AC topologies with stator converters for the further steps of his study [4], mainly due to its worse performance at rated power owing to the high losses in the converter. However, according to the information presented in subsection 5.1, the overall benefits that enhanced control capability provides in turbines with full rated converters and the possibility of operating the rotor near the optimum tip-speed ratio over a wider range of speeds has proven to be more important than a poorer efficiency of those WECSs at low wind speeds.

The results of the evaluation presented in [12] are summarized in figure 12. The 201 MW OWPP is based on sixty seven 3 MW DFIG turbines and three different electric topologies are considered:

- MVAC collector system with a HVAC transmission system.
- MVAC collector system with a VSC-HVDC transmission system.
- A novel proposal where a single-phase medium-frequency collector system is followed by a single-phase VSC-HVDC converter.

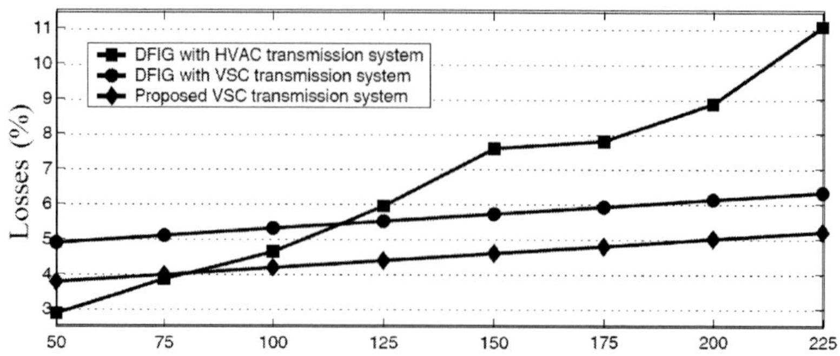

(a) Losses as a function of the transmission distance in km

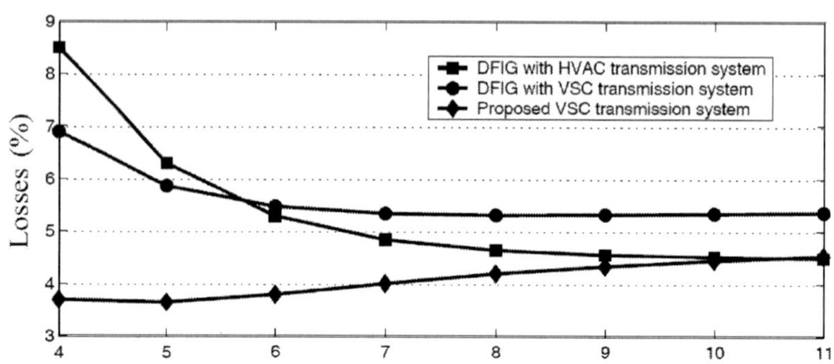

(b) Losses as a function of the average wind speed in m/s

Figure 12. Losses in a 200 MW OWPP with different layouts [12]

According to these results, the break even distance between HVAC and HVDC technologies for the transmission link of a 200 MW OWPP is near 115 km, only taking into account the electric power losses in each alternative.

Finally, it makes sense to mention some recent analyses regarding multiterminal VSC-HVDC schemes in conjunction with fixed speed WECS. A primary design parameter for the CS of an OWPP based on such schemes is the selection of group size, i.e. the number of WECS to be connected to each VSC converter. Jovcic established the main reference in this subject in [44], although other interesting recent works also deal with this issue [45,46].

This group sizing has to start necessarily by fixing the characteristics of the basic structure of the modular CS. Generically, this basic structure is based on several electric generators, a transformer and a VSC converter. Four-pole squirrel-cage 5 MW 4 kV generators are postulated as the most suitable for the proposed OWPP, connected to the TS trough 4/99 kV transformers and $\pm$ 80kV VSC converters connected in parallel to the $\pm$ 80kV HVDC TS [44]. Two directly correlated variables have to be specified to define the CS and the OS, the number of OWT in a group and the rated power of the VSC. Because of the significant cost of electronics, the authors presume that each VSC will be operated at the rated current of the switches (200 A for 5-6 kV IGBTs) which leads to a minimum of 30 MW if the converters are not going to be under-utilized. Then, a range of tests are performed on the PSCAD windfarm model for 2, 4, 6 and 8 turbine group sizes. The energy produced by a generator group is compared with the energy of another group of the same size, under the same wind speed signals, but operated with converters at each individual machine. The results are shown in figure 13. The optimum group size for the proposed OWPP is determined to be six *fixed-speed* turbines per converter.

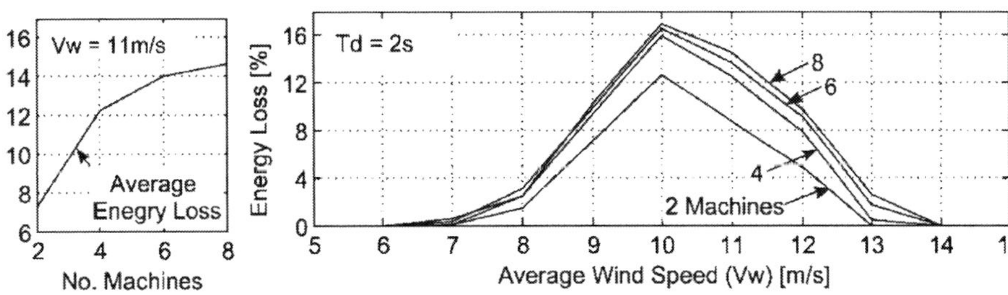

(a) Average loss (b) Average loss as a function of the mean wind speed

Figure 13. Energy losses as a function of the group size in an OWPP based on a multiterminal VSC scheme [44]

5.2. Comparisons Based on the Economic Evaluation of the Alternatives

The results of the evaluation of topologies for 160 MW OWPP presented in [4] are summarized in figure 14(a). The small AC windfarm was best for transmission distances below 10-20 km and the AC/DC windfarm was the best for transmission distances above 130 km. The large AC windfarm is the best option between 20 and 130 km. For the still theoretical DC windfarms, the small DC option is not considered a good solution. Surprisingly, the large DC option is always better than the AC/DC windfarm mainly because of the lower prices of DC cables. The series DC windfarm presents the lowest EPC of all the alternatives for transmission distances above 20 km, although the author shows some uncertainty about how expensive it would be to have the HV insulation in each OWT. For all the alternatives, figure 14(b) shows the influence of the mean average wind speed over the calculated EPC.

In [16], HVAC and VSC-HVDC transmission systems are compared by means of their YZTC. The basic configurations for the analysis of 100, 200 and 500 MW OWPP have been commented on in section 3.. The economic results are summarized in table 15.

On the whole, the HVDC solution is more expensive than the AC solution in all case studies. Among the interesting sensitivity analyses performed on the 100 MW OWF, the one performed on the transmission distance shows a break-even point of costs for AC and DC technologies at a distance of about 90 km. Other sensitivity analyses show that the reduction of the parameters characterizing the converter station (cost, level of losses, and unavailability), which is expected to be given by future technological development, could have a lower but important impact on the economics of the connection. Furthermore, a significant increase in cable installation and protection costs could be an additional reason,

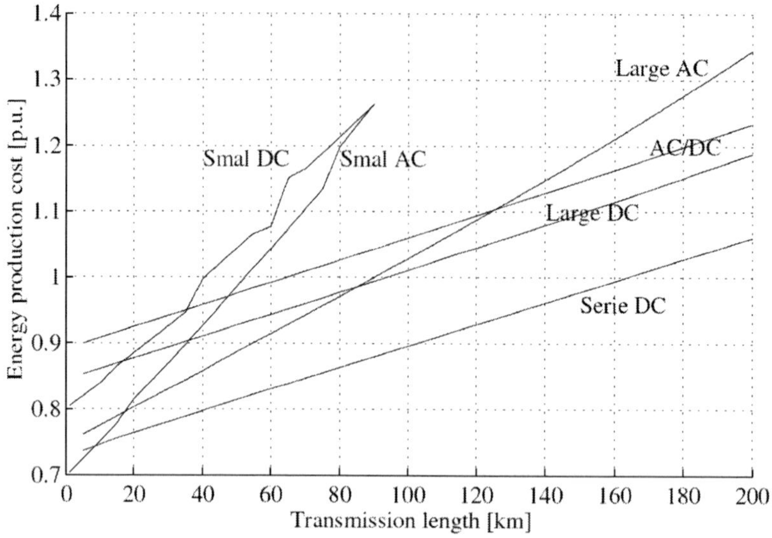

(a) Normalized EPC at an average wind speed of 10 m/s

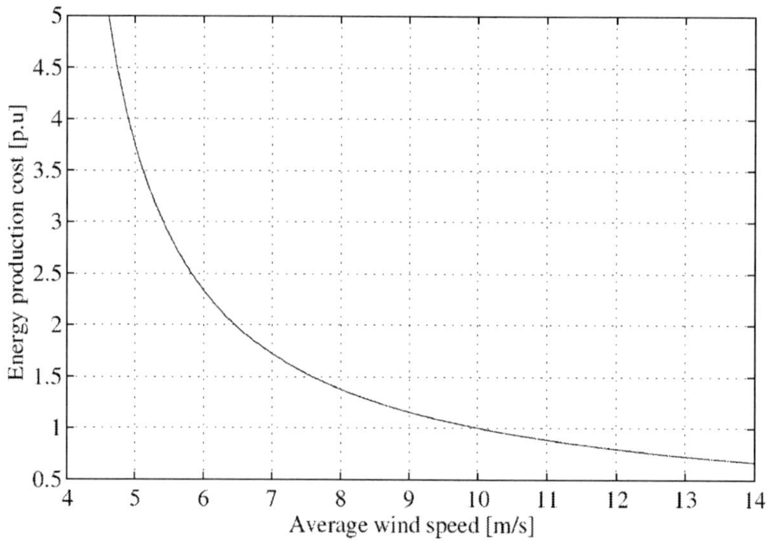

(b) Normalized EPC as a function of the wind speed

Figure 14. COE for alternative layouts in 160 MW OWPPs [4]

which leads to a convergence of total costs for both alternatives. Finally, the authors assert that the comparison becomes more favourable for DC solutions when considering the influence of different types of wind turbines (*fixed-speed* turbines mentioned in subsection 5.1.

Table 15. Project costs for HVAC and HVDC links [16]

		Year Zero Total Costs (MEUR)					
Case	Link	Capex	Losses	OM	ENS	Total	Total/MW
100	AC	58.21	10.03	5.03	1.75	75.02	0.7502
MW	DC	60.39	17.61	5.14	6.91	90.05	0.9005
200	AC	78.86	16.43	5.43	3.49	104.21	0.5211
MW	DC	105.27	33.53	6.75	13.83	159.38	0.7969
500	AC 150 kV	181.66	40.06	11.44	6.55	239.71	0.4794
MW	AC 400 kV	209.36	44.36	10.32	8.73	272.77	0.5455
	DC	260.74	83.59	15.41	22.82	382.56	0.7651

Conclusion

The evaluation methodologies for OWPP electric layouts should consider technical and economical aspects in conjunction. The structure of the techno-economical analyses published during the last decade shows that there has been an evolution in approaches, which leads to the following classification:

1. Methodologies that describe the electric power production assessment in detail but which neglect to consider the time value of money in the economical analysis (2003-2006).
2. Methodologies that consider the time value of money and other economic aspects but do not comment on the procedure followed to assess the electric energy production (2005-2010).
3. Non-analytical stochastic optimization methodologies applied to project planning for OWPP collector and transmission system design (2007-2012).

The main drawbacks of the approaches taken in group 1 is derived from the fact that the analyses are performed on an annual basis. Therefore, neither the influence of the time value of money nor the correlated influence of the variation of the mean wind speed and the electricity price are taken into account. On the contrary, the energy yield assessment procedures are described in detail, including the formulation for the electric power losses assessment.

The approaches taken in group 2 enhance the economic side of the studies taking into consideration the whole lifetime of the OWPP. Therefore, the time value of money is introduced considering a discount factor, as well as other economic variables such as taxes, interests or profit. However, the details of the energy yield assessments in these studies are not disclosed. In some cases the studies of the previous group are cited, whereas in other cases, simulation procedures are just mentioned. Although lifetime analyses are performed, the scenarios derived from the correlated variation of the wind speed and the electricity price are not evaluated. The wake effect within the OWPP is not taken into account.

Regarding the approaches taken in group 3, there is an increasing interest in this field of research, according to the number of papers recently published. However, until actual data from commissioned OWPP is available to check and tune these methodologies properly, I find analytical approaches in conjunction with statistical availability analyses more appropriate for the OWPP electric topology evaluation.

Different methodologies for the assessment of the energy yield and electric losses in pure AC and AC/DC windfarms have been studied. Some of them try to calculate production and losses analytically, whereas others are based on simulations. Among the analytic approaches, Brakelmann's proposal deserves special attention. However, in the case of HVAC transmission systems, the complexity of suggested calculation procedure demands the use of specialized software for its resolution. Among the approaches based on simulation packages, little detail is disclosed about the methodologies, which makes it difficult to replicate the results and to apply the methodology to a different case study. Regarding the inclusion of the losses of individual components in the system level analysis, Lundberg's strategy is a good reference.

The project cost centres generally considered are the initial investment cost, OM costs, electric power losses, ENS, seabed lease, insurance and taxes. The investment cost centres recurrently considered are OWT, foundations, cables, OS structure, compensation equipment, switch gear, transformers and converters. The network connection costs are neglected most of the time.

Some papers present data related to a certain case study, whereas others present cost models applicable for some of the components under certain boundary conditions. However, in most of the latter, the fuzzy definition of the scope of application makes extrapolation of the results difficult. It is also worth noting that for pure AC windfarms there is a number of recent articles disclosing valuable data. On the contrary, for HVDC transmission systems the information disclosed is very limited.

The high cost of the ENS suggests that availability maximization should be one of the main drives in the design of the electric system of OWPP. However, the assessment of the energy yield taking into account the availability of all the systems that make up the OWPP presents, at least, two difficult tasks to cope with. On the one hand, the lack of MTBF and MTTR data published by recognized organizations, and on the other hand, the evaluation of the influence of two stochastic and highly correlated variables: the electric faults and the weather until these faults are solved. This is still an open issue.

The results presented in the most relevant studies published during the last decade have been gathered and commented on. These results seem to have driven the decisions taken in the past, but the scenario is changing as different technical solutions are considered for new projects.

References

[1] "Annual market update 2011", Global Wind Energy Council (GWEC), Tech. Rep., March 2012.

[2] A. Madariaga, I. Martínez de Alegría, J. L. Martín, P. Eguía, and S. Ceballos, "Current facts about offshore wind farms", *Renewable and Sustainable Energy Reviews*, vol. 16, no. 5, pp. 3105–3116, 2012.

[3] K. E. Thomsen, *Offshore Wind: A Comprehensive Guide to Successful Offshore Wind Farm Installation.* Elsever Inc., 2012.

[4] S. Lundberg, "Performance comparison of wind park configurations", Deptartment of Electric Power Engineering, Chalmers University of Technology, Tech. Rep., 2003.

[5] ——, "Evaluation of wind farm layouts", in *Nordic Workshop on Power and Industrial Electronics*, NTNU, Department of Electrical Power Engineering, Trondheim, Norway, Junc 2004.

[6] ——, "Wind Farm Configuration and Energy Efficiency Studies. Series DC versus AC layouts", Ph.D. dissertation, Chalmers University of Technology, Gothenburg, Sweden, December 2006.

[7] H. Brakelmann, "Efficiency of hvac power transmission from offshore-windmills to the grid", in *Proceedings of the 5th conference of the IEEE Power & Energy Society in Europe (PowerTech)*, Bologna, Italy, June 2003.

[8] ——, "Loss determination for long three-phase high-voltage submarine cables", *European Transaction on Electric Power*, vol. 13, no. 3, pp. 193–197, 2003.

[9] T. Ackermann, N. Barberis Negra, J. Todorovic, and L. Lazaridis, "Evaluation of electrical transmission concepts for large offshore wind farms", in *Copenhagen Offshore Wind*, October 2005.

[10] L. Lazaridis, "Economic comparison of HVAC and HVDC solutions for large offshore wind farms under special consideration of reliability", Master's thesis, Royal Institute of Technology, Department of Electrical Engineering, Stockholm, Sweden, 2005.

[11] N. B. Negra, J. Todorovic, and T. Ackermann, "Loss evaluation of HVAC and HVDC transmission solutions for large offshore wind farms", *Electric Power Systems Research*, vol. 76, no. 11, pp. 916–927, 2006.

[12] S. Meier and P. C. Kjaer, "Benchmark of annual energy production for different wind farm topologies", in *Proceedings of the 36th Annual Power Electronic Specialists Conference (PESC)*, Recife, Brazil, June 2005.

[13] R. A. Walling and T. Ruddy, "Economic optimization of offshore substations and collection systems", in *Proceedings of the 5th International Workshop on Large-Scale Integration of Wind Power and Transmission Networks for Offshore Wind Farms*, Glasgow, Scotland, April 2005.

[14] M. Scutariu, "Techno-economical optioneering of offshore wind farms electrical systems", in *Proceedings of the 7th conference of the IEEE Power & Energy Society in Europe (PowerTech)*, Lausane, Switzerland, July 2007.

[15] E. Camm, M. Behnke, O. Bolado, M. Bollen, M. Bradt, C. Brooks, W. Dilling, M. Edds, W. Hejdak, D. Houseman, S. Klein, F. Li, J. Li, P. Maibach, T. Nicolai, J. Patino, S. Pasupulati, N. Samaan, S. Saylors, T. Siebert, T. Smith, M. Starke, and R. Walling, "Wind Power Plant Substation and Collector System Redundancy, Reliability, and Economics (IEEE PES Wind Plant Collector System Design Working Group)", in *Proceedings of the IEEE Power & Energy Society (PES) General Meeting*, Calgary, Canada, July 2009.

[16] P. Bresesti, W. L. Kling, R. L. Hendriks, and R. Vailati, "HVDC connection of offshore wind farms to the transmission system", *IEEE Transactions on Energy Conversion*, vol. 22, no. 1, pp. 37–43, 2007.

[17] B. Van Eeckhout, D. Van Hertem, M. Reza, K. Srivastava, and R. Belmans, "Economic comparison of VSC HVDC and HVAC as transmission system for a 300 MW offshore wind farm", *European Transactions On Electrical Power*, vol. 20, no. 5, pp. 661–671, 2010.

[18] N. Negra, O. Holmstrom, B. Bak-Jensen, and P. Sorensen, "Aspects of relevance in offshore wind farm reliability assessment", *IEEE Transactions on Energy Conversion*, vol. 22, no. 1, pp. 159–166, 2007.

[19] M. Zhao, Z. Chen, and F. Blaabjerg, "Optimisation of electrical system for offshore wind farms via genetic algorithm", *IET Renewable Power Generation*, vol. 3, no. 2, pp. 205–216, 2009.

[20] ——, "Generation ratio availability assessment of electrical systems for offshore wind farms", *IEEE Transaction on Energy Conversion*, vol. 22, no. 3, pp. 755–763, 2007.

[21] M. Zhao, Z. Chen, and J. Hjerrild, "Analysis of the behaviour of genetic algorithm applied in optimization of electrical system design for offshore wind farms", in *Proceedings of the 32th Annual Conference on IEEE Industrial Electronics Society (IECON)*, Paris, France, November 2006.

[22] J. S. GonzÃ¡lez, A. G. G. Rodriguez, J. C. Mora, J. R. Santos, and M. B. Payan, "Optimization of wind farm turbines layout using an evolutive algorithm", *Renewable Energy*, vol. 35, no. 8, pp. 1671–1681, 2010.

[23] J. Serrano, M. Burgos, and J. Riquelme, "An improved evolutive algorithm for large offshore wind farm optimum turbines layout", in *Proceedings of the 9th conference of the IEEE Power & Energy Society in Europe (PowerTech)*, Trondheim, Norway, June 2011.

[24] S. Lumbreras and Ramos, "A benders' descomposition approach for optimizing the electric system of offshore wind farms", in *Proceedings of the 9th conference of the IEEE Power & Energy Society in Europe (PowerTech)*, Trondheim, Norway, June 2011.

[25] L. Fingersh, M. Hand, and A. Laxson, "Wind turbine design cost and scaling model", National Renewable Energy Laboratory of the U.S. (NREL), Tech. Rep. NREL/TP-500-40566, December 2006.

[26] P. Pezzini, O. Gomis-Bellmunt, and A. SudriÃ -Andreu, "Optimization techniques to improve energy efficiency in power systems", *Renewable and Sustainable Energy Reviews*, vol. 15, no. 4, pp. 2028–2041, 2011.

[27] R. C. Dugan, M. F. McGranaghan, S. Santoso, and H. W. Beaty, *Electrical Power Systems Quality*, 2nd ed., S. S. Chapman and S. M. Smith, Eds. Mc Graw - Hill, 2002.

[28] S. Stoft, *Power System Economics. Designing Markets for Electricity*, I. Press, Ed. John Wiley & Sons Ltd, 2002.

[29] "Workshop on the OFTO availability incentive", Office of gas and electricity markets of the U.K. (Ofgem), Tech. Rep., April 2011.

[30] *Std 493-2007 Recommended Practice for the Design of Reliable Industrial and Commercial Power Systems*, The institute of Electrical and Electronics Engineers, Inc (IEEE) Std., 2007.

[31] "The Challenges facing AC offshore substations for wind farms", *Electra (Cigré)*, no. 252, pp. 32–38, 2010.

[32] *Std 493-1980 (Gold Book) Recommended Practice for the Design of Reliable Industrial and Commercial Power Systems*, The institute of Electrical and Electronics Engineers, Inc (IEEE) Std., 1980.

[33] G. Quiñonez Varela, G. W. Ault, O. Anaya-Lara, and J. R. McDonald, "Electrical collector system options for large offshore wind farms", *IET Renewable Power Generation*, vol. 1, no. 2, pp. 107–114, 2007.

[34] "Security standard for offshore transmission networks, by the Offshore Transmission Expert Group of the U.K. (OTEG)", Joint consultation of the Department of trade and industry (DTI) and of the Office of gas and electricity Markets (Ofgem) of the U.K., Tech. Rep., December 2006.

[35] M. Dicorato, G. Forte, M. Pisani, and M. Trovato, "Guidelines for assessment of investment cost for offshore wind generation", *Renewable Energy*, vol. 36, no. 8, pp. 2043–2051, 2011.

[36] "ITC/368/2011 Unit values of reference for the investment and operation and maintenance costs for electricity transport facilities in service as of January 1, 2008", http://www.boe.es/boe/dias/2011/02/25/index.php, Retrieved on July 2012.

[37] E. Stoutenburg and M. Jacobson, "Optimizing offshore transmission links for marine renewable energy farms", in *Proceedings of the OCEANS Conference*, Seattle, USA, September 2010.

[38] M. Banzo and A. Ramos, "Stochastic optimization model for electric power system planning of offshore wind farms", *IEEE Transaction on Power Systems*, vol. 26, no. 3, pp. 1338–1348, 2011.

[39] P. Norgaard and H. Holttinen, "A multi-turbine power curve approach", in *Proceedings of the 4th Nordic Wind Power Conference*, Gothenburg, Sweden, March 2004.

[40] A. Beddard and M. Barnes, "VSC-HVDC Availability Analysis", The University of Manchester, Tech. Rep., December 2011.

[41] J. Babajeva and F. Gronda, "Construction risk in offshore wind farms", Fitch Ratings - Global Infrastructure & Project Finance, Tech. Rep., May 2012.

[42] *XLPE Land Cable Systems - User's guide*, 5th ed., ABB High Voltage Cable Unit, 2010.

[43] *XLPE Submarine Cable Systems. Attachment to the XLPE Land Cable Systems - User's Guide*, 5th ed., ABB High Voltage Cable Unit, 2010.

[44] D. Jovcic and N. Strachan, "Offshore wind farm with centralised power conversion and DC interconnection", *IET Generation, Transmission & Distribution*, vol. 3, no. 6, pp. 586–95, 2009.

[45] O. Gomis-Bellmunt, A. Junyent-Ferre, A. Sumper, and S. Galceran-Arellano, "Maximum generation power evaluation of variable frequency offshore wind farms when connected to a single power converter", *Applied Energy*, vol. 87, no. 10, pp. 3103–3109, 2010.

[46] D. Jovcic, "Offshore wind farm with a series multiterminal CSI HVDC", *Electric Power Systems Research*, vol. 78, no. 4, pp. 747–755, 2008.

In: Renewable Energy
Editor: Andreas Poullikkas

ISBN: 978-1-62618-231-8

Chapter 8

EXPLOITATION OF OLIVE SOLID WASTE FOR ENERGY PURPOSES

Paris A. Fokaides[1,*] ***and Polycarpos Polycarpou***[2]
[1]School of Engineering and Applied Sciences,
Frederick University, Cyprus
[2]Production Division, Agricultural Research Institute, Cyprus

ABSTRACT

The promotion of biomass for energy purposes consists one of the main initiatives of the European energy policy. The Mediterranean region possesses a very promising source of biomass, olive solid waste (OSW) that can potentially be exploited for energy purposes.

In the recent past, many studies have been conducted for the investigation of the performance of OSW as an energy source. Although a lot of knowledge has already been gained with regard to the appropriate conditions for an efficient and „clean" OSW energy conversion, this type of biomass still remains unexploited, resulting to a significant loss of green energy on an annual basis.

This chapter aims to present the current research status on the exploitation of OSW for energy purposes.

Research efforts conducted in the field of OSW direct combustion, pyrolysis and gasification are presented and discussed. European standards in the field of olive solid waste quality classes and specifications are also presented and discussed. The most promising technology solutions that could potentially be applied are identified and further analyzed.

Emphasis is given on the gas emissions resulting from the OSW energy conversion as well as the necessary fuel pretreatment conditions.

Keywords: Biomass, olive solid waste, renewable energy sources, combustion, gasification

* Corresponding Author address: Y. Friderikou 7, 1036, Nicosia, Cyprus, P.O.Box 22101, 1517, Nicosia,Cyprus, Email: eng.fp@frederick.ac.cy.

1. Introduction

Biomass has the highest contribution to the generation of renewable energy addressing all three energy vectors: heating, electricity and transport. The global demand for biomass sources for energy purposes is expected to increase significantly. The use of biomass is currently supported worldwide, because it can significantly reduce greenhouse gas emissions. The European Directive 2009/28 reaffirmed the Community's long-term commitment to the EU-wide development of renewable energies and endorsed a binding minimum target of 10% to be achieved by all Member States for the share of biofuels by 2020. Bioenergy can make an important contribution to a sustainable energy supply worldwide. The Mediterranean region possesses a very promising source of biomass, olive solid waste (OSW) that can potentially be exploited for energy purposes. The olive (Olea europaea) is a species of small evergreen tree in the family Oleaceae, native to the coastal areas of Mediterranean. It is short and squat, and usually does not exceed 10 metres in height. Its fruit, called the olive, is of a significant agricultural importance in the Mediterranean basin as the source of olive oil. The fruit is a small drupe, thinner-fleshed, usually 2 centimeters long, and it contains a seed commonly referred to as a pit or a rock or a stone. Olives are harvested in the Northern hemisphere in the autumn and winter. Blond olives are picked from the middle of October to the end of November and black olives are collected from the middle of November to the end of January or early February.

Olive Production

Over 90% of the total world production of olive oil is situated in the Mediterranean basin. The southern European region represents over 75% of the world olive oil production, the largest producers being Spain, Italy and Greece. About 70% of the world's olive oil is produced in EU countries, the largest producers being the aforementioned three, plus Portugal, whose production is one order of magnitude lower; France, Croatia and Cyprus come next, with a production that is about one order of magnitude lower than that of Portugal's. More than 14.000.000 tons of olives are grown annually within the EU, from which over 2.500.000 tons of olive oil are produced in approximately 12.000 olive mills. The olive and olive oil industry is an important employer in the agro-food-sector in Europe with over 800 000 employees [1].

Olive Oil Extraction

Olive oil extraction can either be performed by olive presses or by decanters. The modern method of olive oil extraction uses industrial decanters to separate all the phases by centrifugation. The decanter centrifuge is commonly used to rapidly separate the solids suspended in a fluid, allowing the phases to be readily separated according to their different densities, namely solids, vegetation water and oil. The working principle of the decanter centrifuge is described in [2]. The separation is enhanced by the use of a rotating bowl. The presence of a screw conveyor concentrically located within this bowl, is the distinctive aspect of a decanter centrifuge.

Olives are typically fed into the bowl through a stationary feed tube. The feed discharges from the feed tube into an accelerating feed zone that distributes the slurry into a bowl. The elevated gravitational level causes sedimentation of the solids against the bowl wall.

At the conical end of the bowl the solids are conveyed out of the liquid and are centrifugally consolidated before being discharged. Solids are discharged to a suitable conveyance located under the frame at the conical end. Meanwhile water, travels through the conical section of the bowl, losing solids as it flows to the adjustable dams which determine the liquid level inside the unit.

With the three phase oil decanter, a portion of the oil polyphenols is washed out due to the higher quantity of added water, producing a larger quantity of vegetation water that needs to be processed. Sacrificing part of its extraction capability, the two phase oil decanter uses less added water thus reducing the phenol washing and the olive paste is separated into oil and wet pomace.

This type of decanter, instead of having three exits has only two.

Table 1. Countries olive production, 2010 [1]

Rank	Country/Region	Production (in tons)	Cultivated area (in hectares)
1	Spain	8,014,000	2,092,800
2	Italy	3,170,700	1,190,800
3	Greece	1,809,800	834,200
4	Morocco	1,483,510	735,400
5	Turkey	1,415,000	826,199
6	Syria	960,400	647,500
7	Tunisia	876,400	1,645,100
8	Egypt	611,900	128,700
9	Algeria	555,200	316,300
10	Portugal	239,600	250,200
—	World	20,578,186	9,398,623

The three phase decanter procedure produce two waste streams: The residual olive solid waste (OSW) and the olive oil wastewater (OOWW). The residual solids produced can be further treated with hexane.

A residual oil is extracted and called olive kernel oil. The rejection of this extraction is the Exhaust Foot Cake.

This by-product is used as a combustible in traditional furnaces of small building material factories (pottery works, brick manufactures, etc.) and greenhouses. [3]

The biggest problem in olive oil process is the Olive oil mill wastewaters, which is high in volume and polluting charge.

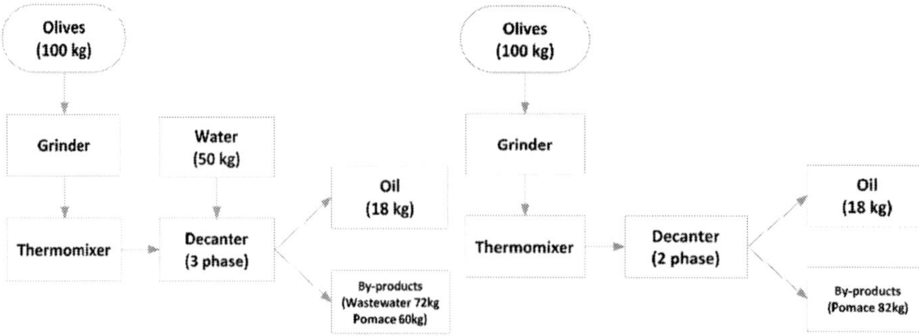

Figure 1. Two (left) and three (right) phase olive oil production process by-products [3].

Olive Solid Waste Energy Value

Olive solid waste is useful for the production of seed oil and kernel wood. The energy content of olive solid waste is defined by the heating power and the results of the proximate and ultimate analyses. The heating power of the biomass can be determined either experimentally or by calculation from the results of the ultimate and/or proximate analyses. Table 2 presents the experimentally determined values of the properties of interest for olive solid waste.

Table 2. Experimentally determined values of olive solid waste properties

	Celma et al. [4]	Olerto et al. [5]	Arvelakis et Koukios [6]	Jenkings et al. [7]	Polycarpou [8]
HHV exp [MJ/kg]	17,8	17,7	21,2	21,59	21,25
LHV exp [MJ/kg]	15,58	16,2	-	-	19,81
Proximate analysis (wt %)					
Fixed carbon	15,87	19,7	19,4	16,28	-
Volatile matter	64,77	66,9	76,0	82,	-
Ash	7,16	5,8	4,6	1,72	2,01
Moisture	12,2	7,6	9,5	-	-
Elemental analysis (wt %)					
C	47,16	50	50,7	52,8	51,65
H	5,62	6,5	5,89	6,69	6,51
O	35,41	36,3	36,97	38,25	38,00
N	4,48	0,8	1,36	0,45	1,84
S	0,17	0,1	0,3	0,05	-
Ash	7,16	5,8	4,6	1,72	2,01

2. EUROPEAN STANDARD 14961 SERIES: SOLID BIOFUEL SPECIFICATIONS AND CLASSES

In order to provide unambiguous and clear classification principles for solid biofuels, the European Committee for Standardization, CEN (TC335) prepared a series of technical specifications for solid biofuels, specifically the EN 14961 standards series. EN 14961 standards determine the fuel quality classes and specifications for solid biofuels. According to the mandate given for the standardisation work, the scope of the CEN/TC 335 includes solid biofuels originating from products from agriculture and forestry as well. The classifications given in this European Standard are provided with the objective of using biomass as a biofuel, and therefore do not deal with all other uses. The biofuels covered by this European Standard are identical to the fuels exempted from the Directive 2000/76/EC on incineration of waste.

Table 3. Specification of properties according to EN 14961, Part 1 [9]

	Master table		
	Origin: According to 6.1 and Table 1		Fruit biomass (3.2.1.2, 3.2.1.4,3.2.2.2,3.2.2.4)
	Traded Form		Grain or seed, kernel
Normative	**Dimensions** (mm)		
	Diameter (*D*) [a], CEN/TS 15149-1, CEN/TS 15149-2		
	D 03	1 mm ≤ *D* ≤ 3,15 mm	
	D 05	1 mm ≤ *D* ≤ 5 mm	
	D 10	1 mm ≤ *D* ≤ 10 mm	
	D 10+	*D* > 10 mm (maximum value to be stated)	
	Moisture, M (w-% as received) EN 14774-1, EN 14774-2		
	M10	M ≤ 10 %	
	M15	M ≤ 15 %	
	Ash, A (w-% of dry basis) EN 14775		
	A1.5	≤ 1,5 %	
	A2.0	≤ 2,0	
	A3.0	≤ 3,0	
	A5.0	≤ 5,0 %	
	A7.0	≤ 7,0 %	
	A10.0	≤ 10,0 %	
	A10.0+	> 10,0 % (maximum value to be stated)	
	Additives (w-%)		Type and amount of additive have to be stated
	Net calorific value as received, Q (MJ/kg) [b], EN 14918		Minimum value to be stated
	Nitrogen, N (w-% of dry basis) CEN/TS 15104		
	N1.0	≤ 1,0 %	
	N1.5	≤ 1,5 %	
	N2.0	≤ 2,0 %	
	N3.0	≤ 3,0 %	
	N3.0+	> 3,0 % (maximum value to be stated)	
Informative	**Amount of fines, F** (w-%, < 1 mm) CEN/TS 15149-2		
	F1.0	≤ 1,0 %	
	F1.0+	> 1,0 % (without additive)	
	Bulk density (BD) as received (kg/m³ loose) EN 15103		Recommended to be stated if traded by volume basis
	Chlorine, Cl (w-% of dry basis) CEN/TS 15289		
	Cl 0.10	≤ 0,10 %	
	Cl 0.15	≤ 0,15 %	
	Cl0.15+	> 0,15 % (maximum value to be stated)	
	Sulphur, S (w-% of dry basis) CEN/TS 15289		
	S0.15	≤ 0,15 %	
	S0.20	≤ 0,20 %	
	S0.20+	> 0,20 % (maximum value to be stated)	
	Ash melting behaviour (°C) CEN/TS 15370-1		Deformation temperature, DT should be stated

[a] 5 w-% may have diameter over the class.

[b] Additives can reduce net calorific value.

Table 4. Specification of elemental composition according to EN 14961, Part 1 [9]

Parameter	Unit	Olive cake		
		Crude olive cake 3.2.1.4	Exhausted olive cake 3.2.2.4	Olive kernels 3.2.1.2
Ash	w-% d	10	3,4 to 11,3	1,2 to 4,4
Gross calorific value $q_{V,gr,d}$	MJ/kg d	19,4 to 21,4	18,1 to 21,6	18,6 to 20,8
Net calorific value $q_{p,net,d}$	MJ/kg d	18,1 to 20,7	13,9 to 19,2	17,3 to 19,3
Carbon, C	w-% d	50	48 to 52	45,7 to 52,3
Hydrogen, H	w-% d	6,9	4,6 to 6,3	6,1 to 6,8
Oxygen, O	w-% d	30	33	38,5 to 42,1
Nitrogen, N	w-% d	1,5	1,4 to 2,7	0,8 to 1,6
Sulphur, S	w-% d	0,2	0,0 to 0,5	0,0 to 0,5
Chlorine, Cl	w-% d	0,2	0,1 to 0,4	0,1 to 0,4
Aluminium	mg/kg d	1250	2 700	559
Calcium, Ca	mg/kg d	6 900	17 200	968
Iron, Fe	mg/kg d	1 000	1 900	391
Potassium, K	mg/kg d	6 000 to 16 000	17 500	6 950
Magnesium, Mg	mg/kg d	3 400	4 000	316
Manganese, Mn	mg/kg d	< 26	17 to 44	12
Sodium, Na	mg/kg d	44 to 1 000	250 to 450	120
Phosphorus, P	mg/kg d	2 450	30 to 1 750	590
Silicate, Si	mg/kg d	14 to 6 600	20 to 11 850	9 to 3 500
Titanium,Ti	mg/kg d	53	145	39
Arsenic, As	mg/kg d	0,4	4	0,8
Cadmium, Cd	mg/kg d	< 0,1	< 0,5	0,2
Chromium, Cr	mg/kg d	3	3 to 13	3
Copper, Cu	mg/kg d	14	10 to 20	9
Mercury, Hg	mg/kg d		0,1	
Nickel, Ni	mg/kg d	2	2 to 17	0,05
Lead, Pb	mg/kg d	2	15	2,1
Vanadium, V	mg/kg d		5	
Zinc, Zn	mg/kg d	19	19	7
Cobalt, Co	mg/kg d		1	
Silver, Ag	mg/kg d		4	
Tin, Sn	mg/kg d		4	

Olive residues from olive pressing are also considered in this standard under the category fruit biomass (Stone/kernel fruits). The most significant characteristics are given in the fuel specification Appendixes. These characteristics vary for different traded form, while the most significant characteristics for all biofuels are moisture content (M), particle size/dimensions (P or D/L) and ash content (A). For example, the average moisture content of fuels is given as a value after the symbol (e.g. M20), which means that the average moisture content of the fuel shall be 20 wt%. Some characteristics, e.g., net calorific value and bulk density, are voluntary, informative. Table 3 provides the required properties, according to the standard, for olive residues and typical values of olive residues elemental analysis, are given in Table 4.

3. Waste to Energy Treatment Technologies: Applications to Olive Solid Waste

The major components of non-food biomass that have a significant heating value and can be exploited for energy production purposes are cellulose, hemicelluloses and lignin. Cellulose is the main constituent of plant cell wall conferring structural support and is also present in bacteria, fungi, and algae. Hemicellulose is the second most abundant polymer (20–50% of Lignocellulose biomass) and differs from cellulose in that it is not chemically homogeneous. Lignin is the third most abundant polymer in nature. It is present in plant cell walls and confers a rigid, impermeable, resistance to microbial attack and oxidative stress [10]. Cellulosic materials are particularly attractive as feedstocks for biofuel production because of their relatively low cost, great abundance and sustainable supply [11]. Cellulose is the most abundant biopolymer on earth and biofuel production from cellulosic biomass has become the major focus of intensive research and development [12]. In Table 5 the biopolymer composition of olive solid residues is provided [13]

Table 5. Biopolymer composition of olive solid waste [13]

Biopolymer composition	wt. %
Cellulose	23.21
Hemicellulose	35.62
Lignin	34.98
Holocellulose	58.83
Organic extractive	9.59
Oil	5.46
Protein	2.16

Pretreatment Procedures

Lignocellulosic material often requires pretreatment to liberate the sugars contained within cellulose fibres embedded in the hetero-matrix of plant cell walls. The process that converts any source of lignocellulosic biomass from its native form, which is recalcitrant to hydrolysis with cellulose enzyme systems, into a form which enzymatic hydrolysis is effective is referred to as 'pretreatment' in bioprocess engineering [11]. An ideal pretreatment step should render lignocellulosics completely susceptible to the action of cellulases. Various types of pretreatment have been developed, but key bottleneck in biofuel production is the initial conversion of biomass to sugars.

Several studies have been conducted [14-23] in the field of solid biomass pretreatment. The production of fermentable sugars by dilution is a method that has been widely applied. Cara et al. (2008) [14], for example studied the production of fermentable sugars from olive tree biomass by dilute acid pretreatment and further saccharification of the pretreated solid residues. Pretreatment was performed at several sulphuric acid concentrations while temperature was in the range 170–210°C. Chen et al. (2012) [15] evaluated two commonly alkaline pretreatment processes based on aqueous ammonia and lime under different enzyme hydrolysis models to provide comparative sugar production performance. The findings of this study suggested that pretreatment methods and enzyme loading model should be considered simultaneously to enhance cellulosic biomass degradation. Weerachanchai et al. (2012) [16] also presented the improvement of biomass properties by pretreatment with ionic liquids. A sequence of the effectiveness of ionic liquid for biomass pretreatment processes was also defined. The increase of pretreatment temperature and the decrease of biomass particle size rendered higher sugar conversion, lignin extraction and lower crystallinity index. Park and Kim (2012) [17] compared the efficacy of soaking and percolation pretreatments with alkaline solutions for lignocellulosic biomass. Various biomasses underwent enzymatic hydrolysis and were pretreated by soaking processes in various alkaline solutions including sodium hydroxide, potassium hydroxide, aqueous ammonia and sodium carbonate. The enzymatic digestibility was found to be relatively low, in the range of 3.2%-8.3%. Nevertheless, the enzymatic digestibility of pretreated solid waste residue was increased by five times compared to that of the initial biomass. The potential of 1-buthyl-3-methylpyridinium chloride, as a pretreatment solvent for lignocellulosic biomasses was investigated by Uju et al. (2012). [18] The yields of regenerated biomasses ranged between 35% and 96%, and varied according to the pretreatment time, type of ionic liquid and biomass. The pretreatment of the biomass with 1-buthyl-3-methylpyridinium chloride resulted in up to 8-fold increase in the cellulose conversion when compared with the untreated biomass.

Fungal pretreatment is also a very promising preatreatment method for solid biomass, as it reduces the recalcitrance of lignocellulosic biomass by lignin-degrading microorganisms and thus potentially provides an environmentally-friendly and energy-efficient pretreatment technology for biofuel production. Wan and Li (2012) [19] provided an overview of the current state of fungal pretreatment by white rot fungi for biofuel production. The specific topics discussed were the enzymes involved in biodegradation during the fungal pretreatment and the operating parameters governing performance of the fungal pretreatment. This study also presented the effect of fungal pretreatment on enzymatic hydrolysis and ethanol production, as well as efforts for improving enzymatic hydrolysis and ethanol production through combinations of fungal pretreatment and physical/chemical pretreatment. The treatment of lignocellulosic biomass with lignin-degrading enzymes isolated from fungal pretreatment, with a comparison to fungal pretreatment was also discussed and modeling, reactor design, and scale-up of solid state fungal pretreatment was presented. With regard to thermal pretreatment techniques, Zhu and Pan (2010) [20] presented a comprehensive discussion of the key technical issues in thermal pretreatment. This study defined the barriers to efficient cellulose saccharification and the pretreatment energy consumption as well as evaluation criteria for the pretreatment performance. An important outcome of this study is the pretreatment energy efficiency (kg/MJ) index, which is based on the total sugar recovery (kg/kg wood) divided by the energy consumption in pretreatment (MJ/kg wood). This efficiency index was used to evaluate the performances of three of the most promising pretreatment technologies: steam explosion, organosolv, and sulfite pretreatment to overcome lignocelluloses recalcitrance (SPORL). It was found that SPORL was the most efficient process and produced the highest sugar yield. Biswas et al. (2011) [21] also presented the steam explosion (SE) pretreatment to separate the main components of woody biomass. In this study, it was found that the SE pretreatment enhanced pellet density, impact resistance, and abrasive resistance of pellet. It was also observed that small degradation in ash fusion characteristics and char reactivity increased. Acharjee et al. (2011) [22] measured the equilibrium moisture content (EMC) of raw lignocellulosic biomass, along with four samples subjected to thermal pretreatment, for relative humidities ranging from 11% to 97%. It was found that pretreated biomass revealed EMC below that of raw biomass, an indication that pretreated biomass, both dry and wet, was more hydrophobic than raw biomass. The EMC results delivered in this study were also correlated with results derived from numerical models, resulting to a physical insight into the water activity in lignocellulosic biomass. Although extrusion based biomass pretreatment procedures remain an attractive option due to their flexibility to carry out chemical reactions under relatively high stresses, temperatures and pressures, they are rarely utilized due to rheological constraints. Senturk-Ozer et al. (2011) [23] demonstrated gelation agents that can render biomass suspensions processable. Specifically, carboxy methyl cellulose, was employed in conjunction with alkaline pretreatment of hardwood-type biomass and enabled separation of lignin from cellulose fibers. Furthermore, recycled black liquor, obtained upon pretreatment, was determined to be as effective as carboxy methyl cellulose for rendering biomass suspensions flowable by again facilitating the concomitant application of high shearing stresses and chemical treatment for the pretreatment of the biomass in the extruder.

Organic Waste to Energy Technologies

Organic waste-to-energy technologies can be broadly classified as either biochemical, chemical or thermal processes.

1) The thermal processes for converting biomass include combustion, gasification and pyrolysis. Examples of the former are pellets-fueled boilers. An example of gasification is biomass partial oxidation to yield a low-energy-value fuel gas, which drives a gas turbine to generate electric power. Example of the rapid pyrolysis is the thermal decomposition of wood wastes to yield liquid fuel oils.
2) Biochemical treatment technologies include anaerobic digestion and fermentation. Anaerobic digestion is the decomposition of wet and green biomass through bacterial action in the absence of oxygen to produce a mixed gas output of methane and carbon dioxide known as biogas. Organic wastes can also be converted to ethanol, through bacterial fermentation, which converts carbohydrates in the feedstock to ethanol.
3) Esterification is the main process of chemical processing. In esterification, biodiesel can be produced from straight vegetable oil, animal oil/fats and waste oils.

A primary factor complicating the description of the thermal waste to energy processing of olive solid waste and in general of biomass is the wide variability of its composition. The elemental composition of the olive solid waste has a widely variable composition. As might be suggested by the heterogeneous nature of biomass and the wide variation in its composition, all olive solid wastes do not thermally react in the same manner.

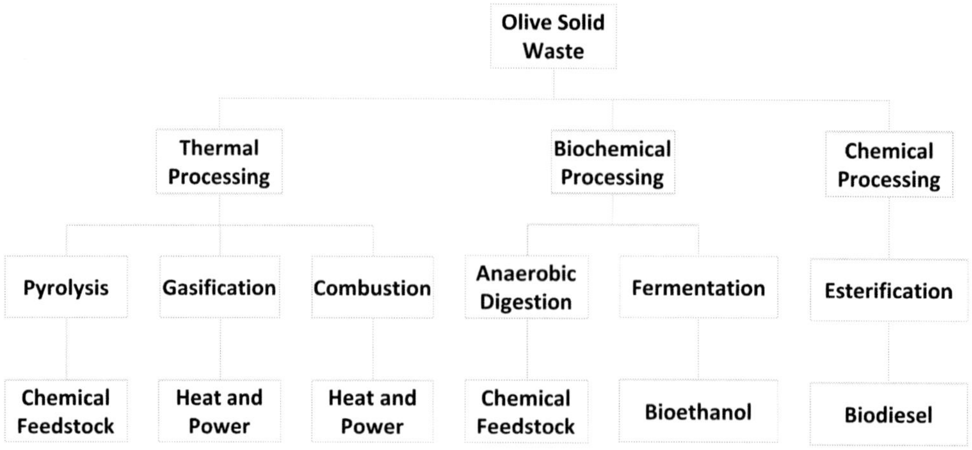

Figure 2. Waste to energy pathways.

Olive Solid Waste Pyrolysis

Pyrolysis reactions offer an alternative way of producing fuels and consequently energy from olive solid waste [24]. Pyrolysis, which is the thermal decomposition of materials in the absence of oxygen, is the simplest and oldest method of processing waste and hydrocarbons in general to produce an utilizable fuel [25]. Pyrolysis according to the operating conditions can be slow or fast. Slow pyrolysis is used rather for carbon materials and solid biofuels production, whereas fast is used for liquid and gaseous biofuels production. Fast pyrolysis of agro-biomass, as for any kind of biomass, requires rapid heating and relatively high temperatures of reactions. These reactions imply a great number of different elementary processes including the thermal decomposition of the macro polymers, the vaporization of liquid components, carbonization reactions and vapor phase cracking. They are of competitive and consecutive types and are strongly influenced by other phenomena, such as sizes of the pyrolysed particles, transfer (heat and mass) efficiencies and temperature and fluid flow characteristics in the reactor. [26].

In the literature, several studies have been reported on the pyrolysis of olive-oil residues. Garcia-Ibanez et al. have studied thermal degradation of olive-oil residue in air atmosphere at three heating rates, drawing the conclusion that increasing the heating rate increases the thermal degradation rate [27]. Blanco Lopez et al. have studied the pyrolysis of olive stones. According to this work, the pyrolysis products with the highest heating value were the liquids. In this study the composition of released pyrolysis gases was also defined, showing that these were a mixture of CO_2, CO,H_2, CH_4, C_2H_4, and C_2H_6. [28]. Zanzi et al. have studied fast pyrolysis of olive waste and wheat straw at high temperatures in a free fall reactor. The higher content of lignin in olive waste resulted in a higher char yield in comparison to other examined forms of biomass. Also higher temperatures were found to favour cracking of the hydrocarbon in the gaseous products and thus increase in hydrogen yield [29]. Putun et al. have studied the slow pyrolysis of olive residue under various pyrolysis atmospheres. They concluded that liquid product yield increased significantly under nitrogen and steam atmosphere compared with the static atmosphere [30].

Olive Solid Waste Gasification

Gasification is a process that converts organic or fossil based carbonaceous materials into carbon monoxide, hydrogen and carbon dioxide. This is achieved by reacting the material at high temperatures (>700 °C), without combustion, with a controlled amount of oxygen and/or steam. The resulting gas mixture is called syngas (from synthesis gas or synthetic gas) or producer gas and is itself a fuel. The power derived from gasification and combustion of the resultant gas is considered to be a source of renewable energy if the gasified compounds were obtained from biomass.

Co-gasification of coal and olive solid waste produces gases that can be used either for energy or chemical purposes, but they may contain undesirable components, such as heavy hydrocarbons, tars, solids and gaseous pollutants, which may interfere with their required applications. Several methods can be used to reduce the undesirable gases components, either by cleaning the gas or by preventing the release of unfavourable gas components during gasification. Catalysis may be used in gasification processes to improve the quality of the product gas by promoting tars and other hydrocarbons further reactions that result from incomplete gasification of pyrolysis products. [31]

Catalysts may be incorporated in the gasifier or may be added to a secondary gas treatment reactor to improve tar cracking, methanation or steam reforming of the produced hydrocarbons and tars [32]. Several catalysts have been tested, either for coal or for biomass gasification: dolomite [33], Fluid Catalytic Cracking Catalysts (FCC) [33], metal catalysts (especially nickel) [34], Y-Zeolites, precious metals [32], alkaline catalysts (LiCl, NaCl and KCl) [35], commercial steam reforming catalysts [36], etc.

However, there is not concise information about catalyst performance and suitability and there is discrepancy in results from different groups. [31]

Olive Solid Waste Combustion

According to Turns (1996) [37], solids combustion differs in a fundamental way from liquid or gaseous fuel combustion in that, heterogeneous chemical reactions assume importance. A heterogeneous reaction is a reaction involving species existing in different physical states, i.e. gas-liquid or gas-solid reactions. The general scheme for olive solid waste combustion is for the moisture to be driven off, followed by the evolution of the volatiles. The volatiles may burn homogeneously in the gas phase or at the biomass particle surface. The solid biomass may swell and become porous as the volatiles are given off. The material left after devolatilization is the char and its associated mineral matter. The char then reacts with oxidants and the mineral matter is transformed into ash, slag, and fine particulate fume in various portions [37].

Combustion applications are generally fuel-flexible when it comes to variations in biomass composition. The main biomass fuel properties that affect the longterm technical and economic success of combustion are moisture content, fixed carbon and volatiles content, impurity (S, N, Cl) concentrations, and ash content. Collectively, these properties affect the energy density of biomass fuels. From an energy perspective, maximum system efficiencies are possible with dry, low-ash biomass fuels. However, drying biomass can be costly and low-ash biomass resources are generally considered premium fuels that tend to be more expensive. [38].

The use and the image of fossil solid fuels can be enhanced in many cases by co-firing it with various biomass fuels. The benefits of co-combustion to the electricity generation community and to the environment are obvious.

The utilisation of olive solid waste together with coal in power plant seems to be a promising technique for the future. Moreover, fluidised bed combustion has been shown to be a versatile technology capable of burning practically any coal and olive solid waste combinations with low emissions. The significant advantages of fluidised bed combustor over conventional combustor include their compact furnace, simple design, effective burning of wide variety of fuels, relatively uniform temperature and the ability to reduction emissions such as SO2 and NOx emissions.[39]

Armesto et al. [39] studied the emissions resulting from fluidised bed co-combustion of coal and olive solid waste. According to their findings the share of olive solid waste in the mixtures had not any significant effect on the combustion efficiency but the type of coal did have great influence on combustion efficiency.

The amount of olive solid wastes in co-combustion mixtures caused a slight decrease of the NOx emissions and a slight increase of the N_2O emissions, due to the decreasing of the flame temperature caused by high moisture content in the olive solid waste. Cliffe and Patumsawad [40] studied the co-combustion of olive cake with coal in a fluidized bed combustor. They determined that the combustion efficiency of 20% olive cake and coal mixture was less than that of the coal combustion by itself, but it was higher than that of 10% olive cake and coal mixture, due to lower elutriation from the bed.

A 20% olive oil waste mixture gave also a higher CO emission than both 100% coal firing and 10% olive cake waste mixture. The co-combustion of olive cake with lignite coal in a circulating fluidized bed combustor was studied by Atimtay and Topal [41]. Various runs were conducted with mixtures of olive cake and lignite, being burned with various excess air ratios (λ). It was found that if λ was less than 40–50%, appreciable amounts of CO and unburned hydrocarbons were formed and the combustion efficiency dropped to 84–87%. CO. CmHn emissions were lower for coal than the emissions for olive cake and coal mixtures at the same λ and temperature.

A secondary air addition in the freeboard was suggested to increase the combustion efficiency. Atimtay and Varol [42] investigated the co-combustion of coal and olive cake in a bubbling fluidized bed with secondary air injection. The results of this study showed that the secondary air injection into the freeboard can be a useful solution to decrease the CO and hydrocarbon emissions and to increase the combustion efficiency. For the setup used in this study, co-combustion of olive cake with 25 wt% coal for excess air ratio of 1.7–1.9, secondary air flowrate of 50–70 L/min, and superficial gas velocity of 0.66–0.71 m/s was found to be the optimal combination both to obtain combustion efficiency higher than 99%, and to obtain NOx and SO2 emissions which are below the limits.

Conclusion

The present global effort to curb greenhouse gases emissions involve substituting fossil fuels with renewable fuels and improving efficiency of energy utilization. Of the various renewable fuels, biomass is in many of the countries the largest largest renewable resource. Hence, there is increasing interest in developing and commercializing innovative biomass energy conversion technologies. In this chapter the main characteristics of olive solid waste as a potential energy source were examined. This review study presented the state of the art knowledge in the field of olive solid waste exploitation for energy purposes, resuming that there is still a lot of progress to be achieved. The European target of 20% by 2020 for the penetration of renewable energy sources in all other energy sectors is limited to the modest 10% for biofuels. The opportunities and challenges for olive solid waste use on the renewable energies national action plans are unique for the southern European states.

Acknowledgment

Authors acknowledge financial support from the European Regional Development Fund (ERDF) and from national sources in terms of the Interreg IVC Programme (Greece-Cyprus) (KEDELEA – K3_02_02).

References

[1] Food and Agriculture Organization of the United Nations - Statistics Division, http://faostat3.fao.org/home/index.html, last visited 01 September 2012.

[2] Letki, A.G., 2007. Centrifucation, Decanters. *Encyclopedia of Separation Science*, pp 1-10.

[3] Masghouni, M., Hassairi, M., 2000. Energy applications of olive-oil industry by-products. *Biomass and Bioenergy* 18:257-262.

[4] Celma, A.R., Rojas, S., Lopez-Rodriguez, F., 2007. Waste-to-energy possibilities for industrial olive and grape by-products in Extremadura. *Biomass and Bioenergy* 31:522–534.

[5] Ollero, P., Serrera, A., Arjona, R., Alcantarilla, S., 2003. The CO2 gasification kinetics of olive residue. *Biomass and Bioenergy* 24:151–61.

[6] Arvelakis, S., Koukios, E.G., 2002. Physicochemical upgrading of agroresidues as feedstocks for energy production via thermochemical conversion methods. *Biomass and Bioenergy* 22:331–348.

[7] Jenkins, B.M., Baxter, L.L., Miles, Jr T.R., Miles T.R., 1998. Combustion properties of biomass. *Fuel Processing Technology 54*:17–46.

[8] Polycarpou, P.,2006. Utilization of Biomass for Energy Production by Direct Burning in Cyprus. *Technical Bulletin* 227. Agricultural Research Institute, Nicosia.

[9] EN 14961-1:2010. *Solid biofuels – Fuel specifications and classes* – Part 1: General requirements.

[10] Agbor, V.B., Cicek, N., Sparling, R., Berlin, A., Levin, D.B., 2011. Biomass pretreatment: Fundamentals toward application. *Biotechnology Advances* 29:675–685.

[11] Lynd, L.R., Weimer, P.J., Zyl, W.H., Pretorius, I.S., 2002. Microbial cellulose utilization: fundamentals and biotechnology. *Microbiology and Molecular Biology Reviews*,66:506–77.

[12] Lynd, L.R., Wyman, C.E., Gerngross, T.U., 1999. Biocommodity engineering. *Biotechnology Progress,*15:777–793.

[13] K.D. Maher, D.C. Bressler, 2007. Pyrolysis of triglycerides for the production of renewable fuels and chemicals. *Bioresour. Technol.* 98:2351–2368.

[14] Cara, C. Ruiz, E., Oliva, J.M., Saez, F., Castro, E., 2008. Conversion of olive tree biomass into fermentable sugars by dilute acid pretreatment and enzymatic saccharification. *Bioresource Technology* 99:1869–1876.

[15] Chen, B.Y., Chen, S.W., Wang, H.T., 2012. Use of different alkaline pretreatments and enzyme models to improve low-cost cellulosic biomass conversion. *Biomass and Bioenergy* 39:182-191.

[16] Weerachanchai, P., Jan Leong, S.S., Wook Chang, M., Ching, C.B., Lee, J.M., 2012. Improvement of biomass properties by pretreatment with ionic liquids for bioconversion process. *Bioresource Technology* 111:453–459.

[17] Park, Y.C., Kim, J.S., 2012. Comparison of various alkaline pretreatment methods of lignocellulosic biomass. Energy, http:// dx.doi.org/10.1016/j.energy.2012.08.010.

[18] Uju, Shoda, Y., Nakamoto, A., Goto, M., Tokuhara, W., Noritake, Y., Katahira, S., Ishida, N., Nakashima, K., Ogino, C., Kamiya, N., 2012. Short time ionic liquids pretreatment on lignocellulosic biomass to enhance enzymatic saccharification. *Bioresource Technology* 103:446–452.

[19] Wan. C., Li, Y., 2012. Fungal pretreatment of lignocellulosic biomass. *Biotechnology Advances* 30:1447–1457.

[20] Zhu, J.Y., Pan, X.J., 2010. Woody biomass pretreatment for cellulosic ethanol production: Technology and energy consumption evaluation. *Bioresource Technology* 101:4992–5002.

[21] Biswas, A.K., Yang, W., Blasiak, W., 2011. Steam pretreatment of Salix to upgrade biomass fuel for wood pellet production. *Fuel Processing Technology* 92:1711–1717.

[22] Acharjee, T.C., Coronella, C.J., Vasquez, V.R., 2011. Effect of thermal pretreatment on equilibrium moisture content of lignocellulosic biomass. *Bioresource Technology* 102:4849–4854.

[23] Senturk-Ozer, S., Gevgilili, H., Kalyon, D.M. 2011. Biomass pretreatment strategies via control of rheological behavior of biomass suspensions and reactive twin screw extrusion processing. *Bioresource Technology* 102:9068–9075.

[24] Burcu Uzun, B, Eren Putun, A , Putun, E, 2007. Composition of products obtained via fast pyrolysis of olive-oil residue: Effect of pyrolysis temperature. *J. Anal. Appl. Pyrolysis* 79:147–153.

[25] Demirbas, A, 2008. Biodiesel—A Realistic Fuel Alternative for Diesel Engines, *Springer*, London, United Kingdom.

[26] Zabaniotou, A., Damartzis, Th., 2007. Modelling the intra-particle transport phenomena and chemical reactions of olive kernel fast pyrolysis. *J. Anal. Appl. Pyrolysis* 80:187–194.

[27] Garcia-Ibanez, M. Sanchez, A. Cabanillas, A., 2006. Thermogravimetric analysis of olive-oil residue in air atmosphere. *Fuel Process. Technol.* 87:103–107.

[28] Blanco Lopez, M C, Blanco, C G, Martinez-Alanso, A, 2002. Composition of gases release during olive stones pyrolysis. J. Anal. Appl. Pyrolysis 65: 313-322.

[29] Zanzi, R, Sjostrom, K, Bjornbom, E, 2003. Rapid pyrolysis of agricultural residues at high temperature. *Biomass Bioenergy* 23:357–366.

[30] Putun, A.E., Uzun, B.B., Apaydın, E., Putun, E., 2005. Bio-oil from olive oil industry wastes: Pyrolysis of olive residue under different conditions. *Fuel Process. Technol.* 87:25–32.

[31] Andre, R.N., Pinto, F, Franco, C, Dias, M, Gulyurtlu, I, Matos, M A A, Cabrita, I, 2005. Fluidised bed co-gasification of coal and olive oil industry wastes. *Fuel* 84:1635–1644.

[32] Bridgwater AV, 1994. Catalysis in thermal biomass conversion. *Appl. Catal. A: Gen.* 1994;116:5–47.

[33] Gil J, Aznar MP, Caballero M-G, Corella J., 1999. *.Ind. Eng. Chem. Res.* 38:4226–35.

[34] Garcia, L, Salvador, ML, Arauzo, J, Bilbao, R., 2001. In: Bridgwater AV, editor. Progress in thermochemical biomas conversion, vol. 1:346–57.

[35] Encinar, JM, Gonzalez F, Rodrıguez, JJ, Ramiro, MJ., 2001. *Fuel* 80: 2025–36.

[36] Caballero MA, Aznar MP, Gil J, Corella J., 1997. *Ind. Eng. Chem. Res.* 36:5227–39.

[37] Turns, S.R., 1996. *An introduction to combustion, concepts and applications*. McGraw Hill, ISBN 0-07-911812-7.

[38] Decker, S.R., Sheehan, J., Dayton, D.C., Bozell, J.J., Adney, W.S., Hames, B., Thomas, S.R., Bain, R.L., Czernik, S., Zhang, M., Himmel, M.E., 2007. Biomass Conversion. Kent and Riegel's Handbook of industrial Chemistry and Biotechnology. *Springer*, ISBN-10: 0387278427.

[39] Armesto, L, Bahillo, A, Cabanillas, A, Veijonen, K, Oteroa, J, Plumedc, A, Salvador, L, 2003. Co-combustion of coal and olive oil industry residues in fluidised bed. *Fuel* 82:993–1000.

[40] Cliffe, K.R., Patumsawad S., 2001. Co-combustion of waste from olive oil production with coal in a fluidised bed. *Waste Manage* 21:49–53.

[41] Atimtay A.T., Topal H., 2004. Co-combustion of olive cake with lignite coal in a circulating fluidized bed. *Fuel* 83:859–67.

[42] Atimtay A.T., Varol M., 2009. Investigation of co-combustion of coal and olive cake in a bubbling fluidized bed with secondary air injection. *Fuel* 2009;88:1000-1008.

In: Renewable Energy
Editor: Andreas Poullikkas

ISBN: 978-1-62618-231-8

Chapter 9

Electrification of Rural Areas Using Renewable Energy Sources: Review of Best Practices Around the World

Antonios Tsikalakis and Terpsihori Tzouliadaki
Department of Electrical Engineering, Technological Educational Institute of Crete, Crete, Greece

Abstract

1.6 billion people live in areas where the electricity grid is not available. One of the electrification options is the use of Renewable Energy Sources (RES) based off-grid systems. This chapter describes the progress in RES technologies utilized for this purpose focusing on small scale RES.

A general outline of the financing mechanism is provided in order to understand potential financing ways. A review of Best practices for rural electrification via RES in various, usually remote areas of the world, is provided. This review shows that off-grid electrification via RES is a viable solution. The examples may show sometimes innovative financing, marketing and technical solutions and try to give the paradigm for imitations.

Keywords: Rural electrification, microgrids, renewable energy sources

Introduction

Most of the readers of this chapter may be able to alter lighting conditions to feel more comfortable to read it. This chapter, however, speaks about people who do not have this ability and feel very lucky if they manage to have a solar lantern to read or make simple house-working. For most of the readers, including the authors of this chapter, access to electricity is taken for granted; nevertheless, 1.6 billion people still lack such access. These people reside in remote, poor and often extreme climate areas of our planet.

Lack of access to electricity grids, has been declared by the United Nations and other humanitarian organizations, as a parameter for deteriorating poverty of the population. Recognizing the importance of energy for sustainable development, the United Nations General Assembly has designated, by their resolution 65/151[1], the year 2012 as the International Year of Sustainable Energy for All.

Simple solutions for providing electricity even to low wattage loads may be enough to improve drastically the living conditions, the entrepreneurship and the medical treatment of the residents of these areas as will be thoroughly described in the following paragraphs with concrete examples.

The scope of this chapter is to show that Renewable Energy Sources (RES) have helped in electrifying such rural areas instead of grid expansion or use of fossil fuel and can help even more to this direction. We do not aim to describe the aid of RES in electrifying telecommunication stations, cathodic protection, buoys and other similar successful applications; neither do we aim to focus on simply remote settlements that due to distance from the grid have remained unelectrified. This chapter aims to described organized efforts for electrifying larger areas mainly in areas with remarkably low electrification rate.

For meeting this purpose, the structure of the chapter is as follows. First, the financing incentives and mechanisms for electrifying rural areas are described. Some of these incentives come from humanitarian or other non-profit organisations, some others focus on improving local entrepreneurship and others are aiming at corporate responsibility or CO_2 reduction via Clean Development Mechanism (CDM) actions.

Then some details regarding technologies used focusing on some details of small scale RES and weather precautions are provided. Some information on load management under those circumstances will be also provided. Best practices on rural electrification all over the world is then provided, focusing on the following:

- Short description of the electrification rate
- Impact of RES in increasing electrification rate
- Financing sources of the implemented projects and if possible the amount of financing
- Presentation of Innovative solutions in terms of financing and marketing of such systems
- Presentation of technical issues encountered, if any
- Potential impact in development of the area
- Plans for the future

The case study will cover the following geographical regions

- Middle East, North Africa Countries (Mahreb)
- Sub-Saharean Africa
- South East Asia
- Cold Climate Areas (Mountainous Areas and Polar regions)
- Physical islands in Oceans and the Mediterranean Sea.

As one can easily discern we have tried to present examples from areas all around the world under various climatic conditions which usually use a variety of RES depending on the RES resource available.

RURAL ELECTRIFICATION PROJECTS FINANCING PRACTICES

The unelectrified customer is the mainly interested stakeholder to gain access to electricity. We will not focus on cases that the customer pays for himself to build a rural electrification system. There are plenty of such cases in the developed countries contributing significantly in the research for rural electrification systems.

We will focus on the cases that the customer is not able to finance totally such an installation and seeks for subsidy. We will focus on organised Rural Electrification projects rather than numerous individual paradigms all over the world.

The diagram below shows the potential funding sources for a Rural Electrification programme. Interested individuals capitals, national funds and international funding may be used for implementation of such a project.

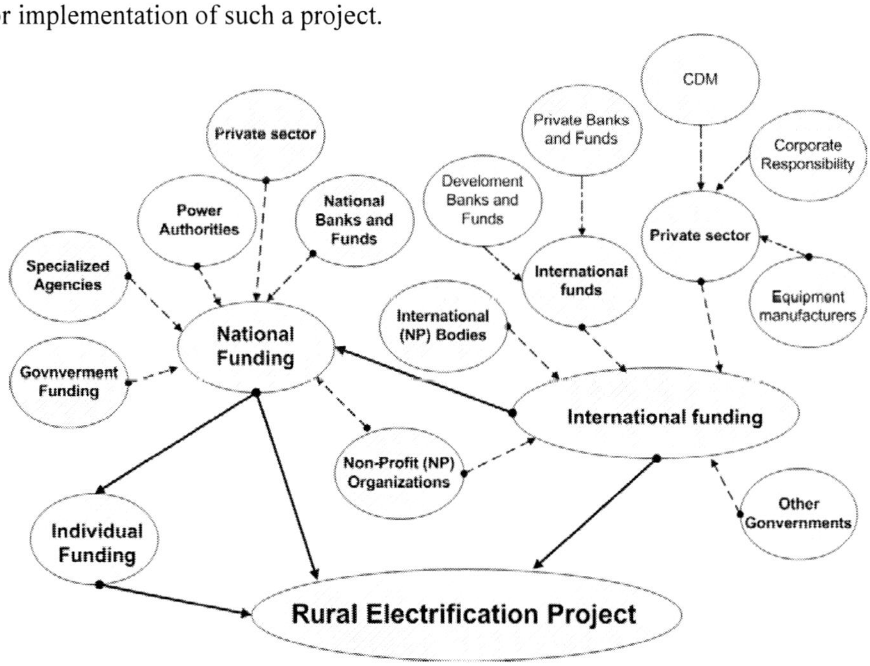

Figure 1. Flow chart of funding sources.

Part of the individual funds may come from national funding (e.g. subsidies) and part of the national funding may be funded by international partnerships. This financial flow is shown with the full line arrows in Figure 1. In the Best Practices section successful implementation of Rural electrification and a large variety of the combinations depicted in Figure 1 are presented.

With regards to national funding, part of the Government budget may be used directly for funding rural electrification projects for improving the population standards of living. Quite often special agencies or funds have been created for this purpose, receiving funds from the government, banking sector, international co-operation and from the partial participation of the beneficiaries themselves. Some of these fund or organisations are part of a greater sustainable development organisation either for the whole country or for some specific isolated areas as will be examined e.g. for Jordan.

The power authorities may play significant role in rural electrification. They may use the option to construct or subsidise the development of RES based rural electrification scheme because of the large distance from the existing grids. Additionally, extended networks over long distances may suffer from voltage deviations or reliability issues. The choice between grid extension and rural electrification should be optimized taking into account available resources and distance from the grid [2]. The power authorities may be funded by national funds directly or seek for international funds and co-operation for implementing rural electrification projects.

National funds and banks may provide attractive financing mechanism for implementing such projects; either with programmes directly aiming the final individual stakeholder or financing organizations or companies tackling with rural electrification. The loans provided may be based on a) financing from other natural funds or b) from international loans at improved interest rate.

The private sector, mainly enterprises selling or constructing equipment for rural electrification, may also provide discounts or attractive payback schemes to the final customers. This may be feasible due to their access to funds at much lower interest compared to individual customers. Some countries subsidize these enterprises in order to increase the electrification rate-since more stakeholders are involved and improve competitiveness.

Last but not least Non Profit organisations such as the Global Village Energy Partnership (GVEP)[3], or religious representatives try to help either by gathering donations for funding rural electrification installation or by significant work of volunteers in the field. As we will see in the Best Practices section organizations in some countries ask for volunteers organizing special programmes to work for rural electrification programmes for some period. These non –profit organization may be locally based, justifying national funding flow, or quite often are international contributing in the international funding of rural electrification projects.

Among the driving forces for funding rural electrification projects is the role of International Non-Profit Bodies such as the United Nations (UN). The United Nations Development Program (UNDP) has been one of the major programs for implementing rural electrification projects. UNDP has activity all over the world,either with direct construction or with financing local partner. The 8 axes of the Millennium Development Goals (MDGs) can be served with the aid of rural electrification as claimed by the UNDP [4] and is proven by various independent surveys [5]. The support of the UNDP helps mitigate and reduce the impact of these threats and provide the rural poor with clean renewable energy and environmentally friendly living. A key part of the support of UNDP in these areas has been the efficient environmental management, risk reduction and adaptation to climate change at national and local level. During the past two decades, the UNDP has implemented over than 2.500 sustainable energy projects around the world, some of them described in the Best Practices section. Key projects will be described here, while numerous reports can be found on the UNDP web-site. Funding comes from own resources of the UNDP, the Global United Nations Development Programme Environment Facility - GEF), the GEF Small Grants Programme, governments and both private and public sectors. The United Nations Industrial Development Organization (UNIDO) [6] also participates in some of these projects.

Based on the above, International funds may fund Rural Electrification Projects to improve the living conditions. These funds can be categorized to a) Special development banks and funds and b) Private banks and funds implementing attractive financial mechanisms. Part of these funds may come from more developed countries in order to foster international partnership.

Indicatively, such international organizations are the World Bank [7] running the Energy Sector Management Assistance Program (ESMAP) [8], the African Development Bank – AfDB, the KfW and the Climate Investment Funds (CIF).

Other governments may subsidize the installation of Rural Electrification Project or even directly finance and implement such projects with their organizations. These organizations may be departments of Foreign Affairs Ministries such as the USAID [9] or the Hellenic Aid [10]. Various such examples can be more found in the Best Practices section.

Last but not least funding source may be the direct contribution of the private sector. Two driving forces may lead private sector stakeholders to finance a rural electrification project in an area thousands of miles away from the main offices of the company. These are a) Reduction of CO_2 footprint and b) Corporate responsibility.

Increased CO_2 emissions, which can be difficulty decreased by changes in the production activities may lead to financing penalty. By implementing a rural electrification project with RES, the interested company can offset the CO_2 emissions providing also sustainable development to other countries helping them to avoid increase of their carbon footprint. This mechanism is generally known as the Clean Development Mechanism (CDM). Therefore rural electrification projects in a host country for CDM projects can be registered as CDM projects and the private company avoid penalties. How could CDM potentially help in increasing sustainability and alleviating poverty has been described in some chapters of textbooks such as[11] or [12].

Corporate responsibility may be another supplement to the CDM needs and may be even linked to marketing reasons. Little amount of the profits, or tiny increase on the price for a product consumed in the developed world can be used to create a "fund" for donating to rural electrification projects[13]. Finally, and more desirable, the humanitarian feeling of the private sector may be the driving force for implementing such projects, especially from managers who work in developed countries but come from countries with low electrification rate.

Among the private sector companies, the equipment providers and manufacturers for RES, cables, switching devices, storage facilities and dedicated loads have additional benefits compared to other companies. Donating, or offering large discounts for equipment for rural electrification apart from corporate responsibility, gives the opportunity to test the equipment in actual and sometimes harsh conditions and improve operating strategies and the equipment itself. It is not uncommon in Rural Electrification Projects Consortia to have active participation of equipment providers as the examples described in this chapter show.

TECHNOLOGIES UTILIZED FOR RURAL ELECTRIFICATION

This section aims to provide a short description of the advances in RES technology utilized. Due to size of the installation, the small capital per installation and difficulties of transportation, RES utilized for rural electrification present some differences compared to large grid connected projects. Moreover, the role of storage is emphasized for the smooth operation of such systems. The converters have been improved to serve various additional operations, whilst the typical loads used for such systems are described along with their role in maintaining secure operation of the equipment. Finally, an overall analysis of the control devices and policies is provided to show how all these components are co-ordinated to maximize benefits.

Production

Production units may be a) PVs, b) Wind turbines c) Micro or Pico Hydro d) Biomass or biogas-fuelled internal combustion engines, microturbines or stoves (if electricity production is not foreseen) e) Fuel Cells with hydrogen being result of electrolysis. f) Other RES such as wave, tidal or geothermal sources but not actually utilized at such low capacity to form grids. Very often these production units should co-operate with storage devices and quite often co-operate with oil-fuel units if hybrid systems are to be built.

Custom or even hand made constructions have been also recorded for power production, especially for wind and hydro.

PVs

Solar energy, actually PVs, because solar thermal electricity units are large enough to be considered for rural applications, is the major RES utilized in terms of number of installations and diversity from extremely hot to extremely cold areas.

No significant changes are required for the typical PV module and similar PV panels with the ones used for installation, e.g. at large interconnected PV plants, can be used. Both the relatively low weight and variety of sizes have helped so that PV panels can be easily transferred and installed almost to any part of the world even using donkeys, camels and small river boats in the absence of access roads. For many years, PVs have helped greatly in off-grid applications including telecommunications, cathodic protection, measurements stations, space applications and so on. Rural electrification is an area where PVs have dominated as examined in the next section.

Wind Turbines

Unlike PVs where a panel can be used either for a large scale interconnected PV park or for a single house installation, wind turbines types and capacity should be different from large scale interconnected projects. Unlike grid-connected applications where wind turbines should follow specific stricter and stricter grid requirements [14] their size gets larger and larger [15] and the off-shore wind power market is blooming [16], small or even micro wind turbines (below 1kW) are required for rural electrification.

Small wind turbines do not need the foundation work of large wind turbines and can be erected with Guyed Tower, Tilt-Up Tower or Self-Supporting Tower made of poles or lattice. Anchors and guys are far more widely used in small rather than large wind turbines. However, their erection is far more difficult than the one of PVs and for larger installations may require cranes. Some smaller wind turbines can be even erected on the buildings with the Vertical Axis Wind Turbines (VAWT) having advantage in such a case.

Small wind turbines may produce either in AC but plenty of models produce in DC, owing their origin to marine applications for battery charging. Some of them use induction generators, some other, the current trend, permanent magnet generators. There is a variety of small wind turbines manufacturers, plenty of them being Small Medium Enterprises. These companies have presented a variety of techniques for yaw and output control, blades control, shape and numbers. Others operate downwind, helping operate with free yaw system, others upwind. Some manufacturers target in marine applications offering such coatings, other manufacturers aim at polar area application while others have focused on urban applications. Even erecting wind turbines may be different among these manufacturers.

Large wind turbines market has been dominated by Horizontal Axis Turbines whilst for small wind turbines there are some manufacturers aiming at Vertical Axis installations. Last but not least the VAWTs due to lower starting torque and the ability to be much easier installed at roofs of buildings could be a potential viable solution to rural electrification [17].

As can be easily discerned there is significant diversity of products to select from. This diversity creates opportunity for development and improvement of small-scale Distributed Generation units. Recent study [18] has shown that the capability of local manufacturers should be utilized for production micro-wind turbines to improve the sustainability of the electrification action. To have an idea of the variety of designs, control and nacelles utilized, small wind turbines manufacturers catalogues have been issued for example by the Nordic Folkecenter for Renewable Energy [19] and the SWIIS consortium [20].

Unlike large wind power projects where wind potential measurements campaign is made before implementing the project, small wind turbines installation may be based on simplified measurements or at the best on wind power atlases over a region. Thus, the following requirements should be met when implementing such a site.

a) To be able to operate at lower wind speeds and provide some electricity. This is the main reason for using more than three blades per wind turbine to increase starting torque. This may be against the efficiency of wind turbine production achieved with larger wind turbines but at such places more operating hours are required. If the developer has sufficient information on the wind potential the most efficient combination can be selected.
b) To be able to withstand hurricanes if the area faces such conditions. Some of them use the furling control action and generally the manufacturers speak of surviving speed rather than for cut-out wind speed. A small wind turbine may use the furling technique to reduce power production as the wind speed increases. There may be significant differences in wind power vs wind speed characteristic curves between a small wind turbine and a larger one with cut-out speed at no more than 30m/s, usually 25m/s.

Their cost per installed kW is quite higher compared to large scale wind parks and varies according to the needs. Components required for erecting the wind turbine can increase significant the cost for building a small wind turbine. Usually the manufacturers issue price list, usually after contact with them or even provide simplified tools (e.g. Bergey [21]) to assess the viability of investing on small wind turbines.

Even though wind turbines are not as widely widespread as PVs, the examples from various countries shows that can be a supplementary solution especially in areas with known high wind power potential like mountainous, coastal and polar regions.

The American Wind Energy Association (AWEA) has dedicated part of their web-site for small wind turbines and frequently issue market reports, roadmaps and various report for the development of this market under the distributed wind power sub-section [22]. More details on small wind turbines issues and updates can be also found in [23].

Mini and Pico Hydro

Water is not available in all areas of the world; to make matters worse, running water or springs with sufficient flow or high hydraulic head are not as widespread as solar or even wind power resource. This is why small or mainly mini and Pico Hydro use is restricted to specific areas of the world, mainly Asia and Latin America with some efforts to be reported in East Africa [24]. Despite this restriction, micro and Pico hydro have greatly contributed to rural electrification in China, Nepal, Vietnam and many South American countries during the past 30 years. A classification of hydro plants with regards to their capacity to small, mini, micro and Pico hydro is shown in Figure 2 and is line with the bibliography on small hydro plants. Although all these technologies could be regarded as small hydro power, they have specific technical characteristics that warrant their own definition. Since the systems we will speak about, aim at early electrification of few houses in a territory we will focus mainly in Micro and Pico hydro in the following sections.

Pico hydro, since are usually in the order of few hundreds Watts are known also as "family Hydro" since they can make a family self-sufficient for electricity needs. In these plants, no dam or reservoir storage is involved and the discharge is made directly from the intake. Mutli-pole engines are used for this purpose for the best co-operation with the hydro-turbine. Larger hydro power installations use salient pole synchronous generators, much easier to directly maintain frequency than other RES, or asynchronous generators. Permanent magnet generators are also used especially for Pico-hydro applications [25].

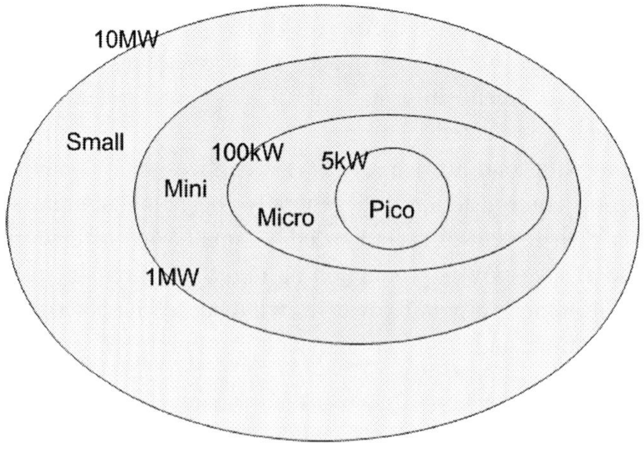

Figure 2. Classification of small Hydro power based on their capacity.

Regarding turbines used, crossflow and pelton turbines can be mush easier manufactured locally; the difficulties with hydraulic head is the one that has led mainly to crossflow turbines. These turbines have two particular benefits: a) wide range of heads and power ratings and b) that can be much easier locally manufactured. Despite the issues of low head with impulse turbines like crossflow, for low capacity, as in our case, the hydraulic head can be as low as 4m for such turbines. Low-head propeller turbines can be also available, and this was the reason for the "family hydro" bloom in China, Vietnam and Laos. Very affordable turbines capable of running with a couple of meters head and costing in the order of 20-$50 (excluding balance of plant e.g. transmission cables and draft tube) dominated the market in the late 1980s and early 1990s with capacity in the order 200-500W. Such turbines were illegally imported and sold for cash in North Vietnam and Laos and the market developed primarily by word of mouth without significant government subsidies but without taxes as well. To improve the quality of these hydro turbines, a more durable design with better bearings and improved manufacturing quality and after-sales service was developed in Vietnam, taking the commercial name PowerPal. The improved quality of the PowerPal systems translates into considerably higher cost, the price of a 200 W unit is US$ 340 and a 2kW double valve turgo costs US$ 2,980. A price list for Power Spout turbines is available at [26]. A different improved version of the Vietnamese low head turbine, which includes a manual flow control mechanism, is produced in Indonesia. The so-called open flume turbines are available as standardised packages for different flows and heads below 6 m. The turbine finally used in a micro or pico hydro is a result of the hydraulic head, the flow rate, the required capacity and the capital available [27]. There is significant research effort on utilizing as low head as possible for relatively high production to take advantage of streams flow [28]. There are also efforts for utilizing the kinetic energy of the stream.

Pico hydro units may incorporate very simple materials and have plenty room of innovation. For example a five gallons bucket hydrogenerator using extremely common materials can be constructed following the descriptions in[29]. This requires a minimum of 30 feet of head and a flow of 26 gallons per minute in order to reach the 13.7 volts needed for 12 volt charging via either permanent magnet alternator or even a DC alternator from a typical car.

Practical action, a registered charity in the UK, has developed a variety of manuals for rural electrification among which for Micro-hydro projects. One simplified but very realistic and practical guide [30] presents guidelines for selecting or not a Micro-hydro project and some questions to be asked by the local communities before implementing such a project. The European Small Hydro Association (ESHA) has issued various documents on implementation of Hydro power among which a document on application of Small Hydropower for Developing Countries [31] and also a Small Hydropower Checklist. The UK Department for the International Development (DfID) has issued a practical guide for pico hydro applications [32]. The manufacturers like PowerSpout issue very descriptive manuals on operating a Low head Hydro turbine [33].

Biomass

Biomass is greatly used in developing countries for meeting heating and cooking needs. Woods, wood chips or anything else available is utilized for the basic needs of the population. This may require much time and effort especially for women, deforestation takes place, biomass is not utilized with high efficiency and finally respiratory problems may be apparent to the residents. Regarding electricity, solid biomass is not actually utilized. If this is to be utilized steam cycles would be required which would be sufficient for large number of final customers. Liquid or gaseous biomass can be used in internal combustion engines. Liquid bio-fuels are used to displace fuel oil (diesel or gasoline) in typical Internal combustion engines. may come from various plants with most famous one the Jatropha oil [34]. Thus cultivation of Jathorpa for fuel increases sustainability of the electrification action provided that food is enough to feed the region.

Gaseous biomass, i.e. biogas can be used instead of other gaseous fuels in ICE. This can be the result of anaerobic digestion converting excreta from various animals or even human into biogas to be directly used as fuel even for ICE. Significant progress on this field has been made in Bangladesh as described in the corresponding action.

Storage Methods

RES production intermittency is the main cause of utilizing storage in such systems. Storage is essential for storing energy at excess periods to use it later.

Even in RES-hybrid systems with Internal Combustion Engines (ICE), storage may be required to smooth the operation of the ICE or even help in avoiding its operation. Efficient storage management can help in avoiding excessive start-ups [35]. Therefore, efficient management of storage devices is an important parameter for the success of RES-based autonomous power systems, as the ones examined here.

Storage methods are mainly based on batteries due to their simplicity, modularity and lack of geographical restrictions. Supercacitors and flywheel applications are very limited, even though they have much higher power capabilities for short time periods compared to batteries.

The vast majority includes lead-acid batteries either sealed (more and more often) or in open cycle. In more sophisticated applications and as costs are reduced Li-Ion batteries are used. Replacing batteries is one of the typical problems of such systems and during the life-time of such a system, batteries may be replaced many times. It is highly recommended that deep discharge batteries are used and not automotive ones. Initial surveys for non-operating PV-battery systems in Kiribati had shown that 43% of the systems had replaced the original deep discharge batteries with automobile batteries having inadequate capacity and a short life expectancy. In the same survey it was stressed that without charge controller, battery life time is reduced. Significant caution should be taken for the self-discharging of the batteries especially in applications where the periods of consumption are much later than periods of production [36], [37]. Moreover, the rating of capacity, the charge time the final content and the depth of discharge is another issue that should be taken into account when designing such systems. Faulty batteries may lead to decreased efficiency and even failure of a Rural Electrification programme.

Converters Used

As examined in the previous sections other sources produce at AC and others in DC. We will see also in the next section that both AC and DC loads may get connected. What if we want DC production to become an AC load? This section briefly describes the converters that may be required.

For the simplest solar light installation, such equipment is not required since all storage, PV and battery require DC. For the connection of additional loads however converters may be required. The typical converters that may exist in a rural electrification project are:

- Inverter for converting DC production to AC of specific frequency. This may (and in our opinion should) be different type for each type of DC source.
- Rectifier for converting AC to DC. This may be used for access to the storage device once the surplus energy is in the form of AC or if DC loads are to be served by an AC source such as a diesel gen-set.
- Maximum power point tracker (MPPT) either for PVs or for wind turbines. This is used since both Voltage and current produced change with the solar radiation or the wind speed and the optimum operating point based on the conditions should be found.
- Charger controller. This is necessary for monitoring the state of the battery and stop serving load or to lead in dumping energy from RES sources.
- Transformer. This may be required as the Rural electrification project has greater power or if distribution of power to the other customers a bit far away from the production is required. This is an electromechanical device and not an electronics (or power electronics) device as the previous ones.

For some manufacturers the inverter includes MPPT or even transformer depending on the connected sources. The co-ordination of all these converters, sources and loads will be described after we examine the potential loads of a Rural Electrification Scheme.

Loads

Unlike large scale RES projects, where the types of loads served is responsibility of the Transmission System Operators (TSOs) and Distributed Network Operators (DNOs), in meeting rural areas needs, efficient load management becomes more and more important, not only for meeting as diverse as possible needs of the population but also for the protection of both the production and the storage equipment. While in grid-connected applications energy efficiency may be desirable, the nature of the loads served in isolated power systems may dictate the type and the technical characteristics of RES and storage devices to be used.

Sometimes loads for rural electrification are merely based on DC, reducing the needs for installing inverters. Some or all of these loads are typically included in rural electrification projects. Even if these loads are not available at the beginning of the project, the project should be scalable so that the rest of the loads are included.

- Basic lighting
- Refrigerators
- Charging of mobile phones
- Computer loads

- Entertainment sets (small TVs or radios)
- Pumps
- Other mechanical loads

Lighting

Lighting is a typical application for rural electrification used mainly to replace kerosene lamps, which are responsible mainly for injuries among end-users and also increase CO_2 emissions. It is estimated that daily use for 4 hours of kerosene lamps lead to 143kg of CO_2 annually [38].

Incandescent lamps (available both at AC and DC) are the most economic lighting solution in terms of initial capital cost. Their main advantage is the fact that being a resistive load, orthogonic pulses output from the inverter or lower voltage than nominal can provide some output, unlike the other electronically driven lamps. However, they are very inefficient and increase the capacity of installation for meeting the same lighting needs. Additionally, they present significantly lower life-time compared to CFL or LED lamps.

Compact Fluorescent Lamps (CFL) are very widespread in such applications since they have significantly higher efficiency and require lower current to operate. Lower current is a significant parameter for sizing especially batteries suffering when high currents are exchanged with the rest of installation. There are both AC and DC CFL lamps depending on the configuration of the rest of the system. At least one manufacturer has designed CFL suitable for use in Solar Homes [39] and DC lamps able also to be dimmed depending on the battery state of charge. Other manufacturer has developed dedicated solar solutions for Africa [40]. As we will see in both Morocco and Bangladesh, CFLs are provided to end-users together with the PV kit installation. LED lighting becomes more and more popular in such installation since low, lower than the CFLs wattage is required and their life-time is much longer than any other artificial lighting source. However, they are still quite expensive per lamp. In Malawi [41], LED lamps are offered as part of the fee for service approach. Many efforts such as D.Light, Millenium Villages and Light Up the World (LUTW) [42] have promoted LED lamps to many developing countries including China and India.

Lanterns are the minimum requirement for lighting. This is a portable lighting device capable of providing light to the end-users Many of them have been till now based on kerosene, creating hazards to end-users as one can watch in a Youtube video [43]. In order to face such danger, solar lanterns have been proposed. In Kenya there is target of replacing 1 million kerosene Lanterns with 1 solar ones. Lanterns may not even require a production facility at the house or an internal cabling installation This lantern may be charged either locally or with the use of sub-central charging stations [44]. These stations may use RES, usually solar for providing electricity for charging the lanterns [45]. Typical costs on solar lanterns, even though we believe they are currently less expensive, can be found in [46]. CFL and LED lanterns are available and an interesting article comparing pros and cons of each solution is provided in [47]. Since LED solar lanterns may be rather expensive for the poor end-users a business model which foresees not buying or donating solar lanterns but renting solar lanterns. This has been designed for India and the Philippines. A latest article presents not only the general idea for renting Solar lanterns via a sub-centralized charging station but also opitimizes the policy followed for maximizing profits for such renting agents [48].

Refrigerators

Preservation of food but more importantly of vaccines and medicine in rural clinics is one of the major needs. Refrigeration improves also the duration of food autonomy and may help in importing food from other regions. Refrigerators are usually operating at AC with induction motors. DC refrigerators are also available but may be more expensive than buying a larger inverter for meeting all AC loads including the refrigerator. Solar refrigerators, using solar power directly to produce cool have started being constructed.

One of the first public loads to be electrified is a rural clinic, which may be simply a small room of a hut. The needs are mainly for keeping vaccines and medicine for future use. In more sophisticated applications, tele-medicine equipment may be used. One of the major programs of this kind is the 300 Photovoltaic Powered Health Clinics: Peru's National Program. The national program is administered by the Ministry of Health, with technical support from the Pan American Health Organization [49]. The management of the program includes planning, procurement, training, implementation, and evaluation. These activities require coordination at the national, regional, and local levels. Similar approach has been used in Indonesia, Zambia, Myanmar, Eritrea with about 1150 rural clinics installed by 1998. NREL has issued guidelines for rural clinics development and sizing. There is also strong-cooperation between the USAId and NREL since USAID utilizes NREL tool in the special dedicated web-site [50].

Mechanical Loads and Pumping

Mechanical power production is one of the major electricity uses all over the world. This is essential for workshops wishing to improve their productivity. The importance of mechanical power on the development of poor areas has been stressed in various reports. There are some interesting ideas [51] on how river stream can be used to drive a pump or how traditional cereal mill and power generation can be combined.

However, electricity becomes the major source for mechanical power, electric Motors of all sizes can be available for a variety of equipment and they can run either on DC or, more often on AC. Increased motor loads may be the main cause of selecting 3 phase configuration of the loads. Additionally when such loads are available there should be special precautions for the starting current reaches up to 5-6 times the nominal current. This creates significant stress to batteries, especially when RES production is very low, since batteries should be able to provide this current even for this short time. Additionally both batteries life-time and capacity is reduced when high current is drawn from them. Pumping is one of the major required activities to improve the access to potable water instead of carrying it over long distances.

Before the electricity becomes a reality, wind pumps were rather common in various areas all over the world. There is still need for such equipment, sometimes met by local manufacturers like Kijito power.

How widespread this equipment can be in Africa can be found in [52]. Now PVs can be used as well to produce electricity to be used with solar pumps as described in [53].

Entertainment, Communication and PC Loads

Providing mobile phones to citizens of rural areas is a relatively inexpensive way to extend the communication networks without erecting communication cables under harsh conditions. A systematic approach for this has been recorded in Morocco and Honduras [54]. One should bear also in mind that there are plenty of communication stations operating with high RES penetration. Solar lanterns, can be combined to provide charging to a mobile phone

Entertainment sets like radios and TVs have also started becoming widespread in these applications. They are usually portable devices that may either operate with DC or via AC sockets. Also for entertainment and communication, internet connection may be required. Such a solution combining Internet access and energy is the ZubaBox [55] a container which can be converted into a 11 seat internet café and can be operated using PVs. A typical ZubaBox is shown in Figure 3. The first applications of this kind have been installed in Kenya and Zambia and the organization asks for any 2nd hand computer equipment for their activities not only for ZubaBox but also for other applications.

Figure 3. A typical Zubabox (courtesy of care of Computer Aid International).

Overall Co-Ordination of Production, Storage and Loads

In the previous paragraphs, the RES production units, the converters utilized and the loads served have been described. All these components should have as smooth as possible co-operation in order to prolong the life-time of the installation, protect the equipment and minimize the fuel consumption, if a diesel gen-set is to be incorporated.

This section presents all the potential combinations of connected devices. The diagram below does not mean that all the components are necessary for hybrid co-operation. Some of the sources may exist in other such networks, some others not.

The Generic electricity and information flow is shown in Figure 4. The units capable of producing DC are interconnected to DC bus, along with any DC loads and storage devices or even electrolysers if Fuel Cells are to be incorporated. If only DC loads and sources exist then no inverter is required and the only existing bus is the DC bus. This is the case of many electrification projects whose aim mainly is to cover some basic electricity needs of the population.

The system becomes more complicated if AC loads and even more complicated if AC sources are to be connected. Then improved co-ordination is required in order to maintain common frequency for the created grid. In common power systems the frequency is maintained via the governor of the production units increasing production when load increases and decreases production as load decreases. In the former case frequency decreases and this is reflected to speed of the axis coupling the generator and the mechanical part of generation.

This is "felt" by the governor which provides signals to increase the fuel or hydro flow and balance production and load.

In the systems we study here, for some periods of operation we do not have the ability of increasing fuel consumption and balance load with production. For this purpose, the role of the storage device may be vital since any deviations between production and demand is smoothed either by charging or discharging the storage device. Deviation in this balance clearly means deviation in the frequency.

Therefore, co-ordination of all these components is essential, especially as the complexity of the equipment increases. This co-ordination or control can be maintained via either expert systems or by a suitable converter circuit.

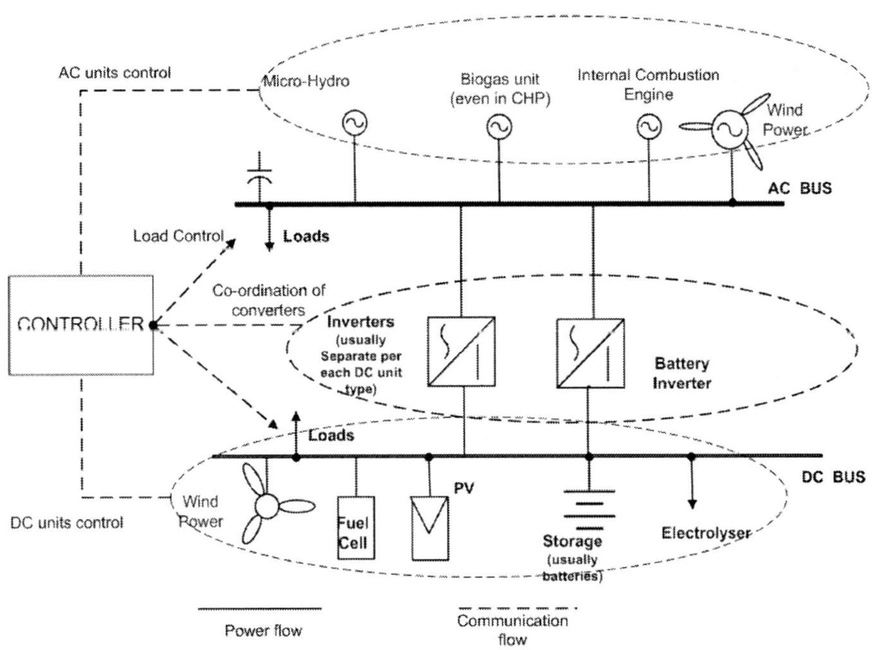

Figure 4. Detailed control scheme for hybrid systems. Some of the components may not be connected depending on the installation.

This controller monitor the system should make decisions on the following:

- What is the State of Charge (SOC) of the batteries? Will they be capable of meeting changes in load balance?
- Should any non-intermittent source (e.g. a diesel gen-set) operate in the AC side and if yes at what operating point (if possible)?

- Should the storage device be charged by starting the operation of a source or should be discharged to avoid paying fuel?
- Should the PVs or Wind turbines operate at their maximum production, as the MPPT indicates for each one of these sources or should they operate at lower operating point due to the full battery and low demand
- Should we connect all the loads, or should we disconnect some of them to maintain the rest of loads served?
- Is addition of the deferrable loads necessary to protect the equipment?
- Should I programme some loads whose operation can be altered to operate? Such an example is the use of pumps and filling a water tank.
- Should I start the electrolyser, if this exists?
- Do we have enough fuel for the next periods?
- Do I need reactive power compensation? (for larger systems)

And above all: *Maintain frequency for AC loads within specific limits regardless the changes in production or consumption.*

Some manufacturers have developed special load controllers for use in such autonomous power systems. Their purpose is twofold: a) The protection of the production equipment and b) the efficient management of the load in order to improve the operation without generator and reduce the periods of not supplying especially critical loads.

Wind turbines should not remain without load connection otherwise this will lead in increasing rotation speed with adverse effect on the stability of the wind turbine itself.

Therefore, in case there is either no load or not sufficient capacity in the batteries, there should be a flexible load like a resistor otherwise the voltage at the end of the wind turbine may be increased significantly. Such devices, a typical commercial terminology is protection box, can be utilized in a variety of permanent magnet units.

Induction machines used as generators should also not remain without load to avoid self-excitation. This is the reason that the manufacturers of wind turbines or hydro turbines offer load to be connected to the generator and the end-users should buy it.

The simplest solution is the use of micro-electronics controller to gather all the signals from the components and make decisions based on specific predetermined rules. An example can be if the SOC is below 60% then start up the generator. These rules may be even the result of agents-co-operation.

Smoother operation can be achieved, if the controller is a converter with the ability of droop operation, which helps in making the appropriate changes in production/demand or storage in order to keep frequency at its nominal value. If the frequency increases, production should be reduced or the charging rate of the battery should be increased. The next solution is to co-operate with the load controllers to connect loads and the outmost solution is to reduce production from RES or even disconnect them.

In larger networks, the controller operation may be aided by a human-machine interface either at the place or via tele-communication. Such is the example of the Huatacondo microgrid SCADA system [56].

BEST PRACTICES ALL OVER THE WORLD

Middle East North Africa countries

Even though these countries have rather high electrification rate, compared to the other cases examined, the vast majority of their desert territories remains unelectrified mainly due to the fact they are both sparsely inhabited and far away from the existing grid. However, in order to improve living conditions in those harsh areas, these countries have put significant efforts on improving electrification rate. In many cases, they have utilized RES, especially PVs and the results as one can see are very promising.

Morocco

The Ministry of Energy and Mines is responsible for the energy planning of the country and for granting any authorizations related to the energy sector [57]. The national utility company, National Office of Electricity (ONE), ensures the planning and realization of the national electrification programs. CDER (Centre de Dèveloppement des Energies Renouvelables) [58] is responsible for RE planning, implementation and follow up of RE programs and projects in Morocco. RES companies like Temasol [59] are used to act as subcontractors of ONE for the rural electrification projects in Morocco. This practice has helped Morocco to make significant progress in solar energy exploitation; 10% of the rural households have been electrified with the use of autonomous PV systems as will be thoroughly described. During the past 10 years, using the PERG initiative[60], the rural electrification rate has risen from 18% in 1995 to 97.5% in 2011. If an energy-related project is to be connected to the grid, an authorization is required from ONE and the Ministry of Energy and Mines. Finally, the Association for Solar and Renewable Energy (Association Marocaine de l' Industrie Solaire - AMISOL) is an organization that supports RES (especially solar energy).

Morocco has long history in rural electrification dating back in 1963 when ONE, with its own resources, began extending the transmission and distribution networks (those in the rural areas included). This first period lasted till 1981. In the meantime, more precisely in 1978, they started planning the PNER (National Program for Rural Electrification) project; by then the total number of rural households supplied with electricity was about 130,000. PNER consisted of two phases. During the implementation of the first one (1982-1986), 287 rural villages representing 68,000 households were electrified. This phase of PNER was financed by 50% by the State and 50% by local authorities. Moreover, a loan by the International Bank for Reconstruction and Development (IBRD) of 30 Million $ was granted. The second phase of PNER started in 1991 and electrified 843 villages representing 155,000 households. PNER II was 100% financed by local authorities and benefited of a loan from the European Investment Bank (EIB) of 30 Million € and a loan of 114 Million $ from IBRD.

The Pilot programme of rural pre-electrification (PPER – Programme Pilote de Pré-Electrification Rurale) was in fact the first attempt to substitute a program logic for a vision of projects in the field of rural electrification off the grid. Built on a Moroccan-French financing scheme, the PPER has been implemented jointly by the Ministry of Energy and Mines (Energy Division), the Ministry of Interior (General Division of the local communities), and CDER. PPER planned the equipment for 240 villages distributed over three provinces: Azilal,

Errachidia and Safi. In 1995, 30 villages (1,500 houses) in these three provinces were equipped with one of the following or combination of decentralised solutions:

- individual solar system;
- collective solar system (battery recharging); and/or
- mini grid

It should be noted that solar solutions have accounted for some 70% of this program.

In parallel, some other actions were implemented, such as "The Village Power" (Moroccan-Spanish co-operation), which has electrified 12 villages in the Northern Provinces (2,500 houses in Chefchaouen, Taounate…), and the "Regional Energy Supply Plan" (SAER - Schéma d'Approvisionnement Energétique Régional), which resulted in equipping over 400 homes during 1989-1992 and creating associations of projects beneficiaries [61].

Despite the huge efforts, the electrification rate through the interconnected grid did not increase considerably the electrification rate of Morocco. In order to phase this issue, the introduction of the decentralised option via the Global Rural Electrification Program (PERG) has followed, starting in 1995. This was a program of the National Power Office (ONE) of Morocco with governmental support. The financing of the project came from various sources. The Equipment Grant of 22.1 mil € from ONE includes a 5 mil € donation from the KfW German Bank, a 5 mil € soft loan from the French Development Agency (AFD), about 1.25 mil € from FFEM (the French Fund for the World Environment), 3.5 mil € from Company's shareholders and another 3.2 mil € via connection fees from customers.

Since 1998, the introduction of PV kits was considered as a means of rural electrification. In 1999 it was decided that the private sector should participate as well. Both these actions accelerated the electrification rate in Morocco. Revamped specifications, in particular with regards to the acceptance procedures and methods of payment, work monitoring decentralization, henceforth provided by Regional Technical Services, simplification of contract management procedures, and quality control of completed works were some of the parameters that lead to this acceleration.

In order to optimally connect areas the use of Geographical Information Systems (GIS) and Distribution Network Planning Software was decisive. The distance of the settlements was the key parameter for either grouping settlements for connecting to the grid or creating mini-grids. In this way, rural electrification can be increased at reasonable cost stating that, if grid electrification cost exceeds 2400€, then a Solar Home System (SHS) should be the option. In order to proceed with the Decentralized option, one operator acts as a delegated manager for ONE. The operator buys and installs SHS - paid by monthly fee and grants from ONE, while the equipment is property of ONE. The operator operates, maintains and renews the equipment for 10 year and the cost is covered by the customer's "Fee for Service".

Table 1. Typical charges for various capacities of PV kits in Morocco

Installed Capacity	ONE participation	Advanced payment for beneficiary	Monthly fee
75 Wp	5500 DH (486€)	900 DH (80€)	65 DH (6€)
200 Wp	11000 DH (972€)	4000 DH (354€)	150 DH (13.5€)

An example of such an operator is Temasol [59]. This company's head office is in Rabat with local backing up agencies close to customers. The communication with people, new application and payments mainly take place on the local weekly markets (souks), about 95% of the installations have been sold there. Thus, the locals do not have to change their habits in order to pay their fee for service.

The operator, once a contract is signed, has to make the installation and supply the equipment within 15 days. Moreover, once per year maintenance of the equipment, such as batteries etc is performed and in case of unexpected problems in the equipment, the operator should be able to respond within 48 hours.

Within the commitments is the replacement and recycling of the batteries. Till now 29,500 batteries have been replaced and 26,000 batteries have been recycled leading to more than 800m^3 of electrolyte and more than 140tons of plastic. Additionally, 7900 control chargers and about 8800 lamps have been replaced from the first installations. The ONE participation amounts to 5100 DH (i.e. 452.8€) on average, excluding tax, and the consortium pre-finances the complement payable by the user.

The participation of the beneficiary household for a 50 Watt kit includes an advance of 700 DH (i.e. 62.2€) (excl. tax) and a monthly fee of 65 DH (below 6€) (excl. tax) for 10 years. Similar structures apply for 75Wp or 200Wp, as Table 1 shows. The simplest electrification system consists of some lamps (usually Compact Fluorescent Lamps) and the availability of one 12V socket operating in DC. Thus, the system is very simple since no inverter is required. New tenders give the ability of increasing capacity and adding a fridge.

Some of the PV kits are installed to form mini-grids and by 2010, 14 such systems with total capacity 120kW had been installed in Ouarzazate.

Within PERG, additionally to the above actions, other RES sources were used as well. More precisely, in 2000 one wind turbine of 15 kW was installed to electrify 71 homes in the village of Moulay Bouzerktoune (Province of Essaouira) and 2 wind turbines of 25 kW each in the village of Sidi Kaouki helped in electrifying 52 additional homes. In 2002, CDER, launched a study in the Provinces of Essaouira and Safi. According to this study 22 villages were eligible to be electrified by wind systems.

Hydropower was also exploited under PERG:

- the micro-hydropower plant of Askaw of 200 kW allows supplying 30 villages in the commune of Iguidi near Agadir, i.e. 593 homes;
- the micro-hydropower plant of Oum -Rbaî (Khenifra province, commissioned in December 2004) of 220 kW can supply 18 villages, i.e. 556 homes and administrative buildings that come under the commune of the Oum Errbia springs, through an isolated grid;
- a third micro-hydropower plant (Maaser plant) that will supply 6 villages (about 344 houses) through isolated grid, is considered.

Within the Micro-Hydropower Plants (MHP) programme, Undertaken by CDER, pilot experiences of MHP projects (up to 100 kW) have been realised in the Haut Atlas (Haouz Region) and in several mountainous regions (Azilal, Asni, Ouneine, Tighdouine.).

All these actions have helped so that, by September 2002, 16,000 Households (150,000 inhabitants) had been electrified in this way.

During the period 2003-2005, 4 tenders for 93,000 households were issued. By the end of 2006, 27,373 villages were supplied with electricity (1,598,441 households).

Only during 2006, 78,286 households were electrified and by the end of 2007 PERG has signified the electrification of more than 35000 villages providing access to electricity for up to 12 million people in the rural areas and a total budget of 20 billion DH.

By the end of 2011, 34070 villages were supplied with electricity (1,938,747 households). It should be mentioned that more than 10% of these households have been electrified using PV kits only-58500 households in 3663 villages.

Additionally to these actions, a demonstration project employing FFEM, TEMASOL and ONEP [62] (National Office for Potable water) for remote villages aims at promoting solar pumping. 10-15 villages will be equipped and operated aiming for providing water to about 6000 people.

The management will be performed by local associations. Currently, 4 solar pumps have been already installed, with 900 persons served. Part of the cost will be covered by ONEP, which brings 0.36 mill €, and FFEM with 0.4 mill €. Moreover, 29 installations of PV pumping systems are foreseen via a number of MEDA/SMAPII projects [63]. Among the activities is the PV installation at about 60 drinking standtaps (Saqayti) for providing water to rural population via pre-paid cards.

Algeria

Some first actions for electrifying remote villages in this vast Southern area have taken place. By 1990 the installed capacity was in the order of 250kWp [64]. Pumping constitutes a significant portion of these installations and experience from the operation under Saharan conditions.

However, some significant more efforts focusing on rural electrification have been made since then. One of these efforts was the electrification of 20 villages in various Wilayas (provinces) in the Sahara desert. 453 dwellings were electrified with installed capacity of 1000 homes achieved, mainly based on the implementation of 108 photovoltaic systems with total capacity of 500kW. The electrified villages were scattered in a region of about 1 million km^2, i.e. about twice the land area of France.

More details on the number of homes electrified and the capacity of PV systems considered are provided in Table 2. Systems of 1.5, 3 and 6 kW were used for this purpose.

Table 2. Electrified villages in Algeria via PVs

Wilaya	Number of Connected homes	Installed capacity (kWp)
Tindouf	156	78
Adrar	45	22.5
Illizi	150	75
Tamanrasset	555	277.5

Other achievements of this action are the construction and operation of a central Hybrid PV-Diesel station of 13 kW_p at Illizi (300 dwellings or about 2000 habitants), the electrification of more than 100 telecommunications sites (power of 650 kW_p) and around 20 security posts (constabulary, communal guards…), and beaconing of 2300 km of paths. .

Egypt

Even though Egypt has significant percentage of desert land very few remote electrification applications exist compared to the other two neighbouring countries.

The New and Renewable Energy Authority (NREA) signed a protocol for co-operation with the Italian Ministry of Environment, Sea and Land to electrify 2 remote settlements in Matrouh Governorate with PV systems of about 43 kW installed capacity.

The project is operating since the end of 2010. This project will provide electricity to 100 households, 80 street poles, 3 mosques, 2 clinics and 1 school. Moreover, a PV system has been used to light one of NREA's remote sites at Matrouh Governorate, with a total capacity of 424 Watt peak for 9 × 11 W (DC) efficient lamps, and a TV set of 60 W (AC). There are also some remote applications such as telecommunication stations, ice making factories, lighting, desalination etc.

Further Rural Electrification of 121 rural villages via PV or PV/Diesel Systems is under investigation and the expected required capacity is about 1.2 MWp.

Thus, access to lighting, potable water and communications can be much easier achieved in remote areas. Moreover, PV irrigation systems are foreseen for remote areas away from electricity infrastructures.

Jordan

4% of the 5.2 million inhabitants of the Kingdom of Jordan live as nomads and about 1% of the population live in desert areas without access to electricity. The vast majority of those areas named Badia [65]. The rural electrification in Jordan is based on international collaboration with various, mainly German, organizations and National funds.

The first period of rural electrification was during 1986-1995, when the majority of installations were for public buildings, like schools, clincs etc. The Royal Scientific Society (RSS) was the main responsible body for the development during this period.

75 PV systems were installed on such buildings, while 22 PV pumps at remote areas in Southern Jordan were installed as well. During 1986-8 clinics in villages Gregra, Mathkour Beer and Reesheh were installed to meet the needs for mainly DC loads for light bulbs, TV systems for educational systems and mainly for refrigerators for keeping vaccines and medicine.

In 1993, 6 similar PV systems were installed by the RSS in the southern villages of Beer Mthkour, Wadi Rum, Finan, Titen and Al-Mizfur. The funding came from the Ministry of education for electrifying mainly school buildings.

The second phase focused on electrifying villages. By 2001 999 villages with 1863 inhabitants had been electrified compared to 883 in 1995.

The third phase currently under development targets in an holistic development of the whole Badia region. More specifically in October 2002, 9 solar home systems were installed in the Rawdat Al-Bindan village for 113 people costing around 50,000€. Similar systems were installed at Thaghrat Aljob at Ma'an Governorate and Al-Faida at Ruwaished area.

Tunisia

According to the current state of development, over 12,000 households and about 200 schools have been electrified. Besides these electrification programmes, some 70 solar water pumping installations have been constructed, mainly in southern Tunisia where there is a good number of surface wells and a need for drinking water. Projects for PV desalination are also being implemented. In sum, the total energy installed, including all uses, is estimated at 1.4 MW of which about 1.2 MW for electrification. The related applications concern: water pumping, electrification, water desalination, agriculture farms etc..

Tunisia had adopted a project to be undertaken by 2007 and that concern PV electrification of 1200 households in 14 governorates. An ongoing project, in co-operation with JBIC (Japan), concerns the decentralised PV rural electrification and water supply within PV. The total cost of the project is estimated at 23.6 MTD. The project targets the electrification of 500 households in the North-West region and 63 water PV pumping and/or desalination projects.

Since 1995, 12 installations have been realised, 7 of which installed in 2005 and 5 in 2006. The power installed ranges between 0.2 and 3 kW for PV and 0.4 and 3 kW for wind. Table 3 summarises the installations in MENA countries.

Table 3. Summary of installations for Rural electrification based on Solar Energy in MENA Countries

Country	Achievements	Future
Morocco	• By the end of 2011: 12 million people in the rural areas (about 2 million households). • During the period 1995-20011 the electrification rate in rural areas has increased from 18% to 98% • 10% of the electrified rural areas have been solely based on PV kits. (more than 50,000) • Use of Private sector as an intermediate for implementing SHS projects via the "Fee-For Service" approach • Small Wind turbines and Micro Hydraulic Projects have been also utilized.	• ONE has expressed the willingness to prolong PERG via PVER • (Plan de Valorisation del' Electrification Rurale). • Take into account Clean Development Mechanism (CDM) in the implementation of rural electrification program.
Algeria	• 20 solar villages till 2001 with around 500kWp. Scattered village in an area of 1 million sq km^2 • 16 solar villages in the period of 2006-2009 –800 dwellings and 500kWp	• Not a concrete plan yet but with strong commitment to promote solar energy. Vast areas due to low population density non-electrified yet.
Egypt	• Favourable wind power conditions additionally to solar conditions • 2 remote settlements in Matrouh Governorate with PV systems of about 43 kW installed capacity, about 100 homes. • Remote applications for scientific rooms	• Rural Electrification of 121 Rural Villages via PV or PV/Diesel Systems is under investigation and the expected required capacity is about 1.2MWp
Jordan	• Badia electrification, 999 villages electrified	• Continuation via Badia development fund

Sub-Saharean Areas

Some of the sub-saharean countries are very poor and the access of people to electricity is very limited. In those areas, both UNDP volunteers and personnel have helped greatly with their actions to eradicate poverty and increase energy access.

The countries analyzed have mainly focused on PV, perhaps without so high electrification rates as other regions, but with high potential. Some of the countries share in common the engagement of the private sector in the form of charity or corporate responsibility.

Nigeria

In Nigeria, 73% of the population has not access to electricity. The Rural Electrification Board -REB was established in 1972 for electrifying areas not served by the power authority, Power Holding Company of Nigeria. However, progress since then has been limited to two villages in Kwalkwalawa of Sokoto State. In 1994 street lighting, 40 residences, 8 shops and 1 school and in 1998, 50 Solar lighting (2-3 per residences) and street lighting were installed. The large RES potential, with the exception of large hydro, has not been exploited [67]. We refer to this country because of an interesting project, the Villasol, being implemented by the Schneider Electric, an electrical distribution manufacturer, as part of the BipBop program [68]. The global aim of the project is to train young residents on rural electrification issues in order to further help in implementing such activities in their own countries. This programme is applied to 27 locations in 15 countries.

Since September 2011, the community of Asore has benefited from reliable, clean and affordable electricity without being connected to the national grid.

Thanks to the "Villasol" electrification solution, the village school and local traders now benefit from continuous electricity supply. At the same time, inhabitants can hire batteries, meaning they can light their homes and recharge their mobile phones thanks to the In-Diya lighting system.

The amortization of the installation Villasol Asore is estimated to be around five years because of the cost and wiring of the business model of a low battery.

Kenya

In Kenya it is estimated that 51% of the urban and only 5% of the rural population have access to electricity due to lack of capitals. 70% of the energy consumed in the country comes from biomass gathered after long time and great effort. In 1973 a rural electrification program established with the co-operation of ''East African Power and Lighting Company (EAPL)'' and the National Electricity Company of Kenya, "Kenya Power and Lighting Company (KPLC)'' being responsible for implementing the project while the Ministry of Energy co-ordinating these efforts. In 2006 the Rural Electrification Authority (REA) was established to accelerate rural electrification for improving socio-economic development of the country. The REA aims to promote RES use, conduct research, develop and manage the Rural electrification program and finally to grant authorization for power production in remote areas.

Another initiative promoting electricity access in the rural areas is "Umeme Pamoja", which translates as "Electricity Together" [69]. This campaign aims to establish a joint group of households, so as to connect them collectively to the grid, thus saving costs. This scheme is financed by the group settlement electrification schemes created by the KPLC. According to them, this scheme is aimed at making electricity connection easier, affordable and faster even without RES.

The KPLC may undertake the construction of off-grid solutions with distributed power systems over the country, i.e. not connected to the transmission system some of which utilized RES. An example is the Marsabit area where the 200kW wind turbine, went out of service in 2006, was replaced by two wind turbines of 250kW each at a total cost of Shs.198 million (1.8million €).

Progress in PVs

The significant cost reduction in PV panels in the mid 80's increased the interest for rural applications. The market has taken off since 1987, when the amorphous silicon panels of 12W were imported into the country for less than $ 100. The government has reduced value added tax tariffs for photovoltaic panels reducing their price by 15-20%.

At the same time, there were some initiatives like the one of Harold Burris, a former Peace Corps volunteer, who set up the first small business, called Solar Shamba, selling small PVs in a coffee producing area outside the capital Nairobi in 1984. By 1990 and had sold about 500 systems. He and Mark Hankins gained funding from the US-AID to create a technical training workshop for PVs. Many of attendants of the above workshop created later successful PV companies advertized even in bumper stickers [70].

At the beginning most buyers belonged to the upper middle class, as the price of materials was very high. The cost reduction and the good practices followed, lead to even greater demand for PVs and a large number of middle class households started buying small PVs for lighting and TV. The new boom phase in the development of the PV was the inauguration of lease and purchase services, which has allowed Kenyans with little income to access PV systems with credit. A review of activities and lessons learnt during the mid 80's to mid 90s can be found in [71] and [72].

Currently, at least 15 private companies selling PV systems operate in Kenya. Since the mid-80s more than 200,000 PV systems have been sold in the country. Since 1990 more than 2.5 MW photovoltaic power systems have been sold to private sector with more than 60% for SHS. By 1999, 3-4% of the rural population had acquired a PV and at least 70% was aware of what was such a system.

The Kibera Community Youth Program (KCYP)[73] started as an assembly unit for PV panels, in the Kibera Community, one of the largest slums in the outskirts of Nairobi. This programme offers job for youth to assembly small and low cost to charge mobile phones or listen to the radio. The KCYP was awarded with International Energy award in 2007.

In September 2009, the Solar Maasai project [74] installed a 5 kW photovoltaic system for a school building called "Empruken Primary" in southwest Kenya. 'The World Centric" supported the project with $ 1,500 sponsorship. The school now has lighting, computers and refrigerators for vaccines. In Mukuri school 25 Solar Lanterns for students and 10 for teachers were donated in 2010.

This was sponsored by the "Rosier Life Trust'' and some parents from the St. Margaret's School in England. This example shows that private charity can aid in rural electrification [75].

The lanterns are rented for reading at home. These lanterns of 5, 7 and 9 W were developed by the Intermediate Technology Consultants (ITC) and have a jack of 9V to power a small radio. Over 1 million solar lanterns have been put so far in Kenya.

Biomass Actions

The Rural Stoves West Kenya (RSWK) Project began in April 1990; it evolved from two earlier projects jointly implemented by ITDG and KENGO (Kenya Energy and Environment Organisation). The Wood-burning Stoves Field Test Project aimed to test five existing stove designs in five different zones of Kenya, and to provide guidelines for choosing an appropriate stove design for dissemination in rural areas of Kenya. The Women Potter Training Project aimed to assist women potters to increase the profitability of their business through improved product quality and diversification.

This included introducing ceramic stoves and stove components as possible products. 13 Women Group were trained to produce the stove named Upesi, and their early signs of success became the pilot project for the RSWK Project. Under this project, a market for improved stoves has been developed in West Kenya, mainly controlled by women's groups, who have mastered the technology.

The current activities are chiefly aimed at commercialisation. Production is estimated at 11 000 stoves annually; the profit generated by the stoves is comparable to wages in rural areas. As a result, the women potters have gained in status, self-confidence, and financial independence.

Under the GEF-UNDP, 6.65million $ project entitled "Market Transformation for Highly Efficient Biomass Stoves for Institutions and Medium-sized Enterprises" [76] was implemented. Under this project, 1,522 stoves in 723 schools were installed with active enrolment of 11,000 students. 550,000 seedlings with survival rates of about 60% (342 hectares) were planted.

Additional 4.35 million $ are linked to actions related to environmental protection in Kenya one of the 6 pilot countries (Ethiopia, Honduras, Kenya, Maldives, Mali and Nepal) to become beneficiaries of the Renewable Energy Program in Low Income Countries (SREP)). This programme is implemented by the Climate Investment Funds (CIF) and is funded by the Department for International Development –(DFID) and the Dutch government with 50 million $ for implementing RES.

Zambia

In Zambia only 2.2% of the rural population can have access to electricity grids. In order to promote rural electrification the Rural Electrification Authority (REA) was established along with the Rural Electricity Fund (REF). The Energy Regulation Act in 1995, involved the private co-operation in the power sector. In 2003 an act was made for Rural Electrification Act focusing on RES.

In 1996, an Energy Service Companies (ESCO) project was established funded by the Swedish International Development Agency (SIDA), -University of Zambia and the Stockholm Environment Institute (SEI) for 1999-2006 in Mbumwae. Under the program Energy Service Companies (ESCO), the Zambian Government buys the PVs, which then are let to energy service companies, for 20 years to repay the loan. Few ESCOs have created since then. These companies install the PV systems in households and shops and the customer will pay a monthly payment for as long as defined in partnership with the company depending on the capacity. Three different options for products were offered:

- 50Wp – 70Wp panel and 100Ah battery without inverter
- 20Wp – 120Wp panel and 150Ah battery with a small inverter
- 120W – 150W panel sufficient to operate a refrigerator

While a typical SHS consists of the components described in Table 4, typical monthly payments are described in Table 5 [77].

In 2006, 400 customers were paying a monthly fee to the ESCOs for solar photovoltaic electricity. Each ESCO has a waiting list of several hundred customers applying for solar installations.

There seems to be no acts of vandalism and very few solar panels were stolen, which may probably be related to a strong social control (the staff of each ESCO know their customers well) and the absence of a local black market for panels. Each client has to sign an agreement by which they take responsibility for the equipment, and can face severe costs if the system is stolen.

There is a good payment record (95% of the clients eventually pay) due to the quality of the service provided by the ESCOs and immediate disconnection in case of non-payment.

Table 4. Typical SHS provided by ESCOs

Component	Description
PV Panel	50Wp Monocrystalline-Si
Battery	90-105 Ah Lead acid with 85% Depth of discharge and 250-300 cycles
Solar Charger	12A Connecting up to 3 panels
4 Lamps	6-13 W, 12V DC Less than 40 lumens/W per lamp
1 Socket	12 V DC For radio or small TV set

Solar Aid is a private company established in July 2003 focusing on PV installation in schools, clinics and community health centres in Lusaka and all over the country, mainly in south with 46 installations.

In July 2009, the Solar Aid launched the "microsolar campaign", training of entrepreneurs, who were called" SunnyMoney Sellers" so as to be able to distribute solar products in all remote rural communities instead of kerosene lamps. Up to date, 15 active entrepreneurs have set up their own independent company and in total more than 1,500 units have been sold.

Table 5. Monthly fee for SHS in Zambia

	Solar Lantern	40Wp SHS	50Wp SHS	75Wp SHS
Monthly fee ZMk	8,000	23,000	25,000	30,000
Monthly fee $	1.6	4.6	5	6

Other UNDP projects in the Area

Tanzania

The UNDP project "Energy and Environment Mitigation to Climate Change"[78] started in April 2004 and ended in April 2009 in Mwanza area. The budget of $100.000, and co-operation with the Ministry of Energy and Minerals (MEM) lead to installation of more than 80 photovoltaic systems in small scale business enterprises to demonstrate the functionality, applicability and the productive use of the technology in the development of rural enterprises and poverty reduction. The demand of solar energy in Mwanza's local dealer's shops increased by 3000 % from the baseline (average 5 enquiries /day), while the number of solar energy merchants increased by 112%. Private sector merchants received business and technical skills training and more than 200 technicians from Mwanza and neighboring regions were trained in solar energy system installation, troubleshooting and maintenance More than 100 public institutions e.g. health facilities and schools were installed with solar photovoltaic systems.

Uganda

The "Bufumira Islands Alternative Energy Demonstration Project" implemented by the UNDP and the "Bufumira Islands Development Association (BIDA)" cost 41.290$. BIDA and the "East African Energy Technology Development Network (EAETDN)" offered technical support to install:

1) A hybrid system at Musoni area with 3 wind turbines of 400W each combined with 4 PV panels for charging batteries at 1$ for each battery charged. 30% of which is used for technicians, 50% for bank loans and 20% for maintenance cost.
2) A PV station at Bufumira with 7 PVs, 3 chargers and 2 100Ah batteries for supplying the BIDA, the local school and the charging needs of the community. It has been recorded that 15 batteries may be charged with this station.
3) A PV station at Bufumira with 8 PVs, 3 chargers, 2 100Ah batteries and one inverter for the secondary school and the charging needs of the community.

South-East Asia Countries

China, Laos [79] and Vietnam have widely used pico-hydro for electrifying great part of rural population with China made pico turbines. Chinese villages have developed the most with 100,000 very small capacity hydro units installed, amounting to 188.5MW [80]. Rural families in Vietnam have installed 130,000 pico-hydro systems (usually 200 Watts) in the last 15 years on a purely commercial basis [81].

Although the cumulative capacity of such smaller hydro plants does not show up in the data, these projects provide essential services to large numbers of citizens in a wide range of countries and local topographies and conditions.Despite these achievements, we will mainly focus on the activities of Bangladesh and India, presenting much greater diversity.

Bangladesh

In this very densely populated country, electrification rate still remains low. PV installation for rural electrification dates back in 1988. Norshingdi project was the first significant PV-based rural electrification programme, initiated with financial support from France. Three battery charging stations with a total capacity of 29.4 kWp and a number of stand alone solar home systems (SHS) with a total capacity of 32.586 kWp were installed. Rural electrification Board (REB) owned the systems and the users paid a monthly fee for the services. In 1995 in order to start RES policy, an independent authority, the Sustainable Energy Development Authority (SEDA) was created. Since 1996, the Infrastructure Development Company Limited (IDCOL)) a non banking financing institution has been managing infrastructure funding along with 10 (currently 30) national Partner Organizations (PO). Siginificant aid has been provided by the Grameen Shakti, as described in our dedicated sub-section. Two major RES sources characterise Bangladeshi projects; solar and biomass. Regarding wind any limited interest focuses on the islands of the coast of Bangladesh while for mini-hydro, few applications exist. The focus has been given all these years on SHS and biogas plants or Improved Cooking Stoves (ICS). ICS help greatly in improving indoor air quality conditions and reduce significantly the cooking time. This has helped greatly the women population who are more free from gathering wood chips or buying other fuels for their cooking. Biogas use along with improved efficiency of ICS help in avoiding the country's deforestation.

The IDCOL started their Solar Energy program in January 2003 and the initial goal was to finance 50.000 Solar Home Systems (SHS) by the end of June 2008. This goal was finally reached in September 2005, three years earlier than the original target, at a cost $ 2 million below the initial budget. The next target set was the addition of additional 200.000 SHSs by the end of 2009, a goal once again achieved earlier in May 2009. Then the target was to reach 1 million SHSs by 2012. By August 2010 645.033 SHSs under the program were installed. This target was surpassed in October 2011 with 1,154,923 SHSs adding 60MW [82]. Latest information from the IDCOL site speaks about 1.45 million SHS by the end of April 2012. Taking into account the installation rhythm, we estimate them now at about 1.8 million SHS. IDCOL continuously revised targets for financing installation, from 2.5million SHS by 2014 to 4 million SHSs by 2016. Bangladesh has set a target to cover 10% of the total energy demand through renewable by 2020.

SHSs are sold (mostly through micro-credit) by POs to the households and business entities in the off-grid areas of Bangladesh. IDCOL provides refinancing facility to the POs and also channels grants to reduce the SHS cost as well as support the institutional development of the POs. In addition, IDCOL provides technical, logistic, promotional and training assistance to the POs. The funds of the programme are gradually reduced as the installation number is increased. The total amount of grant dropped from 90$ to 22€. This is made available by the international funds stated above. For the convenience of the personnel installing SHS, IDCOL has technical information available [83].

Apart from simple SHS, the IDCOL board recently approved financing of a 100-kW solar photovoltaic (PV) based micro-grid by PUROBI Green Energy Limited (PGEL) at Sandip island, Chittagong. A 40-kW diesel generator will be integrated into the proposed power plant in order to ensure adequate power supply during periods of low solar radiation. Total project cost is BDT 55.37 million (i.e. 0.6 million €) and will be financed through mix of debt, equity and grant from KfW. PGEL is a consortium of four NGOs namely Bangladesh Rural

Integrated Development Grub-Street Economy (BRIDGE), Integrated Development Foundation (IDF), Upokolio Bidyut and Mohila Union Samity (UBOMOUS) and Rural Energy and Development Initiative (REDI). Once completed, this Project is expected to supply electricity to adjacent 390 shops, 5 health centres and 5 schools.

A recent study by Mondal and Denich [84] indicates that the selected locations blessed with a considerable annual average global solar radiation (3.81–6.47 kWh/m^2/day), are prospective candidates for the deployment of PV–diesel–battery hybrid power system. Their simulation results for all selected locations indicate that the most economically feasible system for 50 rural off-grid households would be composed of 6 kW PV array together with a 10 kW diesel generator and 10 numbers of batteries of which each has a nominal voltage of 2V and capacity of 800 Ah, and the penetration of solar PV is 43%.

Within the frame of promoting PVs and employment among women, the Coastal Electrification and Women's Development Microenterprise (CEWDM) action was initiated. The project location is Char Montaz, an island with 2000 households, in the southern coastal region of Golachipa Thana of Bangladesh. Today, 35 rural women of Char Montaz are engaged in the operation of a micro-enterprise for construction and sale of DC lamps, which can be used in combination with batteries in Solar Home Systems (SHS). With continuing training from this project, these women learnt lamp construction with quality control, business development and marketing. Today more 1000 lamps are being used with re-chargeable batteries for lighting the rural houses, shops, mosques and fishing boats [85].

The potential of biogas technology is immense in Bangladesh. According to an official estimates there is a cattle population of 24 million and poultry population of 75 million. This can produce about 3 billion m^3 biogas. It is estimated that 4 million biogas plants can be developed.. A total of 24,000 domestic biogas plants had beeminstalled in Bangladesh during 1971 to 2005[86], while there are also over 2000 biogas plants sited on poultry farms [87]. The Bangladeshi government has been heavily involved in the dissemination of biogas plants through the country, with subsidies offered for plant construction. IDCOL has initiated the National Domestic Biogas and Manure Programme (NDBMP) with support from GoB, SNV-Netherlands Development Organization and KfW. Total programme cost is 14.9 million € with farmers contribution in the order of 2.7 million € [88]. Under the project a total of 37,269 domestic sized biogas plants will be financed during the period 2006-2012. By April 2012, 22549 biogas units have been installed within this programme. There are typical sizes for the dome gathering the excreta ranging from 1.2 m^3 to 4.8m^3. Typical equipment specification and required material (bricks, pipes etc) for each one of the typical sizes of the biogas plant are also provided by IDCOL.

At the beginning of the year 2011, IDCOL and the Brac Bank fund larger project to produce energy from biogas. These two have signed an agreement with Paragon Agro Limited (PAL) to create three units producing electricity from biogas, more precisely poultry produced in the PAL farms Paragon Poultry Ltd. (PPL) and the heat would be provided to the farms.

As a side effect, 25,000 tonnes of organic fertilizer will be produced with estimated selling price 15 BDT (0.14€) per kg. One unit will be established in Mymensingh district and the other two in Gazipur, with a total power of 475kW. The total project cost is BDT 146.89 million (1.4million €) and the Brac Bank with IDCOL will jointly provide BDT 88.13million (827 th €).

The Institute of Fuel Research and Development (IFRD) has developed a number of improved stoves. These stoves save 50–70% fuel as compared to the traditional ones. So far, more than 100,000 stoves have been installed in the users kitchens.

The Local Government Engineering Department (LGED) and IFRD are working to install the biogas plants in the rural areas. Khulna University of Engineering and Technology (KUET) has successfully under took a research and development work on biomass briquitting technology with a grant from SIDA. By now, this technology has transferred to private entrepreneurs. Currently, about 60 small private enterprises are producing briquettes from rice husk and saw dust. A summary of other projects running or successfully finalized in Bangladesh is shown in Table 6. The projects supported by the Grameen Shakti are described next and are not included in this table.

Table 6. Various programs in Bangladesh and their achievements

Programme Name	Achievements	Status
Rural Services Foundation	59216 SHS	(April 2010)
BRAC Solar Energy Program for Sustainable Development[89]	72470 solar PV systems, 1000 biogas plants, 10 wind turbines, and 260 Hot Box cookers. 6 PV-Wind hybrid systems	October 2007
Wise Progress Report	23412 (1940kWp)	August 2012
Local Government Engineering Department	40.5 kWp	2006
Rural Electrification Board[89]	20 Solar Powered Irrigation Pump and 1250 SHS (183.7kWp)	On going
Renewable Energy Development and Energy Efficiency Measure of Bangladesh Power Development Board (BPDB) [91]	3 Sets of 10 kW Centralized AC Market Electrification Systems for the electrification of more than 200 shops in each market. • 1000 Solar Home Systems of 120 Wp each • 30 Sets of Solar PV Street Light Systems • 3 Sets of Solar PV Submersible Water Pumps of 50,000 lt per day lifting capacity of each pump. • 9 Sets (360 Wp each) of Solar PV Vaccine Refrigerators for the Health Care Centres	On going

Grameen Shakti

The Grameen Shakti (GS) is a Non-profit Organization, established in 1996 which has revolutionized the local communities as during 1996 to 2010 has installed 520,000 photovoltaic systems in 40,000 villages in Bangladesh, 200,000 of which only during 2010. According to IDCOL estimations, GS had installed 1,020,000 SHS by December 2012. It can be safely said that GS has introduced PVs to Bangladeshi villages.

GS objective is the application of environmentally friendly and affordable energy technologies with emphasis to address the needs of residents without access to electricity or/and natural gas grids. GS is now one of the largest and fastest growing microgeneration programs in the world.

The main two axis refer to SHSs and biogas systems aiming mainly on improved cooking stoves (ICSs). It is estimated that about 3.5 million people are beneficiaries of the GS service. Among the beneficiaries are more than 8500 employees or representatives in 972 branch and 128 regional offices across the country, 100-fold the personnel in 2002. In 2010 46 Grameen Technology Centres were operating in order to reduce the maintenance cost, improve training and expedite after sales services.

To improve access for those unable to afford the low cost solution, there is the option of forming a micro-utility. The interested customer buys the SHS or biogas plant and sells electricity/gas to his/her neighbours. 10,000 micro-utilities have been scattered around the country.

Apart from the obvious environmental, and economic common as in other regions, there is the additional benefit of helping women improve their social status. Women take an active role in the community and in their household finances. GS trains women order to set and repair of photovoltaic panels and electrical sockets in homes and businesses.

SHS

As referred above 520,000 SHS had been installed in the country by the end of 2010, 4-fold the installed systems back in 2007. At the end of 2010, about 20,000 SHS were installed per month. This has led to the millionth SHS on 30th November 2012.

The decentralization of service was one of the main reasons for improving the installation rate. SHSs provide lighting and the ability to charge mobile phone or watch a small TV set. The latter tends to become a social activity. The financial options for paying back a SHS are described in Table 7.

There are plenty of different configurations regarding the size of the panel, the lighting sources and the type and capacity of the storage device offered. There is no need to copy the whole price list here but we will refer to the simplest and the most complicated configuration capable of meeting specific loads for about 4 hours for off grid applications:

a) A 10 Wpanel, 2 or3 LED lights or a 5 watt CFL, a 18AH battery, a charge controller, a frame and cables with price 9800BD Taka (i.e. 92€)
b) A 135 W panel, 11 x 7 W CFL, a 100 AH deep discharge battery, a charge controller, a frame and cables. This system can operate also a TV Set (not included in the package) at 72900 BD Taka (i.e. 690€).

Table 7. Various Options for paying back a SHS to Grameen Shakti

Option	Down payment	Monthly instalments	Service charge (flat rate)
1	15%	36	6%
2	25%	24	4%
3 (forming micro-utility)	10%	42	None
4	100% with 4% discount	N/A	N/A

Urban customer areas can enjoy a PV system coupled for AC loads with 400 W/400 Ah and 1kW inverter at 199000 BD Taka (i.e. 1870€). Lower capacity configurations also exist at lower price [92].

In order to reduce cost, auxiliary equipment is locally manufactured, quite often by women technicians especially trained by the GS for this purpose. It is estimated that about 3000 female technicians have been trained by GS, working either at GS offices or as renewable energy entrepreneurs. Apart from this attractive financing GS has focused on training end-users on how to operate such a system to prolong its life-time.

Biogas

Biogas can be used as local fuel or even to feed a local Internal Combustion Engine unit instead of gas. The financing alternatives are as follows:

a) The buyer pays 25% of the total cost as down payment. The remaining 75% of the cost is to be repaid through 24 monthly instalments with 8% service charge (flat rate) within 2 years.
b) The buyer can construct his plant with his own funds under the supervision of GS engineers. In such a case, half of the technical and supervision fees will be paid in advance and the rest will be paid after the commissioning of the plan.

Typical cost of a 1.6m^3 digester capable of operating for 3-4 hours and capable of using 43kg cows excreta or 24kg poultry ones would cost about 25700BD Takas (680€) and the owner has the ability to pay 2505 BDT down payment and 686 BDT (i.e. 6.5€) monthly instalments. The package includes a) Free after sales service including monthly visits by GS engineers for two to three years b) Option for signing annual maintenance agreement with a small fee during post warranty period. There are the options of more centralized solutions, one unit for neighbouring installations and use of such solutions especially for the agriculture and poultry industry. Even if the quantities of slurry is not enough for running biogas this can be sold to other owners of bio-gas units. Thus, the otherwise polluting excreta can become a trading asset between SMEs. As a by-product organic fertiliser can be produced and used afterwards for agriculture or even aquaculture reducing chemicals fertilizers and their related cost by 30-40%. Only GS had installed 5680 such units during 2010 leading the installed capacity to 14,300 systems [93]. Regarding Improved Cooking Stoves (ICS), 596,000 stoves had been installed by 2012, 147,000 only during 2010. The financing for the end-users is very attractive, especially when one considers western every day life cost. With less than 8€ a domestic single ICS can be bought by a Bangladeshi household and a larger family with slightly more than 10€ can obtain a much larger ICS. Commercial applications can reach 6000 BD Taka or 56€.

GS sees a potential market of at least 2 million ICSs in the first three years of the program. GS plans to depend on two types of local players for expanding ICS - local technicians and local manufacturers. GS has already trained more than 600 local youth especially women to make, sale and repair ICSs. GS has also set up 10 manufacturing units in rural settings for constructing ICS accessories such as metal grates and chimneys.

India

India, the second most populated country in the world, still has some areas unelectrified. There has been significant progress in installing Renewables (Table 8) all over the country to meet increasing demand. Some of these applications are rather innovative or the largest in the whole world, like the largest solar steam cooking or the Bagasse based Cogeneration System, [94]. The Ministry of New and Renewable Energy (MNRE) continues programmes for further grid connection of RES [95]. Three channels of rural electrification exist : The first one via extending Networks, mainly implemented by the Rural Electrification Corporation Ltd(REC)[98], the other by off-grid RES applications managed by[95] and the third one by the SELCO company.The REC is a public sector enterprise under the Ministry of Power, established on July 25, 1969. Its main objective is to finance and promote rural electrification projects all over the country. It provides financial assistance to State Electricity Boards, State Government Departments and Rural Electric Cooperatives for rural electrification projects as are sponsored by them. REC provides loan assistance to SEBs/State Power Utilities for investments in rural electrification schemes through its Corporate Office located at New Delhi and 17 field units (Project Offices), which are located in most of the States. The Project Offices in the States co-ordinate the programmes of REC's financing with the concerned SEBs/State Power Utilities and facilitate in formulation of schemes, loan sanction and disbursement and implementation of schemes by the concerned SEBs/State Power Utilities.

Table 8. Installed Grid Connected Capacity in India,[95]-[97]

Source	Capacity (MW)
Wind	17352
Solar	368
Biomass (including biomass, co-generation etc)	2665
Small Hydro	3252

Various programmes have been implemented for Rural electrification namely [99]:

- Rural electrification under Minimum Needs Programme (MNP)
- Pradhan Mantri Gramodayan Yojana (PMGY)- Prime Minister Village Development Scheme
- Kutir Jyoti Scheme (KJC) (Cottage light scheme)
- Accelerated Rural Electrification Programme (AREP)

We will not focus on these programmes since they have been mainly based on extending grid but we will speak a bit more on the Rajiv Gandhi Grameen Vidyutikaran Yojana (RGGVY)[100] program which aims to electrify most of the India region according to the following definition:

A village is considered electrified if :

1) Basic infrastructure such as Distribution Transformer and Distribution lines are available in the inhabited locality as well as in the Dalit Basti hamlet, where it exists.
2) Electricity is provided to public places like Schools, Panchayat (Local administration) Office, Health Centers, Dispensaries, Community centers etc.
3) The number of households electrified should be at least 10% of the total number of households in the village.

This definition has led to 77.4% electrification by 2009 in terms of villages and only 44% in terms of households. The southern and eastern areas suffer the most.

Under RGGVY, electricity distribution infrastructure is envisaged to establish Rural Electricity Distribution Backbone (REDB) with at least a 33/11KV sub-station in a block, Village Electrification Infrastructure (VEI) with at least a Distribution Transformer in a village or hamlet of more than 100 inhabitants, and standalone grids with generation where grid supply is not feasible. This program also supports electrification using Decentralized Distributed Generation, either RES-based or not. 90% subsidy from Government of India and 10% loan from REC or from own funds of the state/loan from financial institutions is foreseen for such installation. The Monitoring Committee of the RGGVY, coordinates with the MNRE to avoid any overlap in electrification projects.

The target of electrifying 1 lakh (i.e. 100,000) villages and to provide free electricity connections to 17.5 million households Below Poverty Line was achieved by 31st December 2011, instead of March 2012. On July 2012 the actual number was 105,000 dwellings.

MNRE Actions

MNRE with their efforts have helped in great part in the achievements of Table 8. Subsidy programmes and research initiatives aim to keep up this installation rhythm. One of the main programs is the Jawaharlal Nehru National Solar Mission (JNNSM)[101]. Overall objectives for 2022, of this three phased Mission, include: (i) deployment of 20,000 MW of grid connected solar power; (ii) 2,000 MW of off-grid solar applications including 20 million solar lights, (iii) 20 million sq. m. solar thermal collector area; (iv) to create favourable conditions for developing solar manufacturing capability in the country; and (v) support R and D(R&D) and capacity building activities to achieve grid parity. The first phase aims at setting up of 1,100 MW grid connected solar plants, 200 MW capacity equivalent off-grid solar applications and 7 million square meter solar thermal collector area till March 2013. Already by February 2011, the achievements figures are 38.5 MW for off grid PV systems and 120000 sq. meter solar thermal collector area.

Various projects on Biomass are being implemented like Bagasse Cogeneration in Private Sector Sugar Mills, National Biogas and Manure Management Programme (NBMMP) since 1981-82 with 150,000 installations and 440,400 family biogas plants. Also the National Biomass Cookstoves Initiatives (NBCI) launched in December 2009. For rural electrification with biomass, village level projects of 32kWe using Rice husk based gasifier have been proposed.. So far 30 such village level projects have been installed and providing unmet demand of electricity in about 75 villages / hamlets / tolas in East Champaran, West Champaran, Muzaffarpur and Lakhisari of Bihar. It is expected that more than 100 villages / hamlets /tolas would be provided electricity during the current financial year.

There has been significant deployment in setting up of Pico Hydro projects in Karnataka in the last 4 years. Since 2007, till date around 500 Pico Hydro projects have been set up and the numbers are increasing steadily. Many systems have been installed in people living in isolated places near Reserve Forests and National parks where it is not possible to lay transmission lines.

In such places, the Pico Hydro Systems have been a boon to these people. Typically in the hilly areas grid power and solar power is unreliable in monsoon and Pico hydro is a convenient solution due to continuity of supply. Chembu village in Coorg District of Karnataka has seen a remarkable difference due to installations of the Pico hydro systems. The village is surrounded by forests and is typically inhabited by poor farmers who had no access to electricity but had almost perennial streams running through their habitation. Today, around 40 houses in the village have installed these Pico Hydro systems and have access to clean electricity right through the year.

Additionally, a cumulative capacity of 1647 kW of wind-solar hybrid systems has been installed.

Achievements in Rural Electrification via RES through campaigns till 2009 are shown (Table 9). In the same table comparison with previous periods is made. A very detailed description on the achievements currently is provided in the annual report of MNEP, which provides information on initiatives, installation progress, future targets with illustrative examples. Due to the large size of the country and the increased installation rate, the reader is called to advice these reports for having the most updated figures available.

Table 9. Rural Electrification achievements summary in India

	October 1999	October 2003	January 2009
Biogas Units (million)	2.88	3.55	4.09
Improved Wood Stoves (million)	25	35	40
Solar cookers (thousand)			637
Solar Home Systems (thousand)	64	260	434.7
Solar Street lighting	32920	43470	43470
Solar lanterns (thousand)	222	442	697
Solar Pumping	2160	6400	7148

Based on the latest information from the Annual report 2011-12, the Remote Village Electrification Programme has managed to electrify 7113 villages and 1871 hamlets of which in Northern areas, suffering the most, 3186 remote villages and hamlets with 400 remote villages are being implemented in Northern areas. Under the Village Energy Security Test Projects 65 villages were also electrified aiming to protect population from wild animals attacks.

Selco

SELCO Solar Pvt. Ltd, a social enterprise established in 1995, came from the Solar Electric Light Fund and provides sustainable energy solutions and services to under-served households and businesses in India. With an average rural income $ 50 per month it is impossible to make a household install a photovoltaic system, costing 200-500$. The SELCO has leveraged the potential of significant rural banking system to finance sustainable energy systems for poor rural households. Over the years, Selco has forged partnerships with nine regional rural banks (also named as Grameen banks), commercial banks, non-governmental organizations and agricultural cooperatives to develop economical solutions. SELCO also tries to implement business models sufficient for rural population to meet their own needs. Interest rates are based on source of credit ranging from 5% to 14%. Customers pay 10-25% deposit, paying the remaining amount in the period three to five years. Other customers have the ability just to rent equipment according to their needs. Typical services by Selco include custom system design, Installation, Training on proper system use, After-sales maintenance and support. The following loans and financing was received by SELCO:

- By the end of 1996, Selco received $ 5 million loan from the non-profit organization Winrock Indernational U.S.
- The world's first carbon offset funding agreement for solar rural electrification was signed in August, 1999 between SELCO and the Solar Energy trust, providing $500,000 -from PacifiCorp's Klamath Cogeneration Project in Oregon -for SELCO to use over a 30 year period to finance 182,000 SHS in India, China and Sri Lanka. A single 35Wp SHS is expected to displace 7.2 tons of carbon dioxide emissions caused by burning kerosene and recharging car batteries over 20 years. The Klamath Falls project is for 1 million tons or $0.50 per tonCO2.
- The UC International Finance Corporation (IFC). Funded $ 750,000 between 1999 and 2001 and in 2003,provided $ 1 million loan
- 1.7 million $ was made available from three social investors: Good Energies Foundation, Lemelson Foundation, and E and Co. The Selco plans to light up 200,000 rural homes, covering a wider geographical area, over the next four years with this funding.

Apart from this financing the SELCO has established international co-operation with the D-Lab: Development, Design and Dissemination of the Massachusetts Institute of Technology (MIT)[102] and the Engineers Without Borders (EWB) only to mention a few. Pioneering work in rural India of Selco has encouraged many financial institutions to create a special line of credit for solar systems. Between 1997 and 1998 the President of the Malaprabha Grameen Bank, Mr. KM Udupa, convinced of 30.000 $ lending to finance 100 solar lamps. Additionally local people are employed for this purpose.

SELCO Foundation and Labs

SELCO initiated the SELCO Foundation aiming to use the available philanthropic capital to develop robust and field-proven technological and financial models in the field of renewable energy, and generate public awareness about these models. These models can be replicated in other parts of the world, thereby achieving greater leverage on the money spent. Among the activities is the training of the local population also in issues of solar energy, for example by conducting a self-help program for housekeepers from remote areas teaching them basic maintenance actions. One of the major contributions of this foundation is the creation of SELCO Labs. SELCO Labs was created in 2008. It was also recognized that the Labs could contribute to customizing solutions - technological, financial and manpower related - that already existed but were inappropriate for the rural sector. Although, initially started as a department within SELCO, it spun off in 2010 as a separate legal entity under the umbrella of SELCO Foundation, a registered non-profit organization. The main labs are located in the small town of Ujire. Selco labs have made significant work in the field for implementing the targets of SELCO and adapting to the needs of the local population. Some innovative ideas funded and implemented by SELCO including rural electrification with RES are the following.

a) Hawker model

This targets the small entrepreneurs working during the evenings as vendors. They can rent or even buy-depending on their economics- a battery which is charged in a central charging station during the day and is used during their night-time for lighting their outdoor shops.

b) Light for Education

One of the programs funded by SELCO Labs is the Light for Education program [103] hoping to provide good quality light for students, currently using kerosene lamps or candles, and partially resolve the growing problem of indoor air pollution due to kerosene lamps used.

Light for Education program provides a safe, cost-effective and reliable solution already being implemented in several schools successfully. Students are provided an LED powered study light (1W) from the school. The light includes a small detachable battery (2700 mAh NI-MH), which can be recharged at the solar recharge centre at school. Therefore, the child must attend school in order to have light at home. After the end of their school period, students should return the lamps to the school, which are redistributed to the next batch of students, creating a sustainable model for lighting homes of students. The benefits achieved are:

- Independent solar charging system not requiring grid connection.
- Cost effective model because a centralized charging system is used to charge all the battery packs which reduces the high expense of providing individual solar panels for each student

- Technically the performance of the central charging system is much higher than individual solar panels, especially during rainy season. The charging system may become part of a hybrid system and not simply PVs
- Improves studies and attendance in schools as parents also see an added advantage of sending their child to school

c) Integrated Energy Centers (IEC)

As we examined, at the beginning of this paragraph for India, a village can be considered electrified even though a small percent of the population has access to electricity. There are people mainly in temporal settlements not having access to electricity. The IEC aims in central charging stations run with PVs making possible for residents in the proximity of the IEC to access basic energy services on a rental model for lighting, mobile charging, water purification, refrigeration, audio-visual aids and community televisions. The current research till 2013 builds test site IECs aiming to identify the model or models of IEC to be implemented afterwards, either by Selco or in co-operation with other partners. The following systems are currently tested:

a) IEC for floating population with increased seasonal population due to proximity in temples as shown in Figure 5.
b) IEC for population temporally living in area (nomadic way of life). Implementation via tent school
c) IEC for energy centers inside an established accessible public place, like school, hospital etc.
d) IEC for serving residents in a nearby radius (e.g. 2km)

These IECs provide demonstration of lighting appliances, cookstoves, DC fans, TV sets and end-users services with renting services like:

- Provision of lighting appliances, such as solar lanterns or hawker lamps. Typical rental cost in the order of 10Rps (around 12€ct) daily.
- Mobile charge rental, a very popular activity where slots get full quickly. The rental price is in the order of 5Rps (12€ct).
- Rental of mini fridge space
- Provision of purified drinking water
- Laptop charging (especially for IEC in proximity of schools)
- Community TV sets
- Ability to host visit from doctors.

Cold Climate Areas

Renewable Energy Sources have been also applied under very harsh winter conditions. First, some discussions on the precautions to be made for RES projects in cold climate areas

are provided and then some typical examples of RES applications in such areas are described in this section.

Figure 5. A typical Integrated Energy Center (IEC) (Kind Courtesy of Selco Foundation).

Special Precautions to Be Made

In those areas extremely low temperatures occur. This creates additional mechanical and operational stress to the installed equipment. The selected equipment should be able to withstand low temperatures and additionally low maintenance since in such areas it may be rather hard to perform maintenance not only due to the distance to the site but also because of the equipment required to access the installation area. Therefore, the reliability of the equipment should be much improved compared e.g. to urban areas installations.

When cold temperatures coincide with high altitude, e.g. mountainous areas, then lighting may be more often and additional precautions should be made. Additionally, if RES are combined with diesel gen-sets then their efficiency is reduced with altitude according to the manufacturer's data. This is mainly due to lower oxygen content in the air. Low temperatures cause significant capacity decrease so that a special procedure is required to make batteries operational again.

Frozen winds and snowfall create significantly additional mechanical weight either to blades of Wind Turbines or PV panels. Additional weight of Wind turbines increases the losses on the blades and may even lead to their fracture. Icing of the equipment, blades, pylon structure, guys and anchors for small wind turbines, even cabling are a severe threat to the operation of wind turbines and their breakdown may create long time of lack of service. Inefficient de-icing could lead the rotor to stop due to excess vibration or acceleration of the rotor. The erection material should be able to withstand additional weight on the pole. Cylidric poles made of steel should give their position to lattice towers since the latter are less prone to icing; otherwise these poles may bent due to the weight of ice. The standard steel structures become brittle in low temperatures, so some caution is necessary when tubular steel towers are erected during winter. The tubular tower may buckle, so these should not be erected in extreme low temperatures. The lower ends of guys (where they are attached to anchors) need to be protected in severe icing climates, as large ice masses may slide down along the guy wires. Even if the blades have not accumulated ice, low temperatures under high wind speeds may create difficulties in starting torque. This may be resolved with heating and lubrication of the turbines.

Additionally, wind power production is proportional to air density ρ. Air density may be 26.7% higher at –30 deg. Celsius compared to 35 degrees Celsius. This means that the power on the blades is significantly higher compared to other applications in more moderate climates. For instance in Kotzebue, Alaska 66kW were recorded to produce 99kW peak power. All these issues should be taken into account in the dimensioning of the stress and the capacity of the equipment. Relevant research has gradually started in Northern European Countries and Canada, due to the interest in increasing wind power capacity. To face such difficulties the wind task group 13 of the International Energy Association (IEA) has recently issued recommendation issues for cold climates [104]. Some manufacturers like Proven and NorthPowerSystems [105] have tried to apply additional special coatings on the blades, new designs on the blades to withstand arctic environment weather conditions.

PV panels are less vulnerable to such conditions. They have been usually been tested for significant weight of ice which also melts due to sunshine. The manufacturers quite often state what is the capability of withstanding hazel –actually the pressure on the protective glass of the PV. However, special precautions should be made when selecting the inverter for AC loads, even at much more moderate climate areas. Open circuit voltage for PV panels increases as temperature reduces. Thus series connection of PV panels to form an electrical string for connecting to the inverter has higher open circuit voltage as temperature decreases. Unless careful design is made prior to installation, the open circuit voltage may surpass the limit of the inverter and destroy it at fair weather conditions, low temperatures and early morning hours-not unusual for such regions. Since this is a failure caused by inefficient design of the series connection, the inverter manufacturers do not usually remunerate for the loss of the inverter.

On the demand side, heating loads are expected to be higher, lighting hours may last longer since it has been recorded that in order to avoid cold, closets remain close to protect warm leaving the place.

Guidelines in implementing RES projects in cold climate regions can be found either by the Polar power [106] funded by the National Science Foundation, or by external searches [107].

Mountainous Areas-Nepal

Large parts of some countries are located at rather high altitude in mountainous areas like the Himalayas and the Andes. On those areas wind and especially hydro potential may be significantly high; much higher than the rest areas of the world examined.

On those areas long distances, harsh conditions and low population density are barriers for extending electricity networks. Such a country is the Kingdom of Nepal, with 90% of the country located in high mountains with only 15% of the population enjoying grid access. Sometimes, the residents suffer from load shedding [108] with programme downloadable from the Nepal Electricity Authority (NEA). Significant development plans have been applied including rural electrification needs as described below and Hydro potential development with total budget 133,4 million $ [109].

Despite the difficulties, interesting solutions and business models have been proposed in this country as described below.

One of the steps to face this reality was the creation of the Alternative Energy Promotion Centre –AEPC in 1996 [110], aiming to promote and develop RES projects. The Danish government strongly co-operates with this Centre, either in technical terms or even financial. The major achievements of the AEPC are the following:

- Design and implementation of National subsidy policy
- Organization of an Interim Rural Energy Fund (REF) in order to manage subsidies especially from Denmark, Norway, Germany and Nepal
- Financial support from Norway, the World Bank, the UNDP, the Netherlands and the European Commission.

One of the programs implemented under the supervision of the AEPC is the "Energy Sector Assistance Programme (ESAP)".

This program was established by Denmark in 1999 aiming to achieve sustainable development in rural areas for a 20 years period using RES. The first phase of the program created laid the foundation for further beneficial actions for 1.5 million people who live in remote areas of Nepal. This initial phase was of particular interest because it helped in disseminating RES technologies to rural areas, training personnel and improving access to bank products for the development of RES. This phase was financed initially by Denmark with 154 Danish Korons and received further financing from Norway and Nepal leading the total budget to about 31.4 million €. The rest of the phases are summarized in Table 10

AEPC has also organised the Solar Energy Support Programme (SSP) for applying, disseminating and quality assessment of PV system. In 1996, 50% of the cost was subsidized by the APEC and more than 20 private companies were installing PV systems. One of them, the Solar Electricity Company (SEC) [111], is reported to have installed over 15000 PV systems in more than 64 provinces in Nepal. This company is the original designer-supplier and installer of the first pilot project of solar home system in Nepal in 1993. In this project, 68 houses and a community center with electrified PV panels. The Prime Minister of Nepal launched the program, which became the symbol of the successful use of the PV systems and inspired other villages to use this technology and led in installing 45000 homes in Pulimarang Village of Tanahu District.

Additional to installing and selling PVs via resellers in Nepal, SEC has started manufacturing useful electronic peripherals for PV systems such as light devices, battery chargers and DC/DC converters.

Among the AEPC projects is the development of 6 hybrid wind-solar systems of 400W each (150W Solar and 250W wind) in villages Pyuthan, Musgit and Palpa based on the maintenance accessibility. 20 households and 2 schools were beneficiaries of this action.

Table 10. Phases for implementation program in Nepal for the ESAP

Phases	Periods
Phase 1	March 1999 - March 2004
Phase I no cost extension	April 2004 – September 2004
Bridging Phase	October 2004 – June 2005
Bridging no cost extension 1	July 2005 - June 2006
Bridging no cost extension 1	June 2006 – March 2007

Practical action has helped in electrifying many villages in Nepal. Two wind turbines of 100W and 200W were used for providing light and small appliances to the members of the Phakel area.

UNDP has significant presence in Nepal since 1996 through the Rural Development Programme for Energy (REDP). Rural Energy Development Programme (REDP) has served more than 60,000 households living in remote villages of Nepal. Between 1996 and March 2010, the REDP has electrified more than 50,000 households via all type of RES utilizing 9.305 million $.

The Renewable Energy for Rural Livelihood (RERL) program [112] is a successor to the REDP and has been implemented by the AEPC in 26 remote areas of Nepal since 2010. The aim in these selected areas is to connect additional 28,500 households with energy services by 1,65 MWp and installing 250, 100 large and 150 small, power plants in 20 provinces. The total budget is in the order of 1.4 million $ till 2016 and is co-financed by the World Bank, District Development Committees (DDCs), Village Development Committees (VDCs) and local communities.

The Solar and Wind Energy Resource Assessment (SWERA) project was supported by the United Nations Environment Program/Global Environment Facility (UNEP/GEF) project started in 2003 and ended in December 2007 in various locations aiming at households and schools in various countries [113]. In Nepal, the project was jointly implemented with APEC. The number of electrified buildings was not that high but this project was an initial point for systematically collecting wind speed data. The Renewable Energy Project (REP)[114] was a common EU-Nepal (AEPC) project funded by 15 million € from the EC and 675,000€ from the Kingdom of Nepal till August 2011.

The REP focuses on provision of solar energy systems for public services (schools, hospitals, safe water provision, communications, etc) of remote districts, as well as for income generating activities. It is strongly believed by the consortium that quality of the installed system components is one of the key factors that determine the sustainability of the system as a whole.

Therefore adequate attention was paid while drafting the technical specifications of the solar energy system components to ensure that internationally accepted quality products are procured and installed. Like wise, standard sized product package are recommended for installations as described in Table 11.

Himalayan Light Foundation (HLF)

The Himalayan Light Foundation (HLF) [115] is a Non Governmental association aiming at improving the quality of life of the remote population of the Himalayan and South Asia region via the introduction and use of environmentally friendly renewable energy technologies (RETs). In order to achieve their purpose a variety of projects have been proposed.

Solar siblings program aims to bring together students from the US, the UK and Australia with Nepalese students donating solar computers to them [116].

Solar sisters program has the same target, aiming not only at students but also at people around the world. Both projects give the opportunity to participants all over the world to travel in remote areas of Nepal. REP targets mainly districts with low Hydro potential, low income and high solar potential.

Table 11. REP suggested packages

REP Packages	Total Array Power	Appliances
Health Post	600Wp	Vaccine fridge, light, surgery light, HF radio
School 1	1000Wp	Light, Computers, Photocopier
School 2	1900Wp	Light, Computers, Photocopier, Printer
Community Telecom Center	900 Wp	Light, Computers, Photocopier and fax
Computer Literacy Courses	700 Wp	Lights and Computer
Entertainment community	200 Wp	Lights, TV (21"), VCR/DVD
Milling	1500Wp	2HP, motor, milling machine
Water pumping	Case by Case	Motor and pump

Travelling Education Center (TEC) aims in providing multi-media training to Nepalese using solar –powered multimedia kits. The RE-TEC functions under the same principle as the TEC but focuses on disseminating RET information. Among the educational activities HLF has organised the Radio Program on Renewable Energy (RAPORE™) in such a way that every listeners of the "RAPORE™" program could understand the information that has been aired. This is like a radio documentary with 12 pre-recorded episodes. Home Employment Lighting Program (HELP) is a cluster of programs aiming to combine and improve skills of the local population focusing on handicraft with rural electrification program. Thus, the local population have the ability not only to be electrified but also gain a production activity. Renewable Energy System Dissemination Program (RESDP) in 1997 and its successor LEVEL-UP (Lotus Energy Village Electrification and Lighting Utility Project) have helped in electrifying 68 SHS. In conjunction with the Virtual Foundation, USA, the Himalayan Light Foundation installed solar electric lighting systems in several monasteries in the remote Himalayas in 1998 to help the monks also educate the children of the area. During December 1999-September 2001, the Solar Village Electrification Demonstration Project (SOVED), and Paper and Power Project, were implemented. A typical solar system for this project consists of 20Wp panels, batteries, charge controllers mounting and cabling equipment costing 342$. The initial subsidy for the first 75 systems was about 150$ per residence reducing down to 93 $. The local population welcomed this initiative participating actively during the construction phase. In all 106 systems of 20 Watt Solar Home System (SHS) were purchased and installed in 106 beneficiaries of Solar Development Committee in Bongadovan Valley. The installation was followed Installation and Maintenance Training Program conducted by the team from HLF and Lotus Energy Pvt Ltd-the provider of the SHS. Current HELP activities are summarized in the following two cases.

a) Individual villagers produce items for sale allowing them to repay their solar system loan. In Bongadovan valley villages in West Nepal, each villager was responsible for the production of 24 bags monthly. The bags, which are of very high quality, take over 50 hours to be produced. The hours of labor required to make a bag combined with the local hourly labor rate, determined the number of bags produced in order to cover the cost of the solar lighting system after subsidy. Solar lighting helps in maximizing their productivity.
b) The second HELP village, Lekhani, undertook a project called "Paper and Power". There, rather than individual paper makers, a group works to produce a single

> product or products and the village became a community production unit, turning out paper and paper products. The project also helped increase their cash economy by encouraging them to use their skills to produce paper and paper products from Daphne bholua (lokta) and other locally available resources such as Straw (Paral), Agave Americana (Ketuke), Hemp (Gaza), Sabai Grass (Babiyo), Khanyu and bamboo etc. To become participants in the HELP program, the beneficiaries had to meet the following requirements: a) All the recipients should be using Improved Cooking Stove (ICS) at their homes, b) They must plant at least seven trees either at their own periphery or on the community land. Repayment of their SHS loan was based on the number of hours they worked at the community paper production industry. Among the 565 households of Lekhani District, 92 have installed Solar Home Systems.

In this way additional 100$ per household can be saved, due to lower fuel consumption and additional income from selling energy. Solar Powered Ozone Water Treatment Systems or SPOWTS® uses small PV systems to provide completely pathogen free clean drinking water for tourists and villages especially in the Himalayan region. By combining the use of Solar PV with the income generating activity of selling clean drinking water to tourists the solar system sustainability is completely assured along with a new (and substantial) cash income for the system owners maintaining clean treks from empty bottles. (SPOWTS™) program now operating in the Everest trekking region is an excellent example of this kind of HELP™ activity. Revolving Electrification and Income Enterprise Fund (RELIEF) has been commenced in order to fund new project and seek for donations. So far has distributed US$30,000 of systems to over 150 villagers. HLF was seeking RELIEF™ expansion up to US$250,000 by 2008. These model HELP™ projects have proven to be effective methods to bring power to the poorest off-grid villages in remote areas of South Asia and are replicable.

Hydro

Rural electrification through micro-hydro is a credible option for Nepal since the country has exploited only 2 percent of the considerable (due to significant snowfall and rainfall and the altitude difference) potential. Furthermore, micro-hydro provides round-the-clock electricity in close proximity to end-users, with no need for expensive storage or power lines. Under REDP, the predecessor of RERL, 317 sustainable micro-hydro projects have been built with total installed capacity of 5.7 MW, directly benefiting 59,172 households. The success of the solar power programs under RERL, was the basis for the development of the program for micro-Hydro, Micro-Hydro Village Electrification Programme (MHVEP), which is funded by the World Bank with about 12 million $. The aim is to have 4.25 MW installed by 2012, the end of the project. Among the various micro-Hydro projects, Girindikhola micro hydro plant of 75kW has not only provided electricity to 772 households in Kharbang but also to a 100 W FM station since January 2011 [117]. The International Development Association (IDA)-financed Nepal Power Development Project, launched in 2003, featured a Micro-hydro Village Electrification component. Communities formed Micro-Hydro Functional Groups to execute the work, including responsibility for supervising contractors [118]. The project built upon the national strategy launched through the REDP in 1996, which successfully increased rural access to renewable energy sources and formed a solid foundation for scaling up impact. IDA support for the Micro-Hydro Village Electrification Program came through grants of

US$5.5 million in 2003 and US$3.8 million in 2008, covering most of the costs. The remainder comes from cash and labor counterpart provided by participating communities and the Government of Nepal. IDA supports efforts to apply what has been learned thus far to extend micro-hydro applications and improve sustainability. Steady expansion of community-managed micro-hydro systems provided coverage to about 40,000 households in 40 of 51 target districts with increased hydro potential. These systems energize sustainable rural development by creating jobs, protecting the environment, and lighting up households for the first time. In June 2009, IDA approved additional US$12 million to extend implementation through 2012, benefiting additional 36,000 households. IDA will continue to support micro-hydro rural electrification, applying what is learned to refine the effort and improve sustainability. The Khimti Neighbourhood Area Development Project (KiND Project) was a common Himal Power Limited-a private company and the UNDP inaugurated in 2011 with 3.85million$ budget. The project is a kind of Private Public Partnershi (PPP) to provide access to electricity to some 3,900 Households of Dolakha and Ramechhap districts through a 400 kW Haluwa Khola mini hydropower project in Namadi of Ramechhap [119].

The Improved Water Mills (IWM) program for improving traditional water mills is the outcome of co-operation of the Centre for Rural Technology (CRT), with the Directorate General for International Co-operation of the Netherlands (DGIS), the Netherlands Development Organization (SNV), the AEPC, the UNDP and the Government of Nepal. The estimated cost of the project is in the order of 1.3 million €. This project may not aim so much at rural electrification, although there are some examples, as at the improvement of the existing water mills. More efficiently designed kits can be used in water mills, rather widespread in Nepal, improving the income of local population with sustainable production methods. If diesel-fired mills are compared with IWMs, the operating cost would be higher by 400%. One of the options for IWMs is the use of Long Shaft (LS) mills. This allows additional uses to simply milling. With proper design a LS IWM can even provide electricity up to 3kW, depending on the flow and hydraulic head. The installation cost is in the order of 560€ for LS IWM and 234 € for ordinary IWMs. A LS IWM coupled with the ability of 3kW production may cost up to 560,000 NRP (or about 5300€).

The following achievements during the period 2004-2010 have been recorded [120]:

- Installation of 6,349 IWMs including 827 LS IWMs
- Improved livelihood of 330,148 rural households.
- Reduced drudgery of more than 330,000 women by reduction of 2 hours per milling.
- Increased income of IWM owners by 25% in case of SS and 100% in case of LS IWM.
- More than 850 rural households (about 5000 people) are the beneficiaries of the installation of 17 IWM electrification systems.
- Creation of additional employment for more than 1,000 people in IWM sector.
- Well established local service providers network: 16 IWM service centres and 17 IWM kit manufacturers.

Biomass

Next to hydropower, biogas has emerged as a major source of renewable energy in Nepal, providing affordable energy to poor rural households.

AEPC's Biogas programme (Biogas Support Program (BSP)) began in July 1992 [121]. Till then the annual installation rate was only 600 units. One of the phases began to be the first CDM project running in Nepal.

As of July 2011 the following achievements had been recorded:

- 256,662 biogas plants installed, 241,920 under the BSP alone, in over 2800 VDCs and all 75 Districts. 111,000 biogas plants were installed between 1992 and 2003 (Phases I-Phase III) and the rest during the Phase IV, declared as a CDM project.
- Around 95% plants that have crossed guarantee period are operational; 91% of their owners are satisfied.
- 152,410 Household (63%) of biogas plants are connected to toilets.
- About 74% to 89% biogas users use slurry in one or another form and 63% after composting.1.75 Tons slurry and compost fertilizer are produced annually per household.
- Proper usage of bio-slurry and bio-compost in 121500 (90%) households within the Phase IV (July 2003-Dec 2010).
- BSP is ISO 9001-2000 certified in recognition of its strong quality management and subsidy administration systems.
- Around 81 private biogas companies have been strengthened and qualified and 17 biogas appliance manufacturing workshops have been developed and pre-qualified.
- 264 micro-finance institutes are assisting communities with biogas loans.
- Around 9,000 persons got direct or indirect job employment due to BSP.
- Added 371 MW to Nepal's energy production, next to hydro power.
- Reduce global emissions of carbon dioxide as the biogas displace the use of firewood and kerosene, manage animal waste, and produce high quality bio-slurry fertilizer.

Generally, according to the Scaling-Up Renewable Energy Program (SREP)[122] investment plan for the period October 2012 to September 2017, there is high interest in continuing these interesting activities in Nepal.

In the same document the foreseen subsidies have been described. 531.8 million $ is the expected budget allocated as shown in Figure 6. The allocation of capacity is expected to be as follows:

Table 12. SREP allocation of Capacity in Nepal

Small hydropower	50 MW
Mini and micro hydropower	30 MW
Solar home systems	500,000 systems
Biogas, domestic	140,000 plants
Biogas, institutional	10,000 plants

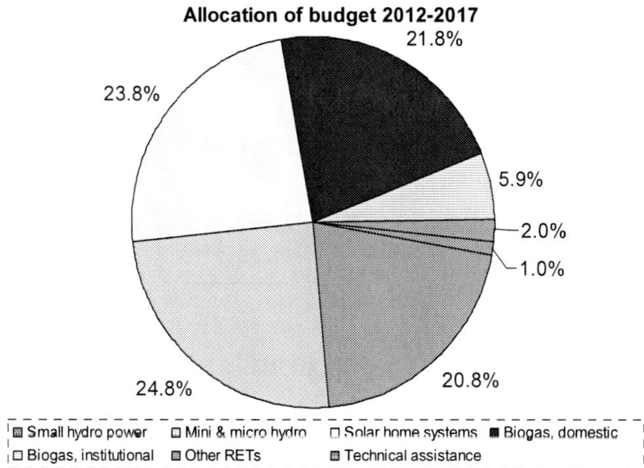

Figure 6. Allocation of budget for Nepal for SREP.

Arctic Areas

The electrification applications focus mainly on electrifying remote scientific stations. Even under those harsh conditions successful examples of RES electrification exist.

In Antarctica (South Pole) the electrification via RES is related to scientific stations or measurement points. These stations may use PVs or wind turbines sometimes combined with diesel units and various storage devices such as the McMurdo Sound Area Hybrid Systems [123] or more sophisticated solutions utilizing Fuel Cells like the one reported in Hope bay area[124].

In the arctic regions more people reside. Thus, apart from the electrification of local scientific stations mainly on Greenland where interesting applications and guidelines can be found [125], there are a few examples based on wind for electrification of remote communities mainly in Alaska [126].

Alaska has over 150 remote, stand-alone electrical grids many of them running on diesel generators serving villages but gradually turning into using RES for reducing operating costs. It is estimated that 134 of them have in small or greater part met their needs based on Wind energy.

The first example comes from Kotzebue [127] where 10 wind turbines of 66kW were installed in 1999 and two additional wind turbines of 50kW and 15 meters diameter not requiring crane to be installed were added in 2002 using freezeback foundation technique. Moreover, 1 Polar 100 turbine, rated at 100kW, with direct drive and not an induction machine was installed in 2002. This co-operates with the existing diesel power plant of 11MW Wind power penetration on the area had reached 39%.

In the isolated village of Wales, Alaska with 160 residents, a Hybrid power system described in Table 13 comprising wind/batteries and diesel generators was created aiming also to provide heat to school.

On St Paul's island, in the Bering Sea a very interesting application for reducing fuel dependency has been installed. One 225kW Wind turbine and two diesel gen-sets of 150kW operating in CHP mode are used to meet the thermal needs of the airport along with the electricity needs of the area. This hybrid system helps in achieving high reliability and average annual wind power penetration in the order of 55-60%.

A unique aspect of the system is that resistive loads are used to maintain load on the wind turbine while, at the same time, provide hot water for heating the facility. Secondary resistive loads coupled with heat storage are used as means of control along with the synchronous condenser instead of batteries. The whole 525 kW wind/diesel CHP system cost approximately $1 million including in-kind support from the TDX Corporation. It is estimated that manages to save $200,000 for electricity fuel and other 50,000$ for heating fuel [128].

Table 13. The Wales, Alaska hybrid system components

Component	Power	Provider
2 Wind Turbines	65 kW (peak continuous)	Atlantic Orient Corp. 15/50
2 Diesel Gen-sets	168 kW	Cummins LTA10
1 Diesel Gen-set	75 kW	Allis-Chalmers 3500
1 Local Load controller	89 kW AC	NREL
1 DC. Load controller Reliance Electric	144 kW 1.2 VDC	NREL
1 Rotary converter	156 kVA	NREL
Battery storage	1.2 VDC /130 Ah /31.2 kWh	SAFT SPH130 NI-Cd Batteries
Auxiliary battery charger 31.2kWh	300 VDC /30 A	NREL

Generally the Alaska Electricity Authority tries to promote high RES penetration to reduce operating costs and fuel dependency. The authority frequently issues renewable energy sources atlases with most recent the one of 2011[129]. The Alaska Center for Energy and Power (ACEP) is an applied energy research program based at the University of Alaska established in January 2008 to meet the energy research needs [130]. For this reason, a specialized Wind-diesel power system working group, the Alaska Wind-Diesel Applications Center (WiDAC) [131], has been created to support the broader deployment of cost-effective wind-diesel technologies for rural communities.

Island Regions

All the above cases refer to electrical "islands" within territories of interconnected power systems. Should these areas were close enough to the Transmission networks, these would not have been islanded electricity networks.

Physical islands in many cases are also electrical islands phasing significant problems and increased cost in meeting their electricity needs. The operational cost may be much higher than the mainland and there are a few examples of high RES penetration levels around the world, summarized in Table 14.

Table 14. Island power systems operating autonomously under Increased penetration (partly information from [132])

	Islands	RES utilized
Nearly 100%	Eigg (Hebriddean) Reao (French Polynesia)	Hydro,Wind and PV Wind, PV
50-75%	Iceland Rathlin Fiji Reunion New Zealand Froya St.Paul (Alaska)	Geothermal Wind Hydroelectric Hydroelectric And bagasse Geothermal and Hydro Wind Wind in Hybrid with CHP
40-50%	Dominica Flores (Azores)	Hydroelectric Hydroelectric
30-40%	Faroe Islands Samoa Sao Miguel St. Vincent and the Grenadines Madeira	Wind and Hydro Hydroelectric Geothermal and Hydroelectric Hydroelectric Hydroelectric
20-30%	St. Pierre and Miquelon, Hawaii Kauai Island(Hawaii) Rarutu Island (French Polynesia) Mauritius	Wind Geothermal and wind Biomass Wind Bagasse
15-20%	Crete Ascension Island (UK) San Clemente Island (USA) King Island Isle of Pines (New Caledonia, France)	Wind, PV Wind Wind Wind Wind Wind

Table 14 contains islands operating autonomously and this is the reason for not including famous RES islands such as Bornholm, Samsoe and Utsira. Many islands face similar problems and try to face these problems creating associations, organizing conferences and in general by creating networks. Clearly increased RES penetration requires diversity of Natural resources. Islands who manage to integrate Hydro power and Geothermal usually can achieve high RES penetration levels much easier than islands utilising more intermittent RES, like PV or Wind. Here we will provide more details on programs which have substantially improved either RES penetration or and more importantly the living conditions of the population of these areas in the Pacific and the Atlantic Ocean and the Mediterranean Islands.

Pacific Ocean Islands

There are numerous of islands, tiny and larger ones in the Pacific Ocean. Many of them are rather vulnerable to climate change and rise of water sea level (e.g. atolls) and the authorities try to find ways to electrify their islands with as minimum as possible environmental impact.

Solomon islands are a state in the Solomon archipelago, the third largest in the South Pacific Ocean. Only Honiara, the capital of the state, has been electrified out of the 347 islands of the state. 90% of the population lives in rural areas with kerosene being their main energy source. Unfortunately, burns and inhale problems due to malfunction of kerosene lamps are not rare in this island state. In 1997, the Solar Electric Light Fund (SELF) [133] has used PV systems to electrify 50 residences in Sukiki village on the island of Guadalcanal in co-operation with Guadalcanal Rural Electrification Agency (GREA), a local organization created for this purpose. Next year, additional 50 residences were electrified in Makaruka village. This project was co-funded by the Council of State Governments, the US Asia Environmental Partnership and the Maryland Department of Business and Economic Development, and the Maryland Energy Administration.

A typical solar system consists of a Solarex 50Wp panel, a Morningstar charging controller, a lead acid battery and 3 CFL. The energy supplied is enough for some hours of lighting and a radio receiver. The GREA is responsible for micro-financing. Each customer interested in installing a solar PV system pays 50$ deposit and 15$ monthly payments for 4 years. Collected payments return to the bank and help in financing other projects too. This type of financing became particularly popular in other villages of the Solomon Islands.

Additional to these actions, vocational activities for training students and village residents in planning and operation of PVs were organized by Solar Energy International (SEI). Personnel of the Ministry of Energy, Mines, and Minerals and the Solomons Islands Electricity Authority (SIEA) [134] were also trained during these activities. Such actions enhance the sustainability of the actions followed.

Micro-hydro was an additional application for rural electrification on this island state. A micro Hydro, of 7kW capacity and minimum production 4kW, was installed by the Appropriate Technology for Community and Environment (APACE) at Vavanga Community in 1993[135]. Originally, 22 houses, 14 kitchens, two church buildings and a bakery building were connected to the supply; however, additions have been made since then. Typically, two 18 W fluorescent lights and a power socket are installed in each household. To enable the river to provide sufficient power for the whole community, current-limiting devices are installed in the households, but unlimited power is supplied at 240 V to the community facilities and businesses. Training activities for implementing and more importantly maintaining the project took place and all the members of the community had some involvement during the construction phase. Similar projects were implemented in villages Iriri, Ghatere Kolombangara, Manawai Harbour community, Malaita, Bulelavata community and Beulah PSS, W P, Raeao and Nariaoa communities.

In Vanuatu, in Efate village, the diesel gen-sets were replaced by Solar home systems of one or two PV panels rated at 50Wp, batteries and LED lighting. The installation cost of 1000$ was met by the Japan International Cooperation Agency (JICA) [136] and the end-users. The latter gave 95$ as a deposit and 10 $ for maintenance and replacing batteries and lamps. There is large room for improvement in this small island state since still 80% of the population lacks access to electricity. By the end of 2012, a 40kW grid-connected solar on Santo is to be also connected along with feasibility studies for 2 hydropower projects.

Kiribati is an island state in the Pacific islands. Only two out of 65 islands are inhabited with around 100,000 residents. PV on those areas has become rather popular for remote electrification. The Kiribati Solar Energy Company (*KSEC*) is the oldest Renewable Energy Sources Company in the Pacific Ocean aiming at electrifying population since 1984[137]. In the 80's apart from the first, not so successful installations, emphasis was put on the training of the citizens and mainly technicians to improve reliability and increase people's awareness on the rural areas on the use of solar PV systems. Training courses were held on Tarawa in 1986 and 1988. Later on, in the early '90s, JICA co-financed 55 solar homes. During 1994-95 EC funded 250 Solar houses and then additional funding of 3.57million $ for PV capacity was provided.

a) During 2001, KSEC installed 310 solar power systems at residences in North Tarawa, Marakei and Nounouti.
b) On August of 2001, the first phase of the project of the European Commission 'Solar Energy for Outer Islands' started. As a result of this project, by 2008 the installed capacity had been increased to 250 kWp compared to 31 kWp and 2100 (Solar Home Systems) were developed. 96 municipal buildings were electrified. As a result, 18% of the rural areas of the Kiribati islands were electrified compared to the low 3% at the beginning of the project.20% of the Gilbert Islands' households were electrified, providing 4 hours/day of basic lighting (100Wp). Furthermore, 46 generators (300-600Wp) have been installed into Maneabas (local meeting houses).
c) In 2009, 4.1 million € were given to increase (actually double) capacity to 500 kWp.

One of the methods to increase capacity is the fee-for-service approach sharing many in common with the related programme in Morocco. KSEC has been applying this approach since 1989. The interested customer pays 50$ deposit and then pays a monthly instalment in the order of 10-50$ not only for paying back the installation but also to have maintenance of the system. This was mainly suggested in order to avoid practices of the past with faulty equipment or bad wiring as the first solar systems by 1990 were improperly operated. It was then also decided that a local technician should be assigned no more than 125 such systems for their regular maintenance and that the villages should have around 50 systems in order to improve viability within the limited income of rural population.

The Pacific ACP including the following countries: Cook Islands, Federated States of Micronesia, Fiji, Kiribati, Nauru, Niue, Palau, Papua New Guinea, Republic of Marshall Islands, Samoa, Solomon Islands, Tonga, Tuvalu and Vanuatu and The European Commission initiated co-operation in 2002 for the period 2002-2007 with funding in the order of €29 million [138]. This is part of the Cotonou Agreement with African, Caribbean and Pacific Ocean state for the co-operation and development of those states [139]. This co-operation continues indicatively with 5 of these Island states, Federated States of Micronesia (FSM), Nauru, Niue, Palau, the Republic of the Marshall Islands (RMI) under the code REP-5 project [140]. Electrification of islands in FSM was one of the major concerns of this project. A successful example was the electrification at the Yap state including the installation of a PV-powered mini-grid on Asor and Fadrai islets in the Ulithi Atoll. The system on Asor has a PV capacity of 19.3 kWp, a battery capacity of 1850 Ah at 48V, and providing electricity to all 18 houses and all public buildings. The system on Fadrai is larger, at 28.1 kWp of PV capacity and 2880 Ah of battery capacity at 48 V, providing electricity to all 31 houses and all public buildings[141]. The systems were installed in March 2009, at a cost of €700,000 allocated to €450,000 for the PV systems and another €250,000 for site preparation, cables, transportation and installation. About 15% of the amount of the PV system is allocated for training, supervision for installation and commissioning.

A recent application in the French Polynesia is the one on the Atoll Reao realized in 2010 serving 40 houses and busineses. The system before its replacement presented fuel costs of 1,000 €/day (250 l of diesel per day at 4 €/lt). Fuel costs of for 100% fuelled power generation have been reduced by 80% through switch to PV/diesel hybrid system consisting of PV park of 140 kWp a battery bank of 1842 kWh an SMA Sunny Island converter of 90kW and the existing diesel gen-sets of 120kVA. This system is expected to save 365 tonnes CO_2 annually and 73,000l. Thus, the investment of 2m € is estimated to be recouped after 6-7 years [142].

Atlantic Ocean Islands

Canary Islands, especially El Hierro [143], Madeira and the Azores have made significant efforts in increasing RES penetration and reducing energy cost. The actions on those islands aim in improving RES penetration, reducing both operating costs and emissions rather than electrifying areas as the cases in the Pacific Ocean.

The Portuguese islands have invested on both Geothermal and hydro energy additional to PV and wind power systems.

In the Azores archipelago RES penetration had already reached 21.5% since 2000 with Sao Miguel Island reaching almost 30%. Madeira has greatly invested in the reversible pump storage unit and EEM is about to replicate the project in Calhetta [144]. However, a very interesting application is on the island of Eigg described below.

Eigg Island

The Eigg Island is one of the Hebridean islands 16km of the western Scottish shore, with $30km^2$ and 90 residents in 37 households. Until 2008, only a diesel station existed on the island [145].

Then, a new power system was created based on RES and reaching 95% penetration. The system serves the same load of the 37 households and 5 malls as before. This hybrid autonomous system incorporates 3 hydroturbines of total capacity 110kW, 4 Proven small wind turbines with total capacity 24kW and a PV rated at 32 kW (after the extension not included in the diagram of Figure 7).

The large hydro unit of 100kW is interconnected to 3.3kV grid while the two smaller ones of 5kW are interconnected to 230V. The Wind Turbines via the Windy Boy inverters feed a transformer rated at 230V/11kV. The 2 already existing diesel generators, 64kW each, are used only during periods of insufficient production.

The operation control of the system is based on 4 Sunny Island inverters of the SMA coupled via a Multicluster – Box 12 with the various producers and consumers. The same company has also provided the required inverters for both PVs and wind turbines.

The Sunny Island inverters rated power 60kW manage about 144kW RES. The battery bank of 212kWh (4400Ah) can provide autonomy of about 12 hours. One part of the electricity network operates at 11kV and the rest at 3.3kV, reducing the ohmic loses.

During operation, the Main Master Sunny Island inverters control the entire grid and balances, demand, consumption, storage and interruptible loads in such a way that the frequency remains stable. Whenever the RES production is higher than the demand (15kW-60kW), batteries are charged, otherwise the slight and temporally increase of the island frequency triggers resistive loads at public buildings.

If necessary, RES production is reduced. This is line with what has been described in the technology section.

In case RES production is not enough, 60kW is drained out of the battery bank. If the State of Charge drops below 60% a diesel generator is activated and maintains frequency (isochronous mode) and power balance. Under those circumstances diesel maintains frequency and any deviation either in production or demand felt by the main controller taking also advantage of the short time overcurrent capacity of the inverter, i.e. 144kW is alleviated.

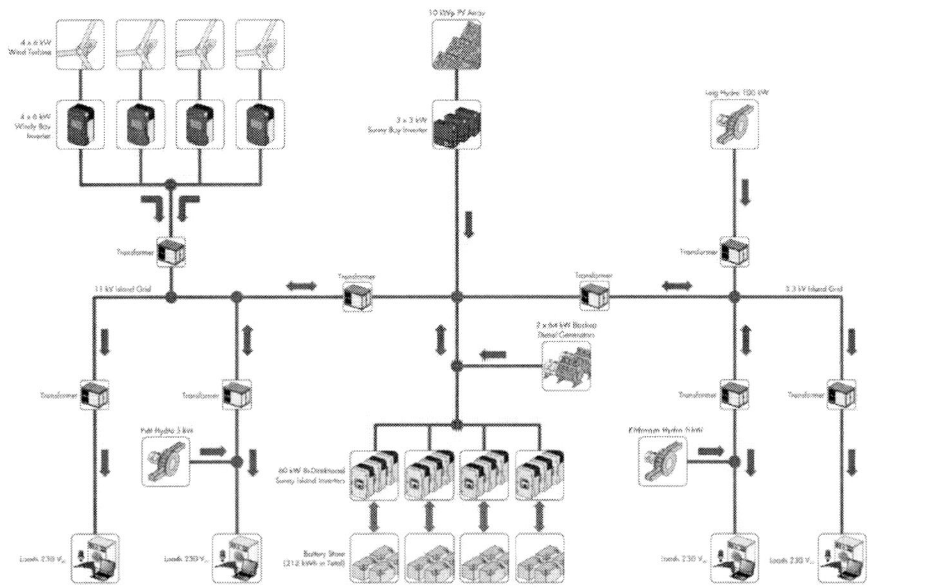

Figure 7. The General control scheme for the Eigg Island System (Courtesy of Dr.Thim-Product Manager Hybrid Energy Solutions at SMA).

Thus, the diesel generator operates at an efficient operating point charging the batteries and optimizing the operation. Thus both detrimental for the gen-set life-time start ups of the generation and operation at low loading, are avoided.

The cost for residents is reduced by 60%. The project manager was the Synergie Scotland Ltd and responsible for the design and installation company Scottish Hydro Contracting Ltd.

The total cost of the project was £1.664 million (2.06million €) and was co-financed by:

- European Regional Development Fund
- Big Lottery Fund
- HIE Lochaber
- Highlands and Islands Community Energy Company
- Scottish Households Renewables Initiative
- Energy Saving Trust
- Highland Council
- Isle of Eigg Heritage Trust
- And resources of the Eigg residents themselves

Mediterranean Islands

Greece and Italy have developed special programmes for improving RES penetration on their islands. Croatia also has shown significant interest in promoting RES on Dalmatian islands [146].

The Italian government to compact high fuel costs has promoted the Sustainable Islands project form increasing RES share in Italian islands [148]. The volcanic island of Stromboli is a typical example of high utilization of RES in hybrid systems PVs. The developed microgrid consists of a diesel gen-set of 160kW, one 100kWp PV and a battery storage device with capacity C_{10}=3000 Ah rated at V=400 V. All these devices are coupled via three inverters, one per phase and static switch in parallel [147]. To better use the production capacity of the HPS, a desalination system has been installed as periodical load to store water during the low occupation period and cover the demand in the summer.

14.8% of the Energy needs of the Hellenic Non-interconnected islands came from RES during 2011, mainly wind. Now the installed capacity has reached 278MW wind power and 100MW PVs. Apart of these applications, in the early 90's some isolated Hybrid PV installations were installed on tiny Greek islands with very few inhabitants. These systems are in the order of multiples of 700Wp and are also equipped with lead acid batteries. 38 such systems were installed in 15 very small islands serving in all less than 100 people. Additionally, on three islands, namely Arkioi[149], Antikythira and Gavdos, larger PV installations of 37.5kW, 25kW and 20 kW respectively have been used to meet large part of the energy needs of the inhabitants since the late 80's. These applications are in addition to 50kWp of Hybrid power systems based on RES on the Holy Mountain region with plenty of monasteries, an electrical island within the Hellenic Interconnection System.

CONCLUSION

Off-grid applications based on RES have become a reality either for simply lighting or for meeting an entire island needs such as the case of Eigg where the population has not left their standard of leaving due to RES installation and the diesel unit operates very rarely. Research on improving these applications is continuous not only in technology but also in social, financial terms as continuous is the need for electrifying people. More flexible and more affordable units are being produced to meet the needs for rural electrification. Operation and ageing of storage devices have been modelled in much more detail compared to the past. More and more loads are developed with view to off-grid applications. Above all the improved capabilities of controlling the various devices so that off-grid application resemble as much as possible the operation of larger power systems have made such systems at great extent competitive with the simplified solution of a diesel gen-set running for providing some electricity. Therefore off-grid electrification based on RES can be a competitive reality either to grid expansion or ICE operation. Research on Smart Grids, distributed generation and microgrids currently flourishing could even more improve the practices in the systems described.

The best practices examples show imperatively that RES can operate more or less everywhere; from the low altitude atolls in the Pacific Ocean to the high altitude areas in Nepal. RES can withstand as high temperatures as the ones in the deserts and, under the precautions described here, already operate in the frozen deserts of Antarctica. Therefore, in terms of technology advance there is not a place where at least one type of RES cannot be installed. The cost, the engineering efforts and the size for equal type of loads may vary but the result can be the same. How much each type of RES has been utilized in the investigated areas is summarised briefly in Table 15.

The competition among the various manufacturers of equipment will either reduce the cost or bring more efficient (in terms of reliability, energy harvested etc) results at the same cost. The market may seem immense, 1.6 billion people without electricity in areas where population is generally increasing, but the companies would like to take a larger portion of it.

Table 15. How often each type of RES for rural electrification is used in various areas investigated

	PV	*Wind*	*Small Hydro*	*Biomass*
MENA Countries	High	Limited	Very Limited	Very Limited
Sub Saharean Areas	Practically Only PVs	Very Limited	Very Limited	Very Limited
S-E Asia	High	Low-medium (India mainly)	High (the highest in our analysis)	High (the highest in our analysis)
Mountainous Area	High	Low	High	Medium-High
Arctic Areas	Medium	High	Small	Small
Pacific Ocean Islands	High	Limited	Low	Medium
Atlantic Ocean Islands	Medium	High	High	Medium
Mediterranean Sea	High	High	Negligible	Very limited

The best practice review has shown various combinations, sometimes very innovative, for the financial implementation of a rural electrification scheme with RES. Participation of the individuals, dedicated National bodies co-operating with international funds and non-governmental bodies. Many governments have realised the potential impact of rural electrification on the development of their country and why not in the peace of the community. Micro-finance products, fee for service, subsidies, rental of equipment, soft loans, donations, participation of International bodies, tenders for Clean Development Mechanism (CDM) etc. All of these financing schemes more or less agree that usually the beneficiary should take part even with amount equal to a western country small coin for the whole installation. As the market matures, this participation should become higher. Even if this is not possible, the solution in Nepal with handicrafts may be an idea capable of creating a market even at the end of the project. This will make the beneficiaries feel that they own the system, they should protect it and operate it in an appropriate way.

Simply connecting a RES based off-grid solution and leaving the place has proven to be inefficient in all terms. The equipment still requires some maintenance, even limited or even precautions to maintain its life-time or have the desirable efficiency. After sales service is important as well. Something may go faulty, if there is not support, the first end-users will get disappointed. Similar results are to be expected if the equipment is of low quality or misused due to inadequate information, as was the case in Kiribati Island in the 80's. The kerosene may create fires but the same danger will exist if improper cables are used. None should forget that electricity may kill instantly if the safety precautions are not taken into account. Therefore, the local community should be trained in a simplified but effective way for the maintenance and the potential hazards of such equipment. Training people on RES installation can accelerate the rhythm of installations year by year as currently happens in Bangladesh, or as has happened in Morocco.

Would RES rural electrification require improved marketing skills? The efficiency is the parameter convincing the end-users. The end-users have got an issue, they require electricity; they need it, as all of us, as reliable as possible with the available capital and operation cost constraints. If the solution suggested is efficient soon or later will be applied. This was the case of pico hydro in Vietnam and China known by mouth. The resellers or promoters of such systems should try to mix with the local people and try to understand their needs. If the Moroccan citizens would expect to apply and pay the service of a PV kit in Rabat instead of their local souk, the achievements described here would not have been achieved. The fact that the SELCO foundation in India is near the places of pray makes people know the product and buy what is affordable and suitable for them. A rural electrification programme designed from an office in a developed country without support in the field is not very promising to succeed.

Summarising the social benefits of rural electrification and especially via RES we can highlight the following:

- Increased productivity especially for craftsmen who can also work at nights
- Increased income closely related to productivity and the fact that kerosene or other fuels replaced may be more expensive rather than an off-grid system based on RES. Such evidence comes from a microgrid in Kenya where the tailors can increase their daily income up to 70% while reducing prices by at least 15% [150].
- Increased free-time for women. Improved cook stoves, available water closer to the houses, increases free-time for women which can be used for other activities that could potential increase income such as knitting or other handicrafts.
- Improved health conditions since:
    - The rural population can much easier have access to well equipped rural health centers with medicine and vaccines readily available. In larger health centers electricity helps in analyzing some exam results locally.
    - Respiratory problems due to fuel oil are reduced (special benefit of RES based solutions)
    - Solar lanterns, the simplest lighting device, present much brighter illumination than the kerosene lamps.

  - Access to cleaner water. The solar or wind pumps and especially the purification devices described in SE Asia section reduce the risk of diseases due to poor quality water.
  - Reduction and proper management of human and animals excreta for example to produce biogas can help in reducing diseases related to excreta.
  - Electrification helps in maintaining food via refrigerators. The benefits are three-fold for human health a) Not eating perishable food, b) Availability to have food at later times and not only whenever there is harvest c) improving nutrition with food that since was difficult to be preserved was not used in the nutrition of the area (e.g. fish on the mountains).
- Improved communication conditions. As we examined in previous chapters, one of the most popular applications is the charging of a mobile phone. This can be made even if the interested citizen does not have electricity access at home via central charging stations at specific low fee. This can help in improving communication penetration in the most remote place without laying cables at all. This is of particular interest for telecommunication utilities who can increase their clients in such away. Some of the schools electrified by SELCO in India have charging station for laptops. This is of particular interest for the citizens themselves since they have a first step to break isolation. Even in the middle of an unelectrified region, a container with PVs, called Zubabox can supply an Internet cafe and provide an open window to the web.
- Additional entertainment activities. Bars and restaurants can prolong their operation time (as proven in Morocco). In many cases, the systems available for rural electrification have foreseen that the end-users have a radio or a TV set or at least foresee a community TV installation. Thus, the residents can be educated, (see for example the HLF educational program on RES in Nepal), entertained and informed about the news.
- Improved educational system : The students can study all the time and not only when there is sun or via kerosene lamps inhaling their smoke. In many of the rural electrification programmes reviewed in this chapter, one of the major targets was the provision of proper lighting to students for their study. Solar lantern applications are the most famous application for the minimum needs of the population. In some countries they are donated to children, in others the children should return the battery to be charged or take the lantern after the every-day class lessons. No matter what the "Business model" is, various sociological studies agree that the efficiency of students in class has been improved, especially in maths. An electrified school means that teachers have the option to use audio-visual aid for their lessons, or that the classes can take place at night and they can be attended even if there are activities outside. This could also help to combat illiteracy quite increased in those areas. Additionally, electrification via RES can be linked to environmental education activities.
- Reduced emissions. This is not simply an electrification advantage (as some would argue for the above). This is a benefit of RES installation which can displace CO_2 from kerosene lamps, poor efficiency of wood burning, or as by –product of an operating diesel gen-set.
- Environmental protection. In many areas of the world using wood for heating, cooking etc, decreased efficiency of the stove leads to increased consumption, which

may result in deforestation. To make matters worse, Internal combustion engines make significant noise and oil may be spilled. A spillage in a fragile environment like the islands described in the best practices may create significant disaster in the eco-system.

- Utilization of local sources. Many communities over the centuries have been trained from the past generations to rely on their own sources. This mentality is preserved since locally available resources are utilized. There is no need to import energy from other areas. To go further, Electric Vehicles transportation is already under discussion in the Himalayas.
- Increased energy efficiency. Clearly off-grid applications most of the time, or at an entry point as most of the markets described here do not have all the appliances that people may have in a house e.g. on the island of Crete where these lines are currently written. In order to increase the continuity of service and reduce installation size, more efficient appliances are used. Even if the village is connected, this way of living will not change.
- Increase of local entrepreneurship in the field of electricity. The examples we reviewed have shown that significant effort is made in many countries to produce locally some of the components required, especially micro-electronics ones. This could decrease unemployment, potentially reduce cost (transportation cost for sure) and why not bring also export of such equipment to other countries.
- Diversity in the economy. Simply laying cables or installing Transmission System towers will bring temporal work and will let the citizens tackle with their ordinary activities. Rural electrification as applied in many of the countries reviewed, can increase the number of companies tackling with various aspects of rural electrification, producing or installing equipment, maintenance, improving products etc. Some of the rural people may become vendors of such equipment therefore diversifying the economy.
- Increased electrification rate. Even in countries where the decision has been made to built extensive transmission or distribution network this can be made simultaneously with providing access to people residing in the other part of the country with an off-grid system. The countries described have not let aside their transmission network plans or interconnection with other power systems. If the reader checks the web-sites and the tenders of power in countries at those areas can see that there are efforts for extending networks. However, for sure this is capital intensive, requires personnel, equipment and for sure time. Parallel implementation of extending grids and RES electrification will much quicker bring about the desired results.

The experience all over the world has shown that there should be political will and co-operation with agencies and International bodies for electrifying rural areas. Based on the above socio-economic findings, one can realize why there should be political will apart from simply taking votes from the electrified villages. Such an effort should motivate the local sources. When the local feel that the equipment is essential for their lives they will do their best to protect it. The rural electrification programme should co-operate with other development programmes in the region and take advantage the one from the other. Following best practices and avoiding mistakes from other countries is the way for improving the quality and increasing the quantity of such systems.

And what if these areas are to be electrified by transmission network, will we drop away these systems ? Our estimations are that parallel implementation could create tremendous possibilities. The existing off-grid applications will diversify to improve local reliability, energy efficiency, RES penetration of the bulk network or even create business opportunities we have not thought yet. If this is not the essence of smart grids, what is it?

REFERENCES

[1] The United Nations (UN) Resolutions adopted by the General Assembly at its 65th session, Resolution A/65/151, 16th Feb 2011. Available: http://www.un.org/Depts/dhl/resguide/r65.shtml. Accessed on 23.10. 2012.

[2] Mahapatra, S. Dasappa, S., (2012). Rural electrification: Optimising the choice between decentralised renewable energy sources and grid extension. *Energy for Sust. Dev.* 16, 146–154.

[3] Global Village Energy Partnership website. Available: http:// www. Gvepinternational.org/en/business/about-gvep. Accessed on 23.10.2012.

[4] The United Nations Development Program (UNDP), The Millenium Development Goals (MDG). Available: http://www.undp.org/ content/undp/en/home/mdgoverview/. Accessed on 23.10.2012.

[5] Kanagawa, M., Nakata T., (2007). Analysis of the energy access improvement and its socio-economic impacts in rural areas of developing countries. *Ecol.l Econ.*, 62, 319–329.

[6] The United Nations Industrial Development Organization (UNIDO)- Section Environment and Energy. Available: http://www.unido. org/index.php?id=905. Accessed on 23.10.2012.

[7] World Bank. 1996. Rural Energy and Development: Improving Energy Supplies for 2 Billion People. Washington, D.C.

[8] Energy Sector Management Assistance Program (ESMAP) homepage. Available: http://www.esmap.org/overview. Accessed on 23.10.2012.

[9] U.S. Agency for International Development, homepage. Available: www.usaid.gov. Accessed on 23.10.2012.

[10] RES and EE development projects of the Hellenic Aid. Available: http://www.hellenicaid.gr/frontoffice/popup.asp?cpage=USERTEXTandcresrc=273andcnode=8andclang=1. Accessed on 23.10.2012.

[11] Karakosta, C., Psarras J., et al, (2011). Stimulating the Sustainable Transfer of Energy Technologies through the Clean Development Mechanism of the Kyoto Protocol. *Handbook of Sust. Energy.* Ch. 15, 573-589.

[12] Lloyd,B.,(2009). Clean Development Mechanism and Sustainable (Economic) Development - Issues and Implications on Poverty Alleviation in Developing World. *The Kyoto Protocol: Economic Assessments, Implementation Mechanisms, and Policy Implications.* Ch. 9, 217-241.

[13] Moon J., (2007). The Contribution of Corporate Social Responsibility to Sustainable Development. Sust. Dev. 15, 296–306.
[14] Tsili, M., Papathanassiou S., (2009). A review of grid code technical requirements for wind farms. *IET Ren. Power Gener.* 3, 308 – 332.
[15] Upwind project report (2011). Design limits and solutions for very large wind turbines - A 20 MW turbine is feasible.
[16] Perkins, S., D., Everett, M. A., (eds) (2011). Offshore Wind Power: Challenges, Economics and Benefits.
[17] Islam, M, Esfahanian V, Ting, D., Fartaj A., (2006). Applications of Vertical Axis Wind Turbines for Remote Areas. *Proc. of* 5th *Nat. Energy Cong.*,Iran.
[18] Leary, J., While, A., Howella, R., (2012). Locally manufactured wind power technology for sustainable rural electrification, *Energy Policy*, 43, 173–183.
[19] Christensen,K., (2006). Catalogue of Small Windmills from 5W to 50kW. Nordic Folkecenter for Renewable Energy.
[20] Small wind industry implementation strategy, SWIIS consortium, project funded by the EU.
[21] Bergey Wind turbines. Payback Calculator web-page. Available: www.bergey. com /Channels/1F2.htm, Accessed on 23.10.2012.
[22] American Wind Power Association, (AWEA) Distributed Wind section of their web-site http://www.awea.org/learnabout/smallwind/index.cfm.
[23] Gipe, P., (2009). Wind Energy Basics Revised: A Guide to Home- and Community-scale Wind Energy Systems, *Chelsea Green Publishing*.
[24] Meier, T., Fischer, G., (2011). Assessment of the Pico and Micro-Hydropower Market in Rwanda. *Report by the GVEP International – Africa Regional Office*. Available: www.gvepinternational.org.
[25] Howey, D.A., (2009). Axial Flux Permanent Magnet Generators for Pico-Hydropower. *EWB-UK Research Conference 2009*.
[26] PowerPal, Micro-Hydro Power products for home and industry prices. http://www. powerpal.co.uk/RPP%2009_01.pdf.
[27] Paish O., (2002). Small hydro power: technology and current status. *Ren. and Sust. Energy Reviews*. 6, 537-556.
[28] Kemp, P., (2011). Very Low Head (VLH) Turbine. *Presented at the OWA 11th Ann. Power of Water Conf.*
[29] Redfield, S., Five Gallon Bucket Hydroelectric Generator Build Manual. Report for the Appropriate Infrastructure Development Group.
[30] Practical Action. Is a community based micro-hydro Electrification Scheme Suitable for your Village? Brochure Published by Practical Publishing Ltd.
[31] ESHA, The European Small Hydro Association. SMALL HYDROPOWER FOR DEVELOPING COUNTRIES, Brochure prepared by ESHA. Available at www.esha.be. Accessed on 23.10.2012
[32] Maher, P., Smith, N., (2001). Pico-Hydro For Village Power. A Practical Manual for Schemes up to 5 kW in Hilly Areas. *Report for the UK Dep. for Intl. Development (DfID)*.
[33] PowerSpout Low Head Turbine Variants LH and LH Pro Installation Manual (Low Head 1500 W maximum output/unit) Domestic install pictures, Available http://www. powerpal.co.uk/pslh1.html Accessed on 23.10.2012

[34] Achten, W.M.J., Verchot, L., Franken, Y.J., Mathijs, E., Singh, V.P., Aerts, R., Muysa, B., (2008). Jatropha bio-diesel production and use. *Biomass and Bioenergy.* 32, 1063–1084.

[35] Tsikalakis, A.G., Hatziargyriou, N.D., (2008). A software for the short-term operation of isolated power systems with high Renewable Energy Sources penetration. *Proc of the 4th European PV-Hybrid and Mini-Grid Conf.*, 350-357.

[36] Nickoletatos, J., Tselepis, S., (2003). Evaluation of literature search and results of survey about lifetime expectancy of components, in particular the energy storage systems in existing RES applications. *Deliverable D1.4, Benchmarking Project, ENK6-CT2001-80576.*

[37] Jossen, A., Garche, J., and Sauer, D. U. (2004). Operation conditions of batteries in PV applications. *Solar Energy*, 76(6), 759-769.

[38] Chaureya, A., Kandpal, E.T.C., (2009). Solar lanterns for domestic lighting in India: Viability of central charging station model. *Energy Policy.* 37, 4910–4918.

[39] Osram. Intelligent energy-saving lamps that do more than just save electricity.· http://el.wikipedia.org/wiki/%CE%94%CE%AF%CE%BF%CE%B4%CE%BF%CF%82_%CE%95%CE%BA%CF%80%CE%BF%CE%BC%CF%80%CE%AE%CF%82_%CE%A6%CF%89%CF%84%CF%8C%CF%82

[40] Royal Philips Electronics (2008). Philips announces new range of solar powered LED lighting solutions for Africa. Available: http:// www.newscenter.philips.com/main/standard/about/news/press/20090216_new_solar_products_backgrounder.wpd.

[41] Adkins, E., Eapen, S., Kaluwile, F., Nair, G., Modi V., (2010). Off-grid energy services for the poor: Introducing LED lighting in the Millennium Villages Project in Malawi. *Energy Policy.* 38, 1087–1097.

[42] Light Up The world foundation web-site http://www.lutw.org/

[43] Solar Electric Light Fund (SELF), YouTube channel. Available At: http://www.youtube.com/user/solarnow2007?feature=watch

[44] Muhopadhyay, K., Sensarma, B., Saha, H., (1993). SolarPV lanterns with centralized charging system - a new concept for rural lighting in the developing nations. *Solar Energy Materials and Solar Cells.* 31, 437–446.

[45] Solanters Initiative. Replacing 1 million kerosene lamps with solar lanterns. Available at: http://www.solanterns.com.

[46] Rubab, S., Kanpal, T.C. (1996). Financial evaluation of SPV lanterns for rural lighting in India. *Solar Energy Materials and Solar Cells.* 44, 261–270

[47] Mukerjee, A.K., (2007). Comparison of CFL-based and LED-based solar lanterns. *Energy for Sust. Dev.* 11, 24-32.

[48] Hong.G.W., Abe, N., (2012). Sustainability assessment of renewable energy projects for off-grid rural electrification: The Pangan-an Island case in the Philippines. *Ren. and Sust. Energy Reviews.* 16,54-64

[49] Jimenez, A.C. K. Olson, (1998). Renewable Energy for Rural Health Clinics. *Report Prepared by NREL.* www.nrel.gov/docs/legosti/ fy98/25233.pdf.

[50] USAID, Powering Health, Electrification Options for Developing Country Health Facilities. Available: http://www.poweringhealth.org/ index.shtml. Accessed 23.10. 2012.

[51] Bates, L. Hunt, S. Khennas, S. Sastrawinata, N. (2009). Expanding Energy Access in Developing Countries The Role of Mechanical Power. *Report by Practical Action Publishing Ltd and UNDP*.

[52] Harries, M., (2002). Disseminating wind pumps in rural Kenya—meeting rural water needs using locally manufactured wind pumps. *Energy Policy*. 30, 1087-1094.

[53] Barlow, R., McNelis, B., Derrick, A. (1993). Solar Pumping: An Introduction and Update on the Technology, Performance, Costs and Economics, *World Bank Techn. paper*.

[54] Lallementa, D. M., Terradob, E. N., Zhanga, Y., (2006). Empowering information and communication technologies in isolated areas: learning from the solar-net villages program in Honduras. *Ren. and Sust. Energy Reviews*. 10, 46–53.

[55] Computer Aid, Zubabox-Solar Internet Hub. Available: http://www. computeraid. org/Zubabox-Solar-Internet-Hub.asp. Accessd on 23.10. 2012.

[56] Reyes, L., (2011). Huatacondo□s Microgrid and the Social SCADA Concept. *Present. at the Jeju 2011 Symp. on microgrids*.

[57] Ministry of Energy and Mines. Available: http://www.mem.gov.ma. Accessed on 23.10.2012.

[58] CDER (Centre de Dèveloppement des Energies Renouvelables, Morocco), Available: www.cder.org.ma. Accessed on 23.10. 2012.

[59] Tenesol (formerly Total Energie). Available: www.total-energie.com. Accessed on 23.10. 2012.

[60] Programme d'Electrification Rurale PERG, Office National d'Electricité (ONE), Available: www.one.org.ma. Accessed on 23.10. 2012.

[61] Komoto, K. (2009). Energy from the desert: very large scale photovoltaic systems: socio-economic, financial, technical and environmental aspects (Vol. 3). *Earthscan /James and James*.

[62] Office National de l'Eau Potable (National Office of Potable Water) Website. Available: www.onep.ma. Accessed on 23.10. 2012.

[63] Labed, S., (2008). PV Large Scale Rural Electrification Programs and the Development of Desert Regions. *Sust. Energy Production and Consumption, NATO Science for Peace and Security Series C: Environmental Security, Springer –Verlag*. 281-292.

[64] Maafi, A., (2000). A survey on photovoltaic activities in Algeria. *J. Ren. Energy*. 20, 9-17.

[65] New and Renewable Energy Authority (NREA) website, www http://www.nrea.gov.eg/ english1.html, Accessed on 30.10.2012.

[66] The Badia Project in Jordan http://www.badia.gov.jo/. Accessed on 30.10.2012.

[67] Oseni, M.O., (2012). Improving households' access to electricity and energy consumption pattern in Nigeria: Renewable energy alternative. *Ren. and Sust. Energy Reviews*, 16, 3967– 3974.

[68] Schneider Electric. The Bipbop program website: Available: schneider-electric. com/bipbop. Accessed on 30.10.2012.

[69] KPLC, 2012. Umeme Pamoja Brochure. *Kenya Power and Lighting Company (KPLC)* Available http://www.kplc.co.ke. Assessed on 26.10.2012.

[70] Hankins, M. (2001). Commercial breaks-building the market for PV in Africa. *Renewable Energy World*, 4(4), 164-175.
[71] Nieuwenhout, F. D. J., Van Dijk, A., Lasschuit, P. E., Van Roekel, G., Van Dijk, V. A. P., Hirsch, D., and Wade, H. (2001). Experience with solar home systems in developing countries: a review. *Progress in Photovoltaics: Research and Applications*, 9(6), 455-474.
[72] Acker, R. H., and Kammen, D. M. (1996). The quiet (energy) revolution: analysing the dissemination of photovoltaic power systems in Kenya. *Energy Policy*, 24(1), 81-111.
[73] Kibera Community Youth Program (KCYP). Available http:// www.kcyp.net/. Accessed on 23.10.2012
[74] The Solar Maasai Project. Available: http://solarmaasai.blogspot.com/. Accessed on 3.10.2012
[75] Ligthing Africa. Inovative Rural Employers Keep Pace with Employees' Needs. Available: http://www.lightingafrica.org/component/ k2/itemlist/ filter.html?all_langs= yesandsearchword=Mukuri. Accessed on 3.10. 2012
[76] UNDP,Market Transformation for Highly Efficient Biomass Stoves for Institutions and Medium-sized Enterprises. Available: http://www.ke. undp.org/index.php/projects/ market-transformation-for-highly-efficient-biomass-stoves-for-institutions-and-medium-sized-enterprise
[77] Lemaire, X., (2009). Fee-for-service companies for rural electrification with photovoltaic systems: The case of Zambia. *Energy for Sust. Dev.* 13, 18–23.
[78] UNDP Tanzania - Transformation of the Rural Photovoltaic Market in Tanzania http://www.tz.undp.org/ee_proj_trpv.htm
[79] Smits, M., (2008). Technography of pico-hydropower in the Lao PDR. *Report for the Lao Institute for Renewable Energy (LIRE).*
[80] Chinese Rural Renewable Energy-CCRE, 2003
[81] Paish and Green, 2002, The Pico-hydro Market in Vietnam.
[82] Progress with SHS's installation up to 29 April (2012). Available at: http://idcol. Org /prjshsm2004.php
[83] Technical Standards Committee, IDCOL. Technical Specifications for Solar Home System (SHS), IDCOL Solar Program.
[84] Mondal, H., Denich, M., (2010). Hybrid Systems for decentralized power generation in Bangladesh. *Energy for Sust. Dev.* 14,48-55.
[85] Opportunities for Women in Renewable Energy Technology Use in Bangladesh (Phase I), (2004). *Joint Report UNDP/World Bank (ESMAP) and Bank Netherlands Water Partnership Program (BNWPP).*
[86] Alam, J., (2008). Biogas energy for rural development: opportunities, challenges and lacuna of implementation. *Nepal: SESAM/ARTES South Asian Regional Workshop.*
[87] Dimpl E., (2010). Small-scale Electricity Generation from Biomass. Part II: Biogas. Deutsche Gesellschaft für Technische Zusammenarbeit (GTZ)-HERA.
[88] IDCOL (2012). Implementation Plan National Domestic Biogas and Manure Programme in Bangladesh. Available http://idcol.org/energy Project.php. Accessed on 23.10.2012.

[89] BRAC, (2012). The Bangladesh Rural Advancement Committee (BRAC) web-page. Available: http://www.lged-rein.org/brac.php, Accessed on 26.10.2012.

[90] Rural Electrification Board, (2012). Available: http://www.reb.gov.bd/intro.htm. Accessed on 23.12.2012.

[91] BPDB (Bangladesh Power Development Board) Renewable Energy Development and Energy Efficiency. Available: http://lged-rein.org/ bpdb.php Accessed on 23.10.2012.

[92] Grameen Shakti price list for PV Systems. Available: http://www. gshakti.org/ index.php?option=com_contentandview=articleandid=115andItemid=124. Accessed on 23.10.2012.

[93] Grameen Shakti Website. Biomass Section. Available: http://www. gshakti.org/index. php?option=com_contentandview=articleandid=60andItemid=64. Accessed on 23.10. 2012.

[94] Chaturvedi, P. (2005). Renewable Energy in India Programmes and Case Studies. *Science and Technology Vision*, 1, 61-64.

[95] Ministry of New and Renewable Energy (MNRE) Website. Available: http:// www. mnre.gov.in/ Accessed on 23.10.2012.

[96] Balijepalli, V. M., Khaparde, S. A., and Dobariya, C. V. (2010)). Deployment of MicroGrids in India. In Power and Energy Society General Meeting, 2010 IEEE, *Session Microgrids*.

[97] Deshmukh, A., (2009). The Role of Decentralized Renewable Energy for Rural Electrification Maharashtra case study, India. *MSc Thesis, Environmental Sciences, Policy and Management, Univ. of Lund.*

[98] Rural Electrification Corporation Ltd (REC) http://recindia.nic.in/ generation_renew. html. Accessed on 23.10.2012.

[99] MoP. (2008). Ministry of Power Annual Report. In M. o. Power (Ed.), (pp. 23-24). New Delhi: The Government of India.

[100] Rajiv Gandhi Grameen Vidyutikaran Yojana (RGGVY) Available: rggvy.gov.in. Accessed on 23.10.2012.

[101] Jawaharlal Nehru National Solar Mission (JNNSM) webpage. Available: http://www. mnre.gov.in/solar-mission/jnnsm/introduction-2/. Accessed on 23-10.2012

[102] D-Lab: Development, Design and Dissemination Lab of the Massachusetts Institute of Technology (MIT). Available http://d-lab.mit.edu. Accessed on 23-10.2012.

[103] SELCO Labs, Light for Education The First 2000- a Report May, 2012a http://www. selcofoundation.org/selcolabs/content/light-education-1

[104] Baring-Gould, I. Cattin, R., Durstewitz, M., Hulkkonen, M. Krenn, A., Laakso, T., Lacroix, A., Peltola, E., Ronsten, G., Wallenius, T.,(2012). Wind energy projects in cold climates. *Report to the Executive Committee of the International Energy Agency Programme.*

[105] NorthPower. *Broch. Northern Power® 100 ARCTIC*. Available: WWW.NORTHERN POWER.COM., Accessed on 23.10.2012.

[106] National Science foundation, Arctic Program, PolarPower website: www. polarpower. Org, Accessed on 23.10.2012.

[107] Lacroix, A., Manwell, J. F., (2000). Wind Energy: Cold Weather Issues. *Renewable Energy Research Laboratory, University of Massachusetts at Amherst.*

[108] Nepal Electricity Authority (NEA), load shedding http:// www.nea.org.np/ load shedding.html, Accessed on 23.10.2012.
[109] Nepal Power Development Project. World Bank website. Available: http://documents.worldbank.org/curated/en/2012/01/15593516/nepal-nepal-power-development-project-p043311-implementation-status-results-report-sequence-17
[110] Alternative Energy Promotion Center website.Available: http://www.aepc.gov.np/. Accessed on 24.10.2012
[111] Solar Electric Company Co (Pvt) Ltd website, http://sec.digital-culture.net/ . Accessed on 24.10.2012
[112] The Renewable Energy for Rural Livelihood (RERL) program website http:// www.rerl.org.np/. Accessed on 24.10.2012
[113] United Nations Environmental Programme /Division of Global Environment Facility Coordination (DGEF). Available www.uncp.org/ dgef Accessed on 24.10.2012
[114] Renewable Energy Project (REP), Available http://www.rep.com.np/. Accessed on 24.10.2012.
[115] The Himalayan Light Foundation (HLF) homepage: http:// www.hlf.org.np/. Accessed on 24.10.2012.
[116] First World Students Bridging The East West Energy Divide for Remote South Asian School Children. http://www.hlf.org.np/index.php?id=12. Accessed on 24.10.2012.
[117] Electricity enables establishment of community FM station, http://www.undp.org/content/nepal/en/home/ourwork/environmentandenergy/successstories/electricity-enables-establishment-of-community-fm-station/. Accessed on 24.10.2012.
[118] International Development Association-IDA Results. Available: web.worldbank.org. Accessed on 24.10.2012.
[119] Khimti power plant http://www.hpl.com.np/khimti.php. Accessed on 24.10.2012.
[120] UNDP,(2011), Turning Tradition to New Ends: Improving Water Mills In Nepal. Report prepared by the APEC and the UNDP, http://asia-pacific.undp.org.
[121] Maharjan, S., (2005). Nepal Biogas Support Project: A Case of Synergy between Rural Development, Environment and Energy Practice. *Presentation at the ESSD Week.* Available: http://www.esmap.org/node/ 1050. Accessed on 24.10.2012.
[122] Scaling-Up Renewable Energy Program Investment Plan for Nepal, https://www.climateinvestmentfunds.org/ cifnet/country-program-info/nepals-srep-programming. Accessed on 24.10.2012.
[123] McMurdo Sound Area Hybrid Systems http://www.polarpower.org/ examples/ mcmurdo/index.html. Accessed 24.10.2012.
[124] Aprea, J.L., (2012). Two years experience in hydrogen production and use in Hope bay, Antarctica. *Intl J. of Hydrogen Energy*. 37, 14773-14780.
[125] Tracy Dahl, Summit Station Wind Power Pilot Project, Polarpower.org Accessed 24.10.2012.
[126] Isherwood, W., Smith, J.R., Aceves, S.M., Berry, G., Clark, W., Johnson, R., Das, D., Goering, D., Seifert, R. (2000). Remote power systems with advanced storage technologies for Alaskan villages. *Energy*. 25, 1005–1020.
[127] Reeve, B.,(2002). Kotzebue Electric Association Wind Projects. *Presentation at the Intl Wind Diesel Conference Anchorage, Alaska.*

[128] CHP Case Studies in the Pacific Northwest 525 kW Wind/Diesel Hybrid CHP System. Available: www.akenergyauthority.org. Accessed 24.10.2012.

[129] Alaska Energy Authority, 2011. Renewable energy Atlas of Alaska. Available: www.akenergyauthority.org. Accessed 24.10.2012.

[130] Alaska Center for Energy and Powe(APEC), Homepage. Available: http:// www.uaf.edu/acep/about/ Accessed 24.10.2012.

[131] Alaska Wind-Diesel Applications Centre, Homepage. Available: http://www.uaf.edu/ acep/alaska-wind-diesel-applic/about-widac/ Accessed 24.10.2012.

[132] Thomas Lynge Jensen (2000),Renewable Energy on Small Islands – 2nd Edition. Report for Forum for Energy and Development.

[133] Solar Electric Light Fund (SELF), Available: www.self.org Accessed 24.10.2012.

[134] Solomon Islands Electricity Authority (SIEA), Available: http:// www.siea.com.sb/. Accessed 24.10.2012.

[135] Electrification of remote villages using micro-hydro, APACE Village First Electrification Group. Available: www.apace.uts.edu.au/ docpublish/vavanga.html, Accessed 24.10.2012.

[136] Japan International Cooperation Agency (JICA), Available: www.jica.go.jp/english/. Accessed 24.10.2012.

[137] Kiribati Solar Energy Company (KSEC), Available http:// ksec.ttechservices.net/ wp/.Accessed 24.10.2012.

[138] Regional Strategy Paperand Regional Indicative Programme. Pacific Regional S Trategy Paper 2002 - 2007, The 9th European Development Fund.

[139] Cotonou Agreement (2000/483/EC). Availlable: http://europa.eu/ legislation_ summaries/development/african_caribbean_pacific_states/r12101_en.htm Accessed 24.10.2012.

[140] REP-5 EU-Pacific project. Available: http://www.rep5.eu/Gallery/ FSM/Outerislands. Accessed 24.10.2012.

[141] Konings,P., (2008). PV Mini-Grids for Yap State: Sustainable Design for Remote Pacific Islands. *Proc of the 4th European PV-Hybrid and Mini-Grid Conf.*

[142] SMA (2010). Off-Grid Power Supply for Reao Atoll - SMA Solar Technology AG. Available, www.sma.de/fileadmin/content/ global/.../ REFREAO-AEN120110.pdf.

[143] Bueno, C., and Carta, J. A. (2005). Technical–economic analysis of wind-powered pumped hydrostorage systems. Part II: model application to the island of El Hierro. *Solar Energy*, 78(3), 396-405.

[144] Empressa Electicidad Madeira, (2010). Madeira Reversible-Hydro System. Presented on the Stories-EU Conference,Estoril,Portugal. Available: www.storiesporject.eu.

[145] Thim, F., Tapanlis, S., Wade, S., (2010) Conception and Operation of a Unique Large-Scale PV Hybrid System on a Hebridean Island. *Presented at the PVSEC 2010.*

[146] CROTOK (2000). Energy development of islands (Croatia national energy program). Zagreb: Energy Institute "Hrvoje Pozˇ ar"; 2000.

[147] Delvecchio, G., Guerra, M., Lofrumento, C., and Neri, F. (2005). A Study for Optimizing a Stand-Alone Hybrid Photovoltaic-Diesel System to Feed Summer Loads. *In Intern. Conf. on Renew. Energy and Power Quality, ICREPQ,* Spain pp167-168.

[148] Silvestrini, G., Gamberale (2005), M. The Italian "Sustainable Island" Programme . A challenge for Island Sustainable Development. In 100% RES, A challenge for Island Sustainable Development.pp117-123,

[149] Korovesis, P. N., Vokas, G. A., Gonos, I. F., Topalis F. V., (2004). Influence of Large – scale Installation of Energy Saving Lamps on the Line Voltage Distortion of a Weak Network Supplied by Photovoltaic station. *IEEE Trans. on Powe Delivery*. 19, 1787 – 1793.

[150] Kirubi, C., Jacobson, A., Kammen, D. M., and Mills, A. (2009). Community-based electric micro-grids can contribute to rural development: evidence from Kenya. *World Development*, 37(7), 1208-1221.

In: Renewable Energy
Editor: Andreas Poullikkas

ISBN: 978-1-62618-231-8

Chapter 10

TECHNICAL, ECONOMIC, AND MITIGATION PERFORMANCE OF SMALL-SCALE (≤1 kW_e) HYDROELECTRIC AND PHOTOVOLTAIC WATER PUMPING SYSTEMS RELATIVE TO NETWORK ELECTRICITY IN RURAL AREAS IN THE SOUTHWEST OF WESTERN AUSTRALIA

Mark P. McHenry*

School of Engineering and Energy, Murdoch University,
Western Australia

ABSTRACT

This chapter describes rural hydroelectric and photovoltaic (PV) water pumping system's technical performance simulations, and economic and greenhouse gas mitigation modelling research. The pico-hydroelectric (for electricity service provision) and PV water pumping (for stock water supply) simulations were selected for a detailed analysis due to their high sensitivity to water resource availability in rural areas. Both observed and projected changes in rainfall and runoff in the in the predominantly agricultural southwest (SW) of Western Australia (WA) will heavily influence land, water, and energy supply decision-making at all levels, including landholders. The hydroelectric systems are also contrasted against similar technical alternatives in the region (≤3 kWe grid-connected PV and wind turbine systems). The results suggest the technical performance of small-scale grid-connected (≤1 kWe) hydroelectric systems have the higher technical potential of all the small-scale grid-connected technical alternatives over the short-to-medium term, despite current drying conditions in the SW of WA. Similarly, the small-scale stand-alone PV water pumping system could not compete against grid-connected electricity when a connection was available, although the systems would be an effective market adaptation and market mitigation measure to

* Corresponding Author address: Murdoch University, 90 South St Murdoch, Western Australia, 6150. Email: mpmchenry@gmail.com.

supply stock water in most cases. This is due to stock watering requirements being generally off-grid, and a stand-alone PV pumping system NPV would likely be greater than a grid-connected pumping system that requires electricity network extension. Furthermore, a farm water supply pipe network extension may be economically less attractive than a stand-alone PV pumping system, although this would require further analysis of water resource quality and availability at the water point, and cost comparisons between the PV option, and the costs associated with extension of the pipe network, including the actual pipe gauge costs, the pipe network extension distance, and whether the pipe was buried or otherwise (etc). However, all of the renewable energy systems were expensive adaptation and mitigation options, and were unable to compete with the low cost of network electricity when a connection is available in terms of the NPV against 2010 market prices, subsidies, and costs.

Keywords: Hydroelectric; photovoltaic; water pumping; agriculture; rural

Introduction

This analysis was developed to clarify both the seasonal hydrological resource availability and performance and cost of small-scale water-related renewable energy systems in the SW of WA relative to network electricity supply. The Australian continent has relatively few large rivers, low topographical relief, yet despite few remaining opportunities for large-scale hydroelectric generation, small-scale opportunities still exist [1]. The variability of rainfall is the principal climatic factor that determines agricultural management in Australia [2]. Rainfall capture, storage, and transport has long been a major limitation to economic development in WA, particularly in agriculture. At the regional scale in the SW of WA, the variability of rainfall and runoff is a major factor in both small-scale hydroelectric generation and water demand, particularly in periods of low rainfall (November-April). Observed changes in summer rainfall since 1950 show extensive drying in the region [3], with around half of the reduction in observed rainfall from 1958-1975 to 1976-2003 due to a reduction in the number of troughs [4]. Winter rainfall in the SW has decreased substantially since 1950, and decreased abruptly in the mid 1970s by around 15-20%. The largest observed decrease is from March to July, while August to October rainfall has increased slightly [5]. Concerningly, rainfall flow into Perth's dams has dropped by about 50% since the mid 1970s from the 1911-74 long-term average [6]. As the SW of WA is less affected by El Niño/Southern Oscillation Index (ENSO/SOI) than the rest of Australia [7], the decrease in rainfall in the SW of WA is believed to be a combination of increased greenhouse gas concentrations, natural climate variability, and land-use change [5].

The increasing trends in atmospheric pressure in mid winter correlate with the rainfall regime changes in the region [5]. Compounding observed rainfall changes are the associations of Australian droughts with increasing records of mean daily maximum temperatures and evaporation [8]. The IPCCs 2007 4AR included a large-scale global projection of relative changes in rainfall runoff by the end of the 21st century representing the median values of 12 climate models using the SRES (Special Report on Emission Scenarios) A1B scenario. In this projection, the SW of WA saw a 90% model agreement on a reduction in runoff, with the median reduction value of between 20 and 40% of 1980-1999 runoff levels [9]. A similar 4AR multi-model (based on the SRES A1B scenario) projection of changes of SW rainfall for the period 2090-2099 relative to 1980-1999, showed a model agreement of close to 90% in a projected rainfall decrease for December to February of between 20% and 10%, and more than 90% model agreement of a decrease of between 30% and 20% for June to August [9]. In a summary of the ranges of uncertainty for rainfall changes simulated by 15 climate models for the IPCC's 4AR, the annual average rainfall is projected to decrease 3 to 22% by 2030 in the extreme SW, and, 0 to 22% in the rest of the SW, and 0 to 15% for southern coastal regions of the SW of WA. By 2070 these rainfall change model simulation uncertainties projected annual average decreases of between 7 to 70% to occur in the extreme SW, zero to 70% in the rest of the SW, and zero to 45% for southern coastal regions [10].

While noting the magnitude, spatial, and temporal uncertainties in projections such as these, climatic changes are expected to further exacerbate observed stresses on water security from population growth, economic growth, land-use change, and also electricity generation [9, 11, 12]. Therefore, scenario information is increasingly being developed at a finer geographical and temporal resolution andapplied to the SRES storylines, producing new regional scenarios of socio-economic conditions, land use and land cover, atmospheric composition, climate, and sea level [13]. In the rural sector, while long-term projections will necessarily guide long-term agricultural planning, they are of debatable applicability for short-to-medium term small investments such as small-scale grid-connected hydroelectric and stand-alone PV water pumping systems. This is particularly so when current agricultural land and water decision-making is sensitive to existing climatic and rainfall variability [14, 15]. Australia has the highest inter-annual natural rainfall variability of all the continents, and the principal cause of this natural variability is ENSO/SOI [2]. As the ENSO/SOI is coupled to the annual cycle, the increased use of such variability indicators can increase the predictability of the season, and thus decrease seasonal farm managerial uncertainties [7].

As periodic dry years are a normal part of the Australian climate regime [16], historical farm management plans have generally taken this variability into account with a number of strategies (de-stocking, delayed re-stocking, feed storage strategies, among many others) [17]. Land and water management inter-relationships are dependent on local conditions and management options, and long-term systematic assessments and comparisons of appropriate scale mitigation and adaptation options are useful to local agricultural decision-makers [15]. However, the greater use of inter and intra-annual climatic variability indicators will advance risk management by the inclusion of both seasonal indicators and long-term water resource projection characteristics, and associated investments [7, 18].

Simulation and Model Input Data, Scenarios, and Assumptions

Agricultural lands in the SW of WA are generally supplied with electricity from the single largest electricity network in WA, the South West Interconnected System (SWIS). The small-scale hydroelectric and PV water pumping systems were simulated to operate in parallel with SWIS connection to a standard two phase rural distribution line (230-240V, 32A), displacing network electricity of an average rural homestead load, and water pumping load for a single paddock, respectively.

Technical simulations used HOMER version 2.68 beta with a 15 minute interval comparison of electricity demand and electricity generation and flows through each component of the system [19]. As a complete annual time series for a homestead or water pumping electricity demand was unavailable in the region, the homestead load was generated in HOMER using a combination of appliance energy audit, 10 day real-time consumption monitoring, and three years of historical electricity billing data. The audit, monitoring and billing data revealed significant inter-year and intra-day load fluctuation, and thus HOMER's day-to-day and time-step-to-time-step variability was set at 50% and 250%, respectively to reflect all expected loads. The load data generated a maximum peak load on a 15 minute basis of around 10.1 kW, and a 12 kWh day^{-1} scaled average demand, which were consistent with parallel operational electricity demands based on the energy audit appliance and the billing data. The high day-to-day energy demand variability was selected to reflect the high irregularity of rural tasks that persist through weekends, and seasonal changes.

The PV water pumping output was designed to provide maximum water demand (in hot summer conditions, December to March) for around 250 dry (no lambs) adult sheep grazing pasture. The hydrological resource was generated by the author as an iterative combination of intermittent site monitoring of local river systems and rainfall data (both informal measurements and government streamflow gauging sites). The available river hydrological resource and stock water requirements were used as broadly representative for the high-rainfall (>800 mm) coastal region in the SW of WA. The annual average available water flow rate was scaled to 10 L s^{-1}. The input data shown in Table 1 includes the hydrological resource, daily solar radiation on a horizontal plane, clearness index, and air temperature input data.

The solar and temperature data was derived from a Bureau of Meteorology (BOM) SW station at Albany Airport, station no.009741, 69 m above sea level, and was transformed in the RETScreen (version 4) climate database with NASA's Surface Meteorology and Solar Energy Dataset [20]. The simulated site's annual average clearness index is 0.512, the annual average horizontal plane solar radiation received is 4.323 kWh m^{-2} day^{-1} with an annual average temperature of 15.4 °C. An economic model was developed in a simple and clear spreadsheet to incorporate all unique attributes of the various renewable energy technology simulations, policies, emission calculations, and market prices (2010) of capital and operational costs, all projected over a 15 year project lifetime.

Table 1. Generated monthly mean available hydrological flows used for the pico-hydroelectric simulations, and the monthly solar resource and temperature inputs. (The annual average flow of 10.3 L s^{-1} was scaled in the HOMER model to 10 L s^{-1})

Month	J	F	M	A	M	J	J	A	S	O	N	D
H_2O Flow (L s^{-1})	0.5	0.1	0.1	0.5	10.0	20.0	25.0	25.0	20.0	13.0	7.0	2.0
Clearness Index	0.57	0.53	0.50	0.46	0.46	0.47	0.48	0.48	0.49	0.50	0.50	0.51
Rad. (kWh m^{-2} day^{-1})	6.83	5.18	4.56	3.19	2.42	2.08	2.28	3.00	4.03	5.14	5.83	6.780
Av. Ambient Temp.(°C)	19.4	19.8	18.8	16.6	14.3	12.4	11.6	11.6	12.6	13.9	15.7	17.9

Each system model involved a number of assumptions detailed in each respective scenario, and a real discount rate of 8%. The SWIS network tariff used in each model was the government-owned electricity retailer, Synergy's 2010 Home Business Plan (K1) tariff (AUD 0.3823 day^{-1} supply charge and AUD0.2083 kWh^{-1} residences consuming less than 20 kWh day^{-1} over the billing period), which is tailored for rural consumers with a homestead and all shed loads consuming electricity off a single electricity meter [21]. A 'Renewable Energy Buyback Scheme' (REBS) rebate was available at the time for exported renewable energy grid-connected systems on the SWIS of capacity between 500 W and 5 kW, calculated on the net total over the billing period at a rate of AUD0.07 kWh^{-1}. Whilst Synergy's REBS values exported renewable electricity at around one-third of the price of the equivalent import price, to be eligible for REBS the consumer must be on the A1 (a flat residential) or SmartPower (a time of use variable) tariff, and thus residences on the K1 tariff are ineligible. There were around 13,000 K1 customers in 2009 [22], and none were able to access the premium WA feed-in tariff (FiT) when it was available (1 Aug 2010 to 1 Aug 2011), or the benefits of the REBS.

A 400 W Grid-Connected Hydroelectric System Technical Simulation Results

The homestead load profile and total consumption were used in the simulation of an efficient pico-hydroelectric reaction turbine installed on a low gross head (2.5 m), with a inlet pipe loss of 12% to operate at a maximum 400 W DC. The simulated DC hydroelectric generator was connected to an over-specification 1.1 kW grid-connected inverter, supplying the homestead load in parallel with the SWIS network. The system schematic of the simulated system is shown in Fig 1. The HOMER power curve equivalent to the 400 W hydroelectric turbine was developed from a range of available efficient pico-hydroelectric turbines of similar nominal outputs. The average turbine efficiency selected for the simulation was 65%, with an inverter efficiency of 95%. Figure 2 shows the annual and monthly electrical simulation results for the 400 W hydroelectric system, and the annual simulation results of the inverter for each 15 minute interval over the simulated year. The inverter output was reflective of the hydrological resource availability, predominantly in the winter and spring months. The constant output of the inverter was a function of the hydrological resource input data and scheduling.

The simulated monthly average hourly electricity exported to the electricity network was almost zero over the summer and autumn months, reflective on the seasonal hydrological resource availability. The simulated monthly average hourly electricity exported to the electricity network from the 400 W hydroelectric system is shown in Figure 3. Note that around half of the average hourly hydroelectric system output in the wettest months was exported to the electricity network. This demonstrates that storage capacity in pico-hydroelectric systems is a fundamental element of system performance, yet in practice building very large dams is not feasible and unadvisable environmentally.

Table 2 summarises the annual average total electricity consumed, exported, inverter production, the percentage of inverter output consumed in the homestead, and the percentage of homestead electricity supplied by the inverter.

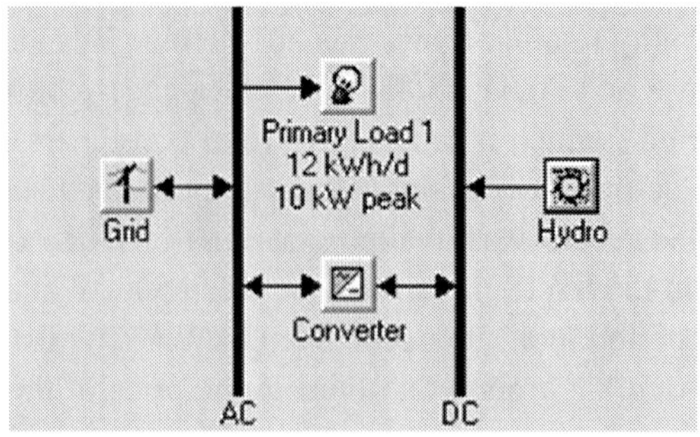

Figure 1. Schematic for the 400 W and the 1 kW hydroelectric system.

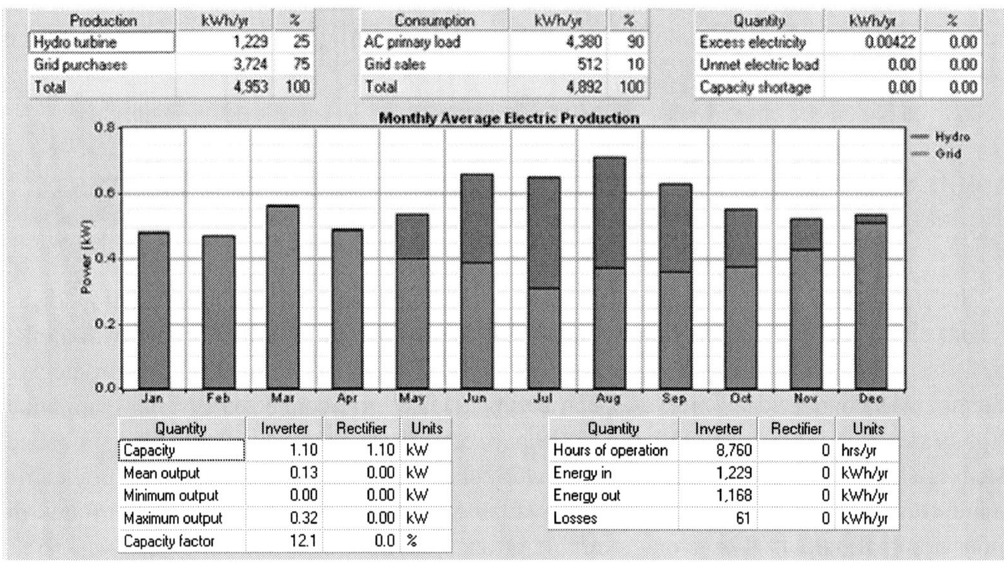

Figure 2. Simulation results for the 400 W hydroelectric system and the inverter output.

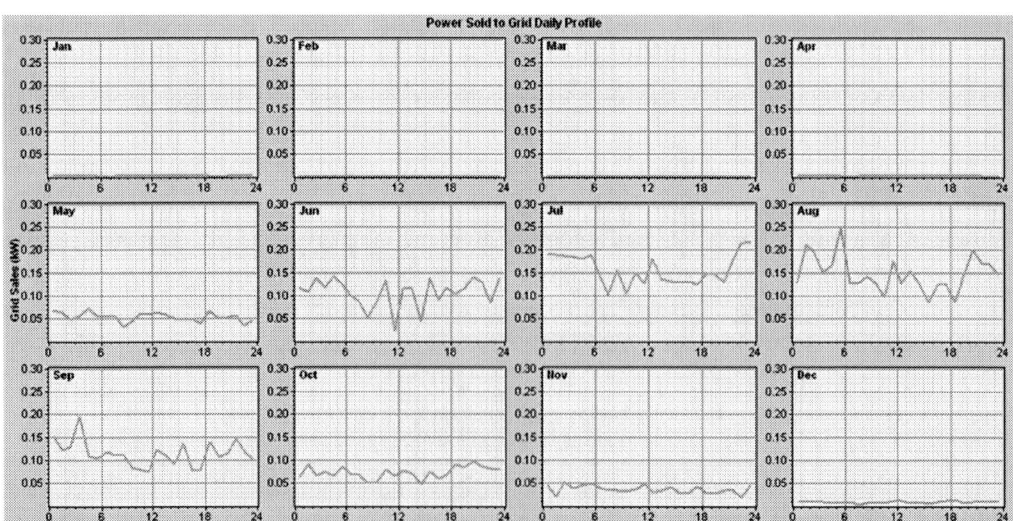

Figure 3. Monthly average hourly homestead electricity exported to the grid.

Table 2. Summary of annual averaged simulated 400 W hydroelectric system outputs

Total homestead electricity consumption from all sources	4,380 kWh $year^{-1}$
Total electricity exported to the electricity network	512 kWh $year^{-1}$
Net electricity production from inverter	1,168 kWh $year^{-1}$
Total inverter production consumed in the homestead	656 kWh $year^{-1}$
% of inverter production consumed by the homestead	56.2 %
% of annual homestead electricity supplied by the inverter	14.8%

The percentage of homestead electricity supplied by the inverter was 56.2% of all electricity requirements throughout the simulated year. The percentage of the 400 W hydroelectric system inverter production consumed by the homestead was 14.8%. The technical performance is comparable to the annual outputs of grid-connected PV and wind systems in the region with similar loads and ratings [23-26].

ECONOMIC AND MITIGATION RESULTS AND DISCUSSION

The market prices for the 400 W hydroelectric system were based on a 2010 average market value for the K1 tariff, as shown in Table 4. The model assumes a zero economic value for electricity exports onto the network from the existing REBS ineligibility of K1 tariff customers. The capital cost of the 400 W hydroelectric system was chosen to reflect an average quality commercially available hydroelectric turbine and enabling equipment (including installation) to meet Australian Standards. The dam construction and cabling to the homestead was not included in the model. The hydroelectric system models does not explicitly include replacement of runners and nozzles, however, the model does include annual costs for general maintenance.

The model also includes a capital subsidy based on the Renewable Energy Certificate (REC). One REC is equivalent to 1 MWh of clean energy generated by a Clean Energy Council accredited renewable energy generator. Small-scale renewable energy rebate structures in Australia are prone to periodic amendments under various guises, yet often remain related to RECs each system is deemed to produce, which are created and sold by consumers who own the generators to Clean Energy Council accredited system installers who onsell the RECs which effectively minimises owner's capital costs. This research included the 2010 Solar Credit Scheme subsidy where small PV, wind, or hydroelectric systems are entitled to a numer of deemed RECs of a fixed value (AUD40.00 each) for both stand-alone and grid-connected power systems [27]. The total REC entitlement for the 400 W system was 38 over the maximum deeming period of five years.

The simulation ignores the value of the total REC entitlement after the period as no installation period multiplier was published post 30 June 2015. The mitigation potential of each system was based on the assumption that electricity exported onto the network does not displace conventional supply, while the direct inverter-supplied load in the homestead would, based on the 2009 (scope 2) SWIS emission factor (0.84 $kgCO_2$-e kWh^{-1}). This factor was assumed to remain constant over the 15 years. The author notes that this is likely an overestimate as the factor has decreased in recent years, and in practice a large number of aggregated small-scale systems will need to influence large-scale generator scheduling to reduce SWIS emissions.

Table 4 shows that the life-cycle market adaptation potential for the 400 W hydroelectric system did not recoup the initial investment, although the market mitigation potential of the system was 8.266 tCO_2-e over the 15 year interval. For each tCO_2-e of mitigation, a cost of almost AUD761 was borne by the system owner. While noting many uncertainties and assumptions, the 400 W hydroelectric grid-connected system was not a commercially attractive option with a NPV of almost AUD-6,500 over the 15 years. The relatively small market mitigation potential offered by the 400 W hydroelectric system investment was not an economically attractive mitigation measure.This was comparable to the market adaptation potential (NPV) and market mitigation potential of both the 1 kW PV and 1 kW wind turbine grid-connected systems (Table 3).

Table 3. A comparison between the 1 kW wind, 1 kW PV, and 400 W hydroelectric grid-connected system's total market adaptation potential (NPV) and market mitigation potential over the 15 year investment cycle. Source: [23-25]

	1 kW PV	1 kW wind	400 W hydro
NPV	AUD-6,436	AUD-5,416	AUD-6,290
Mitigation (tCO_2-e)	9.513	8.467	8.266
Mitigation cost (AUD tCO_2-e^{-1})	677	640	761

Table 4. The DCF and emission results for the 400 W hydroelectric grid-connected system over a 15 year interval, with the NPV in bold (Yr 15 of the 'D. Cumul. $ Flow')

Initial costs & subsidies		Rates/Prices/Factors		Annual costs & production	
Capital equipment	-$9,500.00	Inflation rate	3.00%	Electricity price for import ($/kWh, inc GST)	$0.2083
New meter	-$268.00	Real discount rate	8.00%	Electricity price for export ($/kWh, ext GST)	$0.0000
Capital subsidy	$0.00	Carbon price ($/tCO2-e)	0.00	Annual electricity generated (kWh)	1,168
REC entitlement	38	Carbon price inflation	0.00%	Annual net electricity used (kWh)	656
Total REC value	$1,520.00	REC price	$40.00	Electricity sold	512
		SWIS (kg CO2-e / kWh)	0.84	Total electricity consumed	4,380
		System Insurance	$0.00	Turbine service (TS)	-$30.00
Total	**-$8,248**			Electronics service (ES)	-$400.00

Yr	Cashflow Data	Costs	Income	Net $ Flow	Cumul. $ Flow	D.Net $ Flow	D.Cumul. $ Flow	tCO2 Baseline	tCO2 New	Net tCO2 Mit	Cumul. tCO2 Mit
0	Installation	-$9,768	$1,520	-$8,248	-$8,248	-$8,248	-$8,248	3.679	3.128	0.000	
1	TS	-$30	$137	$107	-$8,141	$99	-$8,149	3.679	3.128	0.551	0.551
2	TS	-$30	$137	$107	-$8,035	$91	-$8,058	3.679	3.128	0.551	1.102
3	TS	-$30	$137	$107	-$7,928	$85	-$7,973	3.679	3.128	0.551	1.653
4	TS	-$30	$137	$107	-$7,821	$78	-$7,895	3.679	3.128	0.551	2.204
5	TS	-$30	$137	$107	-$7,715	$73	-$7,822	3.679	3.128	0.551	2.755
6	TS	-$30	$137	$107	-$7,608	$67	-$7,755	3.679	3.128	0.551	3.306
7	TS	-$30	$137	$107	-$7,501	$62	-$7,693	3.679	3.128	0.551	3.857
8	TS, ES	-$430	$137	-$293	-$7,795	-$158	-$7,851	3.679	3.128	0.551	4.408
9	TS	-$30	$137	$107	-$7,688	$53	-$7,798	3.679	3.128	0.551	4.959
10	TS	-$30	$137	$107	-$7,582	$49	-$7,749	3.679	3.128	0.551	5.510
11	TS	-$30	$137	$107	-$7,475	$46	-$7,703	3.679	3.128	0.551	6.061
12	TS	-$30	$137	$107	-$7,368	$42	-$7,660	3.679	3.128	0.551	6.612
13	TS	-$30	$137	$107	-$7,262	$39	-$7,621	3.679	3.128	0.551	7.164
14	TS	-$30	$137	$107	-$7,155	$36	-$7,585	3.679	3.128	0.551	7.715
15	TS	-$30	$137	$4,107	-$3,048	$1,295	**-$6,290**	3.679	3.128	0.551	8.266
Remaining System Value			$4,000								

A 1 KW GRID-CONNECTED HYDROELECTRIC SYSTEM TECHNICAL SIMULATION RESULTS

The same homestead load profile and total consumption was used in a simulation of a nominal 1 kW DC high-efficiency pico-hydroelectric impulse turbine installed on a medium gross head (6.5 m), with an inlet pipe loss of 12%. The simulated DC hydroelectric system enabling components were identical to the pico-hydroelectric system designed to operate at a maximum of 400 W, although the increased head supports an alternative pico-hydroelectric turbine technology rated up to a maximum of 1 kW in this scenario. The hydroelectric turbine was connected to the same 1.1 kW grid-connected inverter as the 400 W hydroelectric system, and the 1 kW PV and wind systems. The inverter supplied the homestead load in parallel with the SWIS electricity network. The system schematic of the simulated system is identical to Figure 1.

Figure 4 shows the annual and monthly electrical simulation results for the 1 kW hydroelectric system, and the annual simulation results of the inverter for each 15 minute interval over the simulated year. The simulated monthly average hourly electricity exported to the electricity network from the 1 kW hydroelectric system is shown in Figure 5. In a similar manner to the 400 W hydroelectric system, around half of the average hourly hydroelectric system output in the wetter months was exported to the electricity network for the 1 kW system.

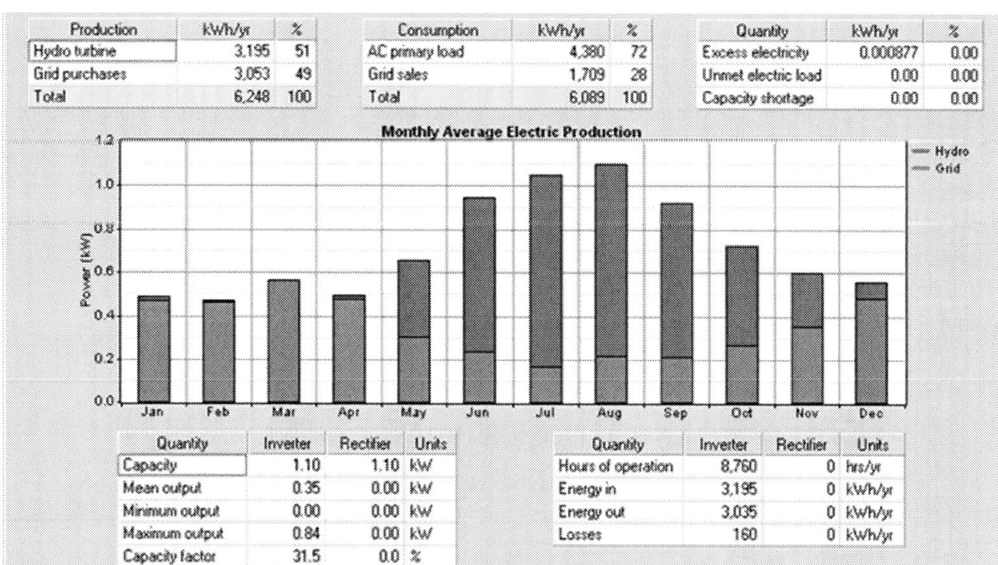

Production	kWh/yr	%
Hydro turbine	3,195	51
Grid purchases	3,053	49
Total	6,248	100

Consumption	kWh/yr	%
AC primary load	4,380	72
Grid sales	1,709	28
Total	6,089	100

Quantity	kWh/yr	%
Excess electricity	0.000877	0.00
Unmet electric load	0.00	0.00
Capacity shortage	0.00	0.00

Quantity	Inverter	Rectifier	Units
Capacity	1.10	1.10	kW
Mean output	0.35	0.00	kW
Minimum output	0.00	0.00	kW
Maximum output	0.84	0.00	kW
Capacity factor	31.5	0.0	%

Quantity	Inverter	Rectifier	Units
Hours of operation	8,760	0	hrs/yr
Energy in	3,195	0	kWh/yr
Energy out	3,035	0	kWh/yr
Losses	160	0	kWh/yr

Figure. 4. Simulation results for the 1 kW hydroelectric system and the inverter output.

Table 5 summarises the annual average total electricity consumed, exported, inverter production, the percentage of inverter output consumed in the homestead, and the percentage of homestead electricity supplied by the inverter. The percentage of homestead electricity supplied by the inverter was 43.7%, and the percentage of inverter production consumed by the homestead was 30.3%. This output data was relatively consistent with both the 3 kW PV and the 3 kW wind rated systems simulated in the region [23-25]. However, the major difference was the larger percentage of electricity produced by the system that was consumed in the homestead in the winter and spring months. This was primarily due to the lower nominal rating of the hydroelectric system which was more matched to the homestead load over wetter months, and resulted in a reduced percentage of exported electricity.

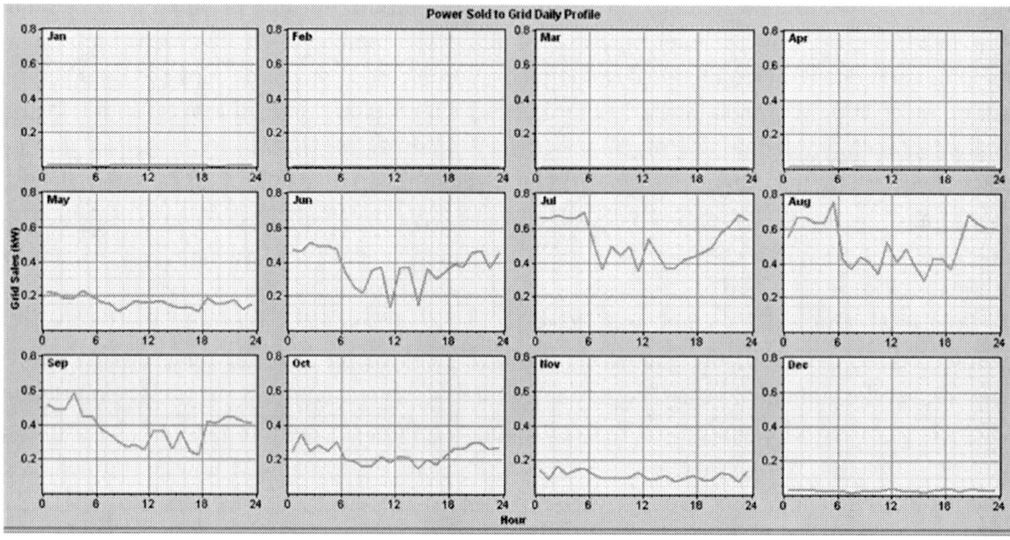

Figure 5. Monthly average hourly homestead electricity exported to the grid.

Table 5. Summary of annual average simulated 1 kW hydroelectric system outputs

Total homestead electricity consumption from all sources	4,380 kWh $year^{-1}$
Total electricity exported to the electricity network	1,709 kWh $year^{-1}$
Net electricity production from inverter	3,035 kWh $year^{-1}$
Total inverter production consumed in the homestead	1,326 kWh $year^{-1}$
% of inverter production consumed by the homestead	43.7 %
% of annual homestead electricity supplied by the inverter	30.3%

Economic and Mitigation Model Results and Discussion

Market prices for the 1 kW hydroelectric system were identical to the 400 W hydroelectric system adaptation and mitigation simulation. The only change in capital costs was the additional capital cost of a longer intake (AUD500), and the larger rebate from the REC entitlement. The total REC entitlement for the system is 95 over the maximum deeming period of five years. In a similar manner to the wind systems and the 400 W hydroelectric system, the simulation ignores the value of the total REC entitlement after the initial period as no installation period multiplier was published for after 30 June 2015. Table 6 shows that the life-cycle market adaptation potential for the 1 kW hydroelectric system did not recoup the initial investment. Table 7 summarises the current market adaptation and market mitigation potentials of the system, with cost of AUD198 per tCO_2-e borne by the system owner. While again noting many uncertainties and assumptions, the 1 kW hydroelectric grid-connected system was not a commercially attractive option with a NPV of around AUD-3,300 over the 15 years. The market mitigation potential of the system was 16.708 tCO_2-e over the 15 year interval, which was less than the comparable 3 kW PV and 3 kW wind systems in the region [23-25]. In a similar manner to the 3 kW PV and wind systems, the relatively small market mitigation potential offered by the 1 kW hydroelectric system was not an economically attractive mitigation measure.

For comparison, the author has undertaken a re-modelling of the 1 kW hydroelectric system with the following assumptions:

- the RECs-related capital subsidy remains the same;
- the 1 kW hydroelectric system outputs and homestead load profiles remain identical;
- electricity import prices remain identical over the period;
- carbon prices remain as zero;
- the K1 tariff is now eligible for Synergy's REBS (AUD0.07 kWh^{-1}), akin to the SmartPower/A1 tariffs, and;
- an additional WA State Government FiT of AUD0.40 kWh^{-1} (net) is available on the K1 tariff for the first 10 years. This creates a real value of AUD0.47 kWh^{-1} for exported electricity in the first ten years, and a real value of AUD0.07 kWh^{-1} after the first ten years.

Table 6. The DCF and emission results for the 1 kW hydroelectric grid-connected system over a 15 year interval, with the NPV in bold (Yr 15 of the 'D. Cumul. $ Flow')

Initial costs & subsidies		Rates/Prices/Factors		Annual costs & production	
Capital equipment	-$10,000.00	Inflation rate	3.00%	Electricity price for import ($/kWh, inc GST)	$0.2083
New meter	-$268.00	Real discount rate	8.00%	Electricity price for export ($/kWh, ext GST)	$0.0000
Capital subsidy	$0.00	Carbon price ($/tCO2-e)	0.00	Annual electricity generated (kWh)	3,035
REC entitlement	95	Carbon price inflation	0.00%	Annual net electricity used (kWh)	1,326
Total REC value	$3,800.00	REC price	$40.00	Electricity sold	1,709
		SWIS (kg CO2-e / kWh)	0.84	Total electricity consumed	4,380
		System insurance	$0.00	Turbine service (TS)	-$30.00
Total	**-$6,468**			Electronics service (ES)	-$400.00

Yr	Cashflow Data	Costs	Income	Net $ Flow	Cumul. $ Flow	D.Net $ Flow	D.Cumul. $ Flow	tCO2 Baseline	tCO2 New	Net tCO2 Mit	Cumul. tCO2 Mit
0	Installation	-$10,268	$3,800	-$6,468	-$6,468	-$6,468	-$6,468	3.679	2.565	0.000	
1	TS	-$30	$276	$246	-$6,222	$228	-$6,240	3.679	2.565	1.114	1.114
2	TS	-$30	$276	$246	-$5,976	$211	-$6,029	3.679	2.565	1.114	2.228
3	TS	-$30	$276	$246	-$5,729	$195	-$5,834	3.679	2.565	1.114	3.342
4	TS	-$30	$276	$246	-$5,483	$181	-$5,653	3.679	2.565	1.114	4.455
5	TS	-$30	$276	$246	-$5,237	$168	-$5,485	3.679	2.565	1.114	5.569
6	TS	-$30	$276	$246	-$4,991	$155	-$5,330	3.679	2.565	1.114	6.683
7	TS	-$30	$276	$246	-$4,745	$144	-$5,186	3.679	2.565	1.114	7.797
8	TS, ES	-$430	$276	-$154	-$4,898	-$83	-$5,269	3.679	2.565	1.114	8.911
9	TS	-$30	$276	$246	-$4,652	$123	-$5,146	3.679	2.565	1.114	10.025
10	TS	-$30	$276	$246	-$4,406	$114	-$5,032	3.679	2.565	1.114	11.138
11	TS	-$30	$276	$246	-$4,160	$106	-$4,926	3.679	2.565	1.114	12.252
12	TS	-$30	$276	$246	-$3,914	$98	-$4,829	3.679	2.565	1.114	13.366
13	TS	-$30	$276	$246	-$3,668	$90	-$4,738	3.679	2.565	1.114	14.480
14	TS	-$30	$276	$246	-$3,422	$84	-$4,655	3.679	2.565	1.114	15.594
15	TS	-$30	$276	$4,246	$824	$1,339	-$3,316	3.679	2.505	1.114	16.708
	Remaining System Value		$4,000								

Table 7. A comparison between the 3 kW wind, 3 kW PV, and 1 kW hydroelectric grid-connected system's total market adaptation potential (NPV) and market mitigation potential over the 15 year investment cycle. Source: [23-25]

	3 kW PV	3 kW wind	1 kW hydro
NPV	AUD-15,015	AUD-13,387	AUD-3,316
Mitigation (tCO_2-e)	20.954	17.728	16.708
Mitigation cost (AUD $tCO_2\text{-}e^{-1}$)	716	755	198

The results of the new modeled scenario are shown in Table 8. The NPV of the 1 kW hydroelectric system with a FiT was an attractive investment from the perspective of the owner at AUD2,074 over the 15 year model, with a strong influence of the 10 year FiT, with a distinct downturn in cashflow after year 10. Based on the new scenario assumptions the 1 kW hydroelectric system has both a positive market mitigation potential of 16.708 tCO_2-e, and a positive market adaptation potential of AUD2,074. The positive market adaptation potential (or a negative abatement cost) of AUD124 tCO_2^{-1} was an attractive investment to both adapt to electricity price increases and reduce any carbon liability of a private investor. However, the author points out that the unsubsidised system (excluding the capital SGU rebate and the FiT) would result in a NPV of AUD-7,116 for the private owner (Table 9).

Table 8. The DCF and emission results for the 1 kW hydroelectric grid-connected system with a 10-year AUD0.47 kWh-1 FiT for exported electricity, and standard Synergy REBS values after the 10 year period. The NPV is in bold (Yr 15 of the 'D. Cumul. $ Flow')

Initial costs & subsidies	
Capital equipment	-$10,000.00
New meter	-$268.00
Capital subsidy	$0.00
REC entitlement	95
Total REC value	$3,800.00
Total	**-$6,468**

Rates/Prices/Factors	
Inflation rate	3.00%
Real discount rate	8.00%
Carbon price ($/tCO2-e)	0.00
Carbon price inflation	0.00%
REC price	$40.00
SWIS (kg CO2-e / kWh)	0.84
System insurance	$0.00

Annual costs & production	
Electricity price for import ($/kWh, inc GST)	$0.2083
Electricity price for export ($/kWh, ext GST)	$0.4700
Annual electricity generated (kWh)	3,035
Annual net electricity used (kWh)	1,326
Electricity sold	1,709
Total electricity consumed	4,380
Turbine service (TS)	-$30.00
Electronics service (ES)	-$400.00

Yr	Cashflow Data	Costs	Income	Net $ Flow	Cumul. $ Flow	D.Net $ Flow	D.Cumul. $ Flow	tCO2 Baseline	tCO2 New	Net tCO2 Mit	Cumul. tCO2 Mit
0	Installation	-$10,268	$3,800	-$6,468	-$6,468	-$6,468	-$6,468	3.679	2.565	0.000	
1	TS	-$30	$1,079	$1,049	-$5,419	$972	-$5,496	3.679	2.565	1.114	1.114
2	TS	-$30	$1,079	$1,049	-$4,369	$900	-$4,597	3.679	2.565	1.114	2.228
3	TS	-$30	$1,079	$1,049	-$3,320	$833	-$3,764	3.679	2.565	1.114	3.342
4	TS	-$30	$1,079	$1,049	-$2,270	$771	-$2,992	3.679	2.565	1.114	4.455
5	TS	-$30	$1,079	$1,049	-$1,221	$714	-$2,278	3.679	2.565	1.114	5.569
6	TS	-$30	$1,079	$1,049	-$171	$661	-$1,617	3.679	2.565	1.114	6.683
7	TS	-$30	$1,079	$1,049	$878	$612	-$1,004	3.679	2.565	1.114	7.797
8	TS, ES	-$430	$1,079	$649	$1,527	$351	-$653	3.679	2.565	1.114	8.911
9	TS	-$30	$1,079	$1,049	$2,577	$526	-$128	3.679	2.565	1.114	10.025
10	TS	-$30	$1,079	$1,049	$3,626	$486	$358	3.679	2.565	1.114	11.138
11	TS	-$30	$276	$246	$3,873	$106	$463	3.679	2.565	1.114	12.252
12	TS	-$30	$276	$246	$4,119	$98	$561	3.679	2.565	1.114	13.366
13	TS	-$30	$276	$246	$4,365	$90	$651	3.679	2.565	1.114	14.480
14	TS	-$30	$276	$246	$4,611	$84	$735	3.679	2.565	1.114	15.594
15	TS	-$30	$276	$4,246	$8,857	$1,330	**$2,074**	3.679	2.565	1.114	16.708
Remaining System Value			$4,000								

Table 9. The DCF and emission results for the 1 kW hydroelectric grid-connected system, minus any government subsidy or FiT mechanism over a 15 year interval, with the NPV in bold (Yr 15 of the 'D. Cumul. $ Flow')

Initial costs & subsidies		Rates/Prices/Factors		Annual costs & production	
Capital equipment	-$10,000.00	Inflation rate	3.00%	Electricity price for import ($/kWh, inc GST)	$0.2083
New meter	-$268.00	Real discount rate	8.00%	Electricity price for export ($/kWh, ext GST)	$0.0000
Capital subsidy	$0.00	Carbon price ($/tCO2-e)	0.00	Annual electricity generated (kWh)	3,035
REC entitlement	95	Carbon price inflation	0.00%	Annual net electricity used (kWh)	1,326
Total REC value	$0.00	REC price	$0.00	Electricity sold	1,709
		SWIS (kg CO2-e / kWh)	0.84	Total electricity consumed	4,380
		System insurance	$0.00	Turbine service (TS)	-$30.00
Total	**-$10,268**			Electronics service (ES)	-$400.00

Yr	Cashflow Data	Costs	Income	Net $ Flow	Cumul. $ Flow	D.Net $ Flow	D.Cumul. $ Flow	tCO2 Baseline	tCO2 New	Net. tCO2 Mit	Cumul. tCO2 Mit
0	Installation	-$10,268	$0	-$10,268	-$10,268	-$10,268	-$10,268	3.679	2.566	0.000	
1	TS	-$30	$276	$246	-$10,022	$228	-$10,040	3.679	2.565	1.114	1.114
2	TS	-$30	$276	$246	-$9,776	$211	-$9,829	3.679	2.565	1.114	2.228
3	TS	-$30	$276	$246	-$9,529	$195	-$9,634	3.679	2.565	1.114	3.342
4	TS	-$30	$276	$246	-$9,283	$181	-$9,453	3.679	2.565	1.114	4.455
5	TS	-$30	$276	$246	-$9,037	$168	-$9,285	3.679	2.565	1.114	5.569
6	TS	-$30	$276	$246	-$8,791	$155	-$9,130	3.679	2.565	1.114	6.683
7	TS	-$30	$276	$246	-$8,545	$144	-$8,986	3.679	2.565	1.114	7.797
8	TS, ES	-$430	$276	-$154	-$8,698	-$83	-$9,069	3.679	2.565	1.114	8.911
9	TS	-$30	$276	$246	-$8,452	$123	-$8,946	3.679	2.565	1.114	10.025
10	TS	-$30	$276	$246	-$8,206	$114	-$8,832	3.679	2.565	1.114	11.138
11	TS	-$30	$276	$246	-$7,960	$106	-$8,726	3.679	2.565	1.114	12.252
12	TS	$30	$276	$246	$7,714	$98	$8,629	3.679	2.565	1.114	13.366
13	TS	-$30	$276	$246	-$7,468	$90	-$8,538	3.679	2.565	1.114	14.480
14	TS	$30	$276	$246	-$7,222	$84	-$8,455	3.679	2.565	1.114	15.594
15	TS	-$30	$276	$4,246	-$2,976	$1,339	**-$7,116**	3.670	2.566	1.114	16.708
Remaining System Value			$4,000								

A 120 W PV STAND-ALONE 60 W WATER PUMPING SYSTEM

There are an estimated 1000 PV water pumping systems in WA alone [28]. The primary reason that WA agriculturalists are installing PV water pumping systems is to improve the reliability of stock water supplies, particularly reducing the need for windmill pumping and associated annual servicing requirements [29]. Water pumping systems are the most common renewable energy system in agricultural regions of WA [28]. However, most water pumping systems are small in terms of rated output and are used when the electricity network, or owner constructed reticulated water supply systems are uneconomic or unavailable.

This feasibility study compares the economics between a small (60 W) water pumping system for stock watering using a 120 W PV array, relative to SWIS network grid-connected pump of identical capacity. The small 60 W pump was designed to provide 1000 L day^{-1} in winter and also 1500 L day^{-1} for stock-watering in summer at a total dynamic head of 16 m. The simulations assumed a PV temperature coefficient of power of -0.5% $^{o}C^{-1}$, a nominal operating temperature of 47°C, an efficiency at standard test conditions of 13%, a derating factor of 85%, a ground reflectance of 20%, and was a non-tracking system orientated with an azimuth of 180 (degrees West of South), and a slope of 35°, (measured from the horizontal plane).

The analysis assumes an average working time per day over a year of 2 hrs and 10 mins, with both the stand-alone and the AC equivalent system electricity consumption based on the same conditions. Therefore, in both cases the pump consumes an estimated 47.45 kWh per year. This simplified calculation was obtained by multiplying the 130 minutes of operation per day by 60 seconds per minute by 60 W, multiplied again by 365 days per year, then divided by 3,600,000 to obtain 47.45 kWh.

The model assumes the same performance and costs of the pumping component, the piping system and associated components, including filters, tank, etc. In terms of operational requirements, the model assumes an AC water pump was identical to its simulated DC-equivalent, and both pumping systems require the same servicing. Figure 6 is a schematic diagram of the simulated system.

The capital costs for all system components including PV module, freight, pump, and balance of system component prices was based on the actual costs in 2010. The system does not include installation, a water storage tank, connection pipes from the water source, bore construction (or maintenance), or dam floats, as these costs are highly variable and many of the components may be available on-site at next-to-zero cost.

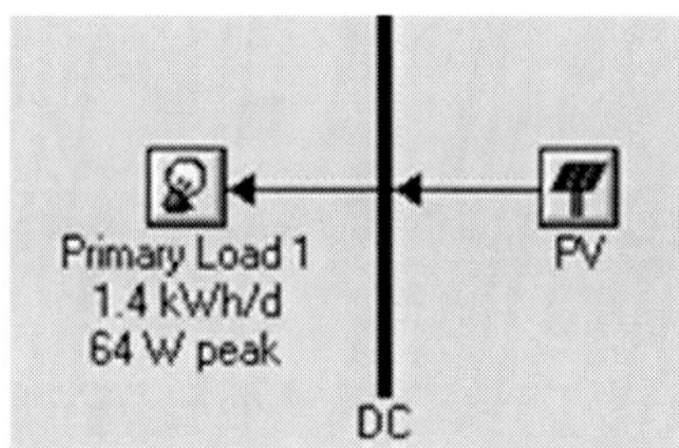

Figure 6. PV pumping system schematic.

The total cost of the PV pumping system (AUD3,700) includes the 120 W PV array and balance of system components, the DC pump, and a float valve and pressure switch to turn the pump off when the water storage tank is full. The cost of the grid-connected system (AUD2,600) was assumed to be identical minus the PV array and balance of system cost (AUD1,100), including the 60 W AC pump.

The economic model includes two scenarios: a stand-alone pumping system when the location does have access to the electricity network, but chooses not to connect, and; a grid-connected system (as the baseline). As the only difference between the two systems in terms of the model was the PV array, the grid-connected pumping system was chosen as the baseline, and only the cost difference of AUD1,100 was modelled.

Similarly, no replacement costs for any of the components were included, as it was assumed that both system costs would be identical, except for maintenance of the PV array and balance of system components.

This was represented by an annual difference of AUD10 per annum to approximate labour costs of a swift seasonal inspection and cleaning the system. No subsidies were included in the analysis of either water pumping system. The K1 electricity tariff supply charge of 38.23 ¢ day^{-1} (including GST), was not included in the grid-connected system economic model as an equivalent average annual daily load cost. This was because pumping loads as small as 60 W would likely be connected to an existing network connection with a much greater load demand to render the supply charge component negligible, and only a unit charge of AUD0.2083 kWh^{-1} was included.

TECHNICAL SIMULATION, ADAPTATION AND MITIGATION RESULTS AND DISCUSSION

Figure 7 shows the monthly electrical simulation results of the output from the 120 W PV array. The total annual output of the array was 168 kWh for the simulated year. The assumed electricity consumption of 47.45 kWh per year from the 60 W pump corresponds to 28% of the total PV output used to supply the pump load (Table 10). The low percentage was a result of both the over production of electricity above the approximate 60 W requirement by the PV array, and intervals where the array was generating some electricity, although was insufficient to run the pump (Figure 8). Table 11 shows that the PV-powered pumping system does not recoup the initial investment, and the owner's net present cost remains AUD1,100 higher than the grid-connected system over the 15 year interval. Note that no salvage value of the PV array and balance of system components were added in year 15 in the model. The total life-cycle market mitigation potential of the system was 0.52 tCO_2-e, which is the maximum possible for the pumping system over the 15 years, based on the simulation assumptions. The market and mitigation potentials of the system, represented in terms of a carbon price the system owner would pay was AUD2,115 $tCO_2\text{-}e^{-1}$. While noting simulation and modelling assumptions and uncertainties, the 120 W PV array pumping system was not commercially attractive compared with the grid-connected water pumping option. The additional cost of the PV pumping system demonstrates that PV could compete on a level playing field if grid-connection was an inexpensive option when commissioning either of these systems. Either a water pipe network extension over a few hundred metres, or any length of electricity network extension would compare poorly against the economics of a stand-alone power supply system, even without subsidy mechanisms. Water pumping for stock watering, however, is often required in areas where there is no electricity network or piping infrastructure to provide water. In the scenario of a stand-alone PV pumping system, the NPV would likely be greater than a grid-connected pumping system that requires electricity network extension. Similarly, a farm water supply pipe network extension from the electricity connection point is likely to be economically less attractive than a stand-alone PV pumping system. However, this would be dependent on the water resource available at the water point and the electricity network connection point, and would require cost comparisons between the PV option, electricity network extension, and extension of the pipe network, including the actual pipe gauge required, the extension distance, and whether the pipe was buried or not (etc). Therefore, stand-alone PV pumping systems would be an effective market adaptation and market mitigation measure to supply stock water in most cases when there is no access to network priced electricity.

Table 10. Summary of annual average simulated technical output for a 60 W pump

Total electricity consumed by the 60 W pump	47 kWh $year^{-1}$
Total electricity produced by the 120 W PV array system	168 kWh $year^{-1}$
Total electricity consumed by the grid-connected system	47 kWh $year^{-1}$
% of PV array electricity consumed by the pump	28 %
% of electricity consumed by the grid-connected pump	100%

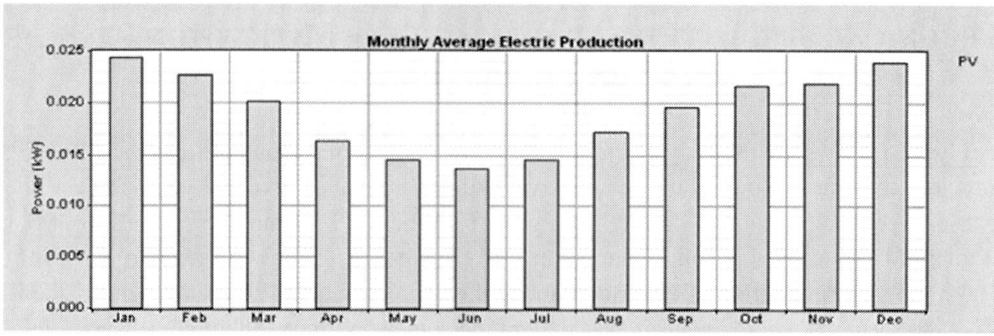

Figure 7. PV simulation results for the pumping system 120 W PV array.

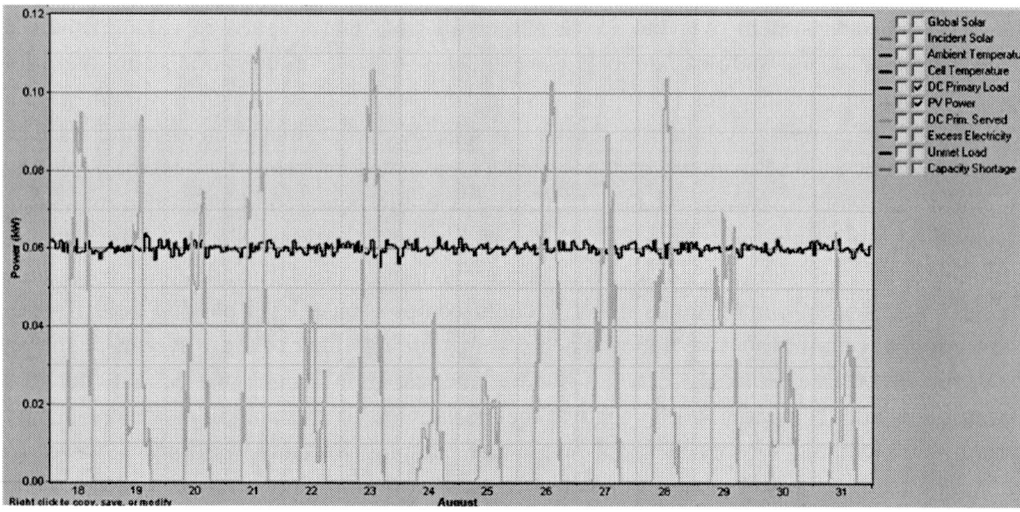

Figure 8. Selected 14 day PV output and pumping loads for the system.

Table 11. The DCF and emission results for the 60 W PV stand-alone water pumping system over the 15 year interval. The NPV is in bold (Yr 15 of the 'D. Cumul. $ Flow')

Initial costs & subsidies	
Capital equipment	-$1,100.00
New meter	$0.00
Capital subsidy	$0.00
REC entitlement	0
Total REC value	$0.00
Total	**-$1,100**

Rates/Prices/Factors	
Inflation rate	3.00%
Real discount rate	8.00%
Carbon price ($/tCO2-e)	$0.00
Carbon price inflation	0.00%
REC price	$40.00
SWIS (kg CO2-e / kWh)	0.84
System insurance	$0.00

Annual costs & production	
Electricity price for import ($/kWh, inc GST)	$0.2083
Electricity price for export ($/kWh, ext GST)	$0.0000
Annual electricity generated (kWh)	168
Annual net electricity used (kWh)	47
Electricity sold	0
Total electricity consumed	47
Array service (AS)	-$10.00
Pump service (PS)	$0.00

Yr	Cashflow Data	Costs	Income	Net $ Flow	Cumul. $ Flow	D Net $ Flow	D Cumul. $ Flow	tCO2 Baseline	tCO2 New	Net tCO2 Mit	Cumul. tCO2 Mit
0	Installation	-$1,100	$0	-$1,100	-$1,100	-$1,100	-$1,100	0.039	0.000	0.000	
1	AS	-$10	$10	$0	-$1,100	$0	-$1,100	0.039	0.000	0.039	0.039
2	AS	-$10	$10	$0	-$1,100	$0	-$1,100	0.039	0.000	0.039	0.079
3	AS	-$10	$10	$0	-$1,101	$0	-$1,101	0.039	0.000	0.039	0.118
4	AS	-$10	$10	$0	-$1,101	$0	-$1,101	0.039	0.000	0.039	0.158
5	AS	-$10	$10	$0	-$1,101	$0	-$1,101	0.039	0.000	0.039	0.197
6	AS	-$10	$10	$0	-$1,101	$0	-$1,101	0.039	0.000	0.039	0.237
7	AS	-$10	$10	$0	-$1,101	$0	-$1,101	0.039	0.000	0.039	0.276
8	AS	-$10	$10	$0	-$1,102	$0	-$1,101	0.039	0.000	0.039	0.316
9	AS	-$10	$10	$0	-$1,102	$0	-$1,101	0.039	0.000	0.039	0.355
10	AS	-$10	$10	$0	-$1,102	$0	-$1,101	0.039	0.000	0.039	0.395
11	AS	-$10	$10	$0	-$1,102	$0	-$1,101	0.039	0.000	0.039	0.434
12	AS	-$10	$10	$0	-$1,103	$0	-$1,102	0.039	0.000	0.039	0.474
13	AS	-$10	$10	$0	-$1,103	$0	-$1,102	0.039	0.000	0.039	0.513
14	AS	-$10	$10	$0	-$1,103	$0	-$1,102	0.039	0.000	0.039	0.553
15	AS	-$10	$10	$10	-$1,093	$3	**-$1,099**	0.039	0.000	0.039	0.592
Remaining System Value			$10								

Conclusion

The results suggest the technical performance of small-scale grid-connected (≤1 kW_e) hydroelectric systems have the higher technical potential than many comparable small-scale grid-connected technical alternatives over the short-to-medium term in the region, despite current drying conditions in the SW of WA. However, the limited number of locations that are allowed to be converted into even very small hydroelectric intake dams limit this type of technology considerably in WA. Conversely, whilst the small-scale stand-alone PV water pumping system could not compete against grid-connected electricity when a connection was available, the technology is an effective market adaptation and market mitigation measure to supply stock water in practice. This is because stock watering requirements are generally off-grid, and a stand-alone PV pumping system NPV would likely be greater than a grid-connected pumping system that requires electricity network extension. This work demonstrates that the local considerations and external factors to technologies, costs, and any available subsidies are generally overshadowed by practical concerns in rural areas. Nonetheless, all of the renewable energy systems assessed were expensive adaptation and mitigation options when competing with the currently low cost of network electricity service provision when a suitable connection is available (in terms of the discounted net present cost against 2010 system market prices and subsidies).

References

[1] Harries, D., McHenry, M., Jennings, P., Thomas, C., (2006). Hydro, tidal and wave energy in Australia. *Int. J. Environ. Stud.* 63(6), 803 814.

[2] Risbey, J., Kandlikar, M., Dowlatabadi, H., Graetz, D., (1999). Scale, context, and decision making in agricultural adaptation to climate variability and change. *Mitig. Adapt. Strat. Global Change*. 4, 137-165.

[3] Whetton, P.H., McInnes, K.L., Jones, R.N., Hennessy, K.J., Suppiah, R., Page, C.M., Bathols, J., Durack, P.J., (2005). Australian climate change projections for impact assessment and policy application: A review. in *CSIRO Marine and Atmospheric Research,* CSIRO.

[4] Hope, P.K., Drosdowsky, W., Nicholls, N., (2006). Shifts in the synoptic systems influencing southwest Western Australia. *Clim. Dynam.* 26, 751-764.

[5] Australian Greenhouse Office. (2007). Detecting, understanding and attributing climate change, Department of the Environment and Water Resources, Editor. 2007, Australian Greenhouse Office.

[6] Power, S., Sadler, B., Nicholls, N., (2005). The influence of climate science on water management in Western Australia: Lessons for climate scientists. *B. Am. Meteorol. Soc.* June 2005.

[7] Nicholls, N., (1991). The El Nino / Southern Oscillation and Australian vegetation. *Plant Ecol.* 91, 23-36.

[8] Nicholls, N., (2004). The changing nature of Australian droughts. *Climatic Change*. 63, 323-336.

[9] Intergovernmental Panel on Climate Change. (2007). Longer Report. In: Climate Change 2007: Synthesis Report. in *Intergovernmental Panel on Climate Change Fourth Assessment Report.* Cambridge, United Kingdom: Cambridge University Press.

[10] Suppiah, R., Hennessy, K.J., Whetton, P.H., McInnes, K., Macadam, I., Bathols, J., Ricketts, J., Page, C.M., (2007). Australian climate change projections derived from simulations performed for the IPCC 4th Assessment Report. *Aust. Meteorol. Mag.* 131, 131-152.

[11] Hightower, M., (2009). Energy and water issues and challenges - completing the energy sustainability puzzle, in *REUSE 09.* 2009, Sandia National Laboratories, New Mexico, USA: Brisbane, Australia.

[12] U.S. Department of Energy. (2006). Energy demands on water resources: report to Congress on the interdependency of energy and water. http://www.ssandia.gov/energy-water/docs/121-RptToCongress-EWwEIAcomments-FINAL.pdf.

[13] Carter, T.R., Jones, R.N., Lu, X., Bhadwal, S., Conde, C., Mearns, L.O., O'Neill, B.C., Rounsevell, M.D.A., Zurek, M.B., (2007). New assessment methods and the characterisation of future conditions. in *Climate Change 2007: Impacts, Adaptation and Vulnerability. Contribution of Working Group II to the Fourth Assessment Report of the Intergovernmental Panel on Climate Change.* Cambridge, United Kingdom: Cambridge University Press.

[14] Barker, T., Bashmakov, I., Alharthi, A., Amann, M., Cifuentes, L., Drexhage, J., Duan, M., Edenhofer, O., Flannery, B., Grubb, M., Hoogwijk, M., Ibitoye, F.I., Jepma, C.J., Pizer, W.A., Yamaji, K., (2007). 2007: Mitigation from a cross-sectoral perspective. in *In Climate Change 2007: Mitigation. Contribution of Working Group III to the Fourth Assessment Report of the Intergovernmental Panel on Climate Change.* Cambridge, United Kingdom and New York, USA: Cambridge University Press.

[15] Klein, R.J.T., Huq, S., Denton, F., Downing, T.E., Richels, R.G., Robinson, J.B., Toth, F.L., (2007). Interrelationships between adaptation and mitigation. in *Climate Change 2007: Impacts, Adaptation and Vulnerability. Contribution of Working Group II to the Fourth Assessment Report of the Intergovernmental Panel on Climate Change.* Cambridge, United Kingdom: Cambridge University Press.

[16] Cullen, L.E., Grierson, P.F., (2009). Multi-decadal scale variability in autumn-winter rainfall in south-western Australia since 1655 AD as reconstructed from tree rings of *Callitris columellaris. Clim. Dynam.* 33(2-3), 433-444.

[17] Harle, K.J., Howden, S.M., Hunt, L.P., Dunlop, M., (2007). The potential impact of climate change on the Australian wool industry by 2030. *Agr. Syst.* 93, 61-89.

[18] Kirono, D.G.C., Chiew, F.H.S., Kent, D.M., (2010). Identification of best predictors for forecasting seasonal rainfall and runnoff in Australia *Hydrol. Process.* 24(10), 1237-1247.

[19] HOMER Energy LLC. (2010). HOMER Version 2.68 beta. 2010.

[20] Department of Natural Resources Canada. (2010). RETScreen Version 4. 2010.

[21] Synergy. (2010). Tariff prices. 2010, 27 October www.synergy.net.au.

[22] Office of Energy. (2009). Electricity retail market review. Final recommendations report review of electricity tariff arrangements. Office of Energy report to the Minister for Energy. Perth: Government of Western Australia.

[23] McHenry, M.P., (2012). A technical, economic, and greenhouse gas emission analysis of a homestead-scale grid-connected and stand-alone photovoltaic and diesel systems, against electricity network extension. *Renew. Energ.* 38, 126-135.
[24] McHenry, M.P., (2012). Small-scale (≤6 kWe) stand-alone and grid-connected photovoltaic, wind, hydroelectric, biodiesel, and wood gasification system's simulated technical, economic, and mitigation analyses for rural regions in Western Australia. *Renew. Energ.* 38, 195-205.
[25] McHenry, M.P., (2012). Are small-scale grid-connected photovoltaic systems a cost-effective policy for lowering electricity bills and reducing carbon emissions? A technical, economic, and carbon emission analysis. *Energ. Policy.* 45, 64-72.
[26] McHenry, M.P., (2011). Integrating climate change mitigation and adaptation: Refining theory for a mathematical framework to quantify private and public cost-effectiveness, and carbon emissions for energy and development projects. *Renew. Energ.* 36(4), 1166-1176.
[27] Office of the Renewable Energy Regulator. (2010). Website. 2010, 19 August www.orer.gov.au.
[28] McHenry, M.P., (2009). Remote area power supply system technologies in Western Australia: New developments in 30 years of slow progress. *Renew. Energ.* 34, 1348-1353.
[29] McHenry, M.P., (2009). Why are remote Western Australians installing renewable energy technologies in stand-alone power supply systems? *Renew. Energ.* 34, 1252-1256.

In: Renewable Energy
Editor: Andreas Poullikkas ISBN: 978-1-62618-231-8

Chapter 11

MODELING THE CARBONIZATION PROCESS TO DEVELOP A COST-EFFECTIVE, SMOKELESS, CONTINUOUS, DOWN-DRAFT RICE HUSK CARBONIZER SUITABLE FOR RICE GROWING REGIONS

R. F. Orge[1,*], M. P. McHenry[2] and R. L. de Leon[3]
[1]Philippine Rice Research Institute (PhilRice), Philippines;
[2]School of Engineering and Energy, Murdoch University, Western Australia;
[3]College of Engineering, University of the Philippines, Diliman, Philippines

ABSTRACT

This chapter discusses modeling of the operational processes occurring in a small-scale, down-draft, continuous rice husk (40 kg h^{-1}) carbonizer suitable for application in rice growing regions. The model was used as a tool to optimize the performance of a constructed carbonizer using material and heat balances. The carbonizer technology operating principles are discussed in terms of four operational 'zones' and the possible reactions occurring in each zone. The material balance model was used to determine the amounts of each participating material at each zone, and the energy balance was generated using the material balance solutions. The final output of the model for O_2, CO, and CO_2 was reconciled with testing performance of the constructed carbonizer. The results suggested that 99.2% (weight basis) of the total CO produced during carbonization was burnt at the ignition chamber zone, resulting in only 0.8% CO emission from the chimney. The energy balance determined there was a high potential for the carbonizer to produce useful heat, for rice farm activities, with flue gasses calculated

* R. F. Orge, Philippine Rice Research Institute (PhilRice), Maligaya, Munoz Science City, 3119 Nueva Ecija, Philippines Tel.: (+63) (44) 456-0354. Email: rforge@gmail.com.

at 724°C. The material and heat balance models were successfully verified by prototype testing.

Keywords: Carbonizer; rice husk; energy; model; biochar

NOMENCLATURE

RH	rice husk
CHR	carbonized rice husk
R_p	the matrix of the elemental composition (%) of the RH based on ultimate analyses
A_p	the matrix of the elemental composition of ambient air
C_{crh}	C content (%) of the carbonized RH produced by the carbonizer
Q_r	the input capacity of the carbonizer, (kg RH h^{-1})
Q_a	the flow rate of air entering the side opening of the carbonizer, (kg h^{-1})
Q_b	the flow rate of air entering the bottom opening of the carbonizer, (kg h^{-1}).
R_w	matrix of the kg of RH components entering the carbonizer
A_{ws}	matrix of the kg of air components entering the side opening
A_{wb}	matrix of the kg of air components entering the bottom opening
t	accounting period considered (one hour)
C_{crh}	portion of the C present in RH remaining in the charcoal after carbonization
C_a	the amount of gasified C taking part in the reaction
$R_w(1)$	the first element of the R_w matrix (%C)
y	yield of CRH (a ratio of the weight of CRH to input RH, in decimal form)
X_{air}	X characteristic of the intake air
X_{rh}	X characteristic of the RH
X_{crh}	X characteristic of the CRH
X_{fg}	X characteristic of the flue gas
m	mass of the material
Cp	specific heat capacity, (J $mole^{-1}$ K^{-1})
ΔT	difference between reference and final temperature, (°K)

INTRODUCTION

Carbonization is a complex process [1, 2] with some published literature equating carbonization with pyrolysis [3-5], while others consider pyrolysis a distinct process to accomplish carbonization [6-11], or as a sub-processes similar to gasification [12]. Several models describing the carbonization process have been developed [13-16], although no literature is currently available that sufficiently describes how the chemical elements in rice husk (RH) and air interact in the numerous reactions involved in the carbonization process. This work incorporates Hien's [17] research on RH combustion systems, both Vinluan's [18], and Rezaiyan and Cheremisinoff's [5] research on gasification technologies, Basu's [19] work on combustion and gasification, and Reed and Das's [12] work on downdraft combustion systems to simplify the problem to equations to predict the behavior of a thermal system like a carbonizer [20]. Reed and Das [12] reported that adjusting the equivalence ratio of a gasifier at a level lower than 0.25 favors the production of carbonized rice husk (CRH). This means a carbonizer is able to operate as a gasifier if an optimal equivalence ratio for gasification is achieved by influencing air intakes.

This chapter uses a combination of gasification and combustion theory to model the various chemical reactions taking place in the operation of a smokeless RH carbonizer, and how these elements interact during the process. The predictive model produced suitable quantitative inputs for assisting the design and optimization of the carbonization process and carbonizer. The results were tested by constructing a small-scale, motor-less downdraft RH carbonizer in the Philippines, designed to enabled future retrofitting of heat recovery components. The resultant carbonizer overcame many limitations of the batch carbonizers currently used in the region, including low heat recovery capability, and in particular the unacceptable operator emission/smoke exposures.

CARBONIZER OPERATING PRINCIPLES

RH is initially fed into the top of the carbonizer, filling the hopper, which is then tightly closed (Figure. 1). The model assumes that hopper cover and all of the hopper joints and connections are perfectly sealed so ambient air taking part in the carbonization process enters only through the side opening where a portion of the RH bed is exposed. Once the carbonizer is ignited and starts to operate by itself, the model assumes the carbonizer operates with the same processes and reactions as that of a gasifier, since both systems operate with limited (below stoichiometric) levels of air. The flow of materials (RH, air, and CRH) entering and leaving the carbonizer is represented in Figure. 2.

The RHs inside the hopper undergo changes in various zones, the boundaries of which overlap in practice [17], but are clearly divided diagrammatically for the discussion. Furthermore, the size of each zone will vary for each operation, primarily due to the changes in RH agitation frequency which is undertaken to allow the CRH to collect in the bottom chamber. In the drying zone the RHs are heated from the ambient temperature to pyrolysis temperatures of around 230°C [17]. In the immediate vicinity of the chimney's base and the upper portion of the ignition chamber is the pyrolysis zone. The RHs are exposed to high temperatures in the absence of air as the air does not flow through this region of the carbonizer. Thus, the RHs lying in this region are 'cooked', transforming them into three products - solid (CRH), liquid (tar), and gas. The region in the immediate vicinity of the ignition chamber is the reduction zone. This zone lies along the path of the air wherein the following reactions are expected to occur in a similar manner as that of a gasifier [12]:

$$C + 0.5\,O_2 \rightarrow CO + 110.52\ \text{kJ mol}^{-1} \quad (1)$$
$$C + O_2 \rightarrow CO_2 + 393.77\ \text{kJ mol}^{-1} \quad (2)$$
$$H_2 + 0.5O_2 \rightarrow H_2O + 285.83\ \text{kJ mol}^{-1} \quad (3)$$
$$C + H_2O \rightarrow CO + H_2 - 131.40\ \text{kJ mol}^{-1} \quad (4)$$
$$C + 2H_2O \rightarrow CO_2 + 2H_2 - 89.87\ \text{kJ mol}^{-1} \quad (5)$$
$$C + CO_2 \rightarrow 2CO - 172.73\ \text{kJ mol}^{-1} \quad (6)$$
$$C + 2H_2 \rightarrow CH_4 + 74.87\ \text{kJ mol}^{-1} \quad (7)$$

$$CO + 3H_2 \rightarrow CH_4 + H_2O + 205.82 \text{ kJ mol}^{-1} \quad (8)$$
$$CO + H_2O \rightarrow H_2 + CO_2 - 41.43 \text{ kJ mol}^{-1} \quad (9)$$

Reactions four and five are the water-gas reactions, reaction six is the Boudouard reaction, reaction seven is the methanation or hydro-gasification, and reaction nine is the shift conversion or water-gas shift [5,19]. The combustion zone is bound by the walls of the ignition chamber. It is where secondary air entering through the bottom opening is introduced to the combustible gases.

The reactions can be described as follows:

$$CO + 0.5O_2 \rightarrow CO_2 + 283.25 \text{ kJ mol}^{-1} \quad (10)$$
$$CH_4 + 2O_2 \rightarrow CO_2 + 2H_2O + 802.89 \text{ kJ mol}^{-1} \quad (11)$$
$$H_2 + 0.5O_2 \rightarrow H_2O + 285.83 \text{ kJ mol}^{-1} \quad (12)$$

In addition to the N_2 component of air which did not play a reactionary role, gases entering into the combustion zone could be any or all of the following products of reactions one to nine: H_2, CO, CO_2, H_2O, and CH_4. As some of these products are reactants of the other reactions, some of the combustible gases may be consumed while in the reduction zone.

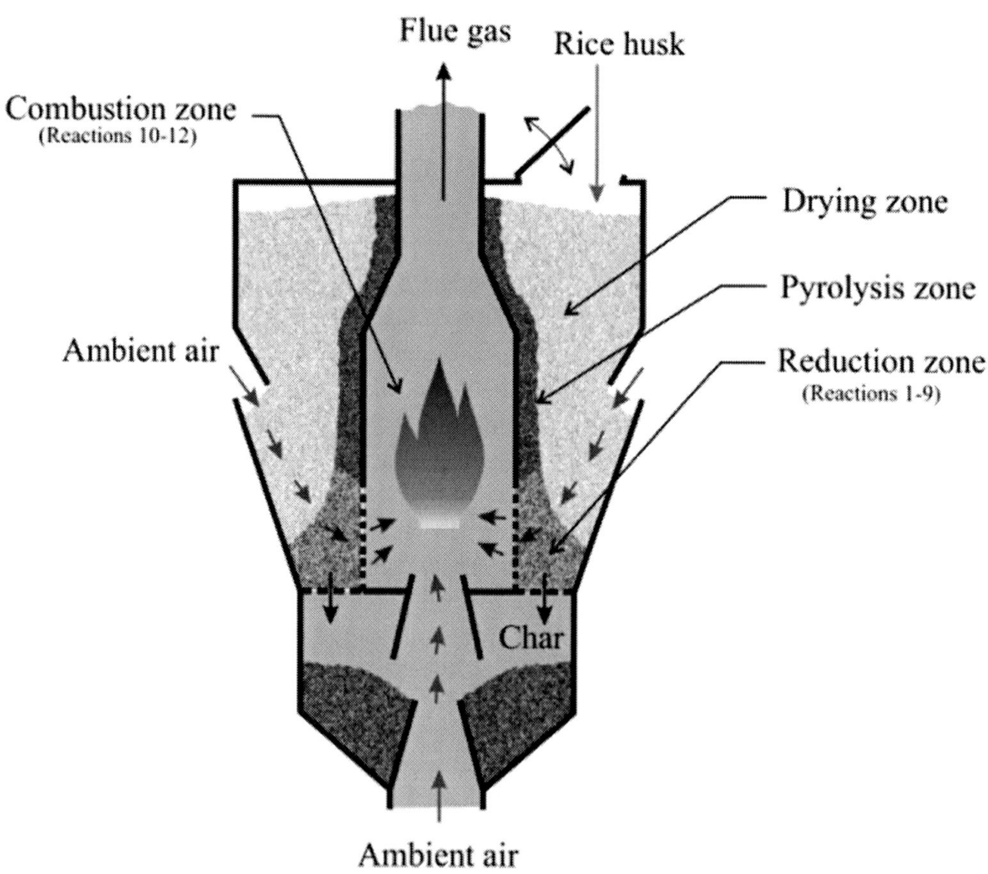

Figure 1. Cross section representing material flows into and out of the carbonizer.

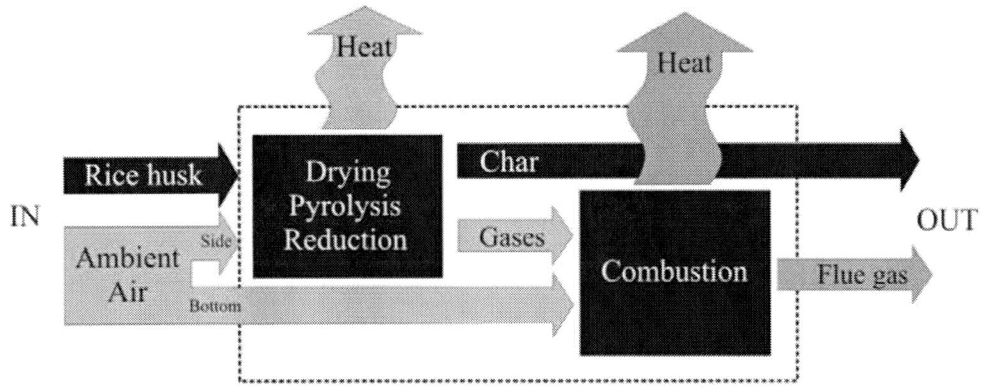

Figure 2. Flow of materials into and out of the carbonizer in a process diagram.

METHODS

The carbonization process was modeled based on the concept of material balance. As matter cannot be created nor destroyed, the total materials that constitute the RH and the ambient air taking place in the carbonization process remains the same at the point of entry and exit of the carbonizer. MATLAB® was used to develop a program to model the carbonization process to accounted for reactions one to twelve to determine a condition wherein a 'material balance' exists, based on the zone boundary conditions. The program output displays a list of the materials and their relative amounts at the point of entry and exit in the carbonizer, and for each zone. Since the carbonizer is a continuous design, the program uses an accounting period of one hour. Thus, the amount of RH considered is equal to the carbonizer's input capacity (kg RH h^{-1}) multiplied by one hour.

The model included input data from the input RHs and actual operation of the carbonizer. For the materials entering the carbonizer, the data were derived from elemental (ultimate) and proximate analyses of the RHs (Table 1). The data for the exiting materials were obtained during testing of the carbonizer (see Table 2 for testing results). The calculations in program were based on the following assumptions:

1) The processes involved do not change with time (steady state) and are carried out at constant pressure (isobaric);
2) Each reaction from one to nine took place during the carbonization process, and each of the carbon-consuming reactions (such as reactions one, two, four, five, six, and seven) receive equal chance of obtaining carbon (C) from RH (as the main source), although each reaction may obtain unequal amounts depending on where material balance is attained. This assumption also applies to chemical substances/compounds which are produced in one reaction but are also utilized as reactants in another reaction;
3) CO, a product of reactions one, four, and six, and reactant of reactions eight and nine, was allowed to exit the chimney unburnt. This was introduced to find a material balance solution to the measured CO gas at the chimney, in addition to satisfying the C allocation flexibility described in assumption 2 above;

4) Both H_2 and CH_4, being more combustible than CO, were assumed to be completely burned before leaving the chimney. The H_2 was assumed to be totally combusted in the reduction zone as it's auto-ignition temperature is 400°C [21], well above reduction zone temperatures obtained in the performance tests results. The CH_4 was assumed to be totally burnt by the time it exits the ignition chamber to pass through the carbonizer's chimney.

Figure 3 provides a visual representation of the sources and destinations of the materials involved in the carbonization process. The nine rows in each of the two grids represent the reactions occurring at the reduction zone (with the number designations indicated at the left).

Table 1. RH data used in the material and heat balance analysis.

Parameter	Value
Proximate Analysis	
Moisture, %	7.8
Ash, %	18.1
Volatile, %	55.5
Fixed C, %	18.6
Ultimate Analysis	
C, %	38.2
Hydrogen, %	5.5
Oxygen, %	35.2
Nitrogen, %	0.9
Ash, %	20.2

Table 2. Performance of the carbonizer

Performance Parameter	Prototype
Ignition time, min	3.2
Total amount of RH consumed, kg	164.5
Total operating time, h	4.1
Input capacity, kg RH h^{-1}	40.12
Char yield, %	40.3
Maximum temperature attained, °C	898.7
Attendance time, % of total operating time	21.3
Reloading with RH, %	8.6
Agitation, %	8.6
Collection of CRH, %	4.1
Purity of CRH output, %	99.1
Emission	
O_2, %	8.9
CO, ppm	431
CO_2, %	7.4

The columns in Figure 3 represent the participating elements and compounds including N_2. A shaded box where columns and rows intersect indicate the presence of an element or compound in a particular reaction. The main sources of these elements/compounds are RH and ambient air which are displayed at the leftmost side as two independent sources. The only source of C was the RH, which were allocated to the six C-consuming reactions. This was the net amount of C after subtracting the C content of the CRH. In the model, one of the two reactants from each row was an independent reactant, and one considered as the independent reactant (the leftmost end of the row in Figure. 3). The calculated dependent reactant was subtracted from the input RH or ambient air material flows, and also the measured output products. Any excess reactant will proceed to the combustion zone, or if not utilized was considered as carbonizer emission gas.

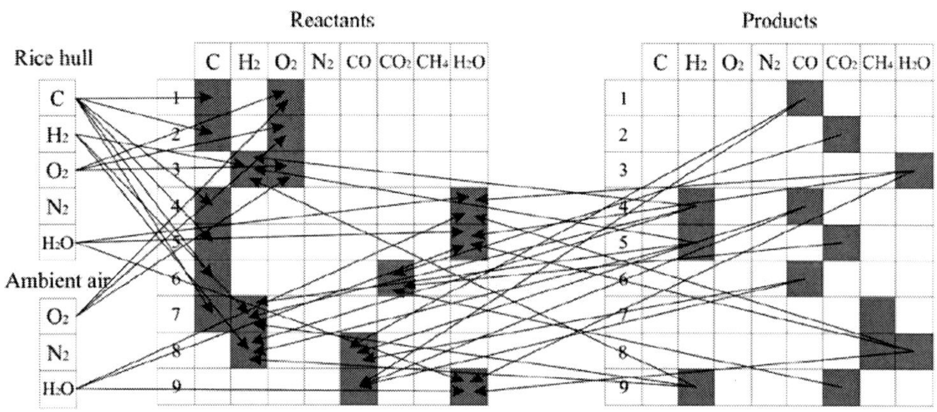

Figure 3. Sources and possible destinations of reduction zone elements and compounds.

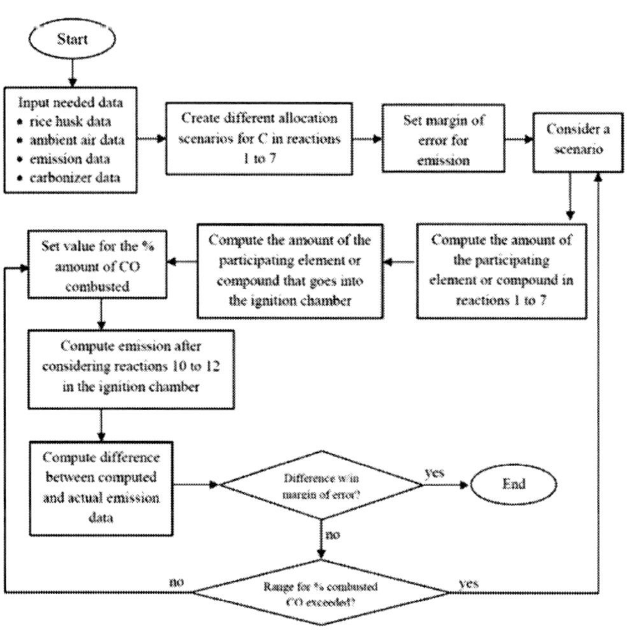

Figure 4. Algorithm for the computer model for the carbonization process.

The development of the model for the carbonization process followed the general algorithm shown in Figure 4. The 'balanced' condition was determined as when the modeled carbonizer emissions matched the measured emission data within a margin of error. The two loops created to filter the possible solution included one that varied the percentage of CO completely burnt at the ignition chamber, and the other was the allocation of C for the C-consuming reactions. The calculations in the computer model are described in this section. Let:

R_p = the matrix of the elemental composition (%) of the RH based from the results of the ultimate analysis = [%C, %H_2, %O_2, %N_2, %H_2O, %Ash];
A_p = the matrix of the elemental composition of ambient air = [%O_2, N_2, %H_2O];
C_{crh} = C content (%) of the carbonized RH produced by the carbonizer;
Q_r = the input capacity of the carbonizer, (kg RH h^{-1});
Q_a = the flow rate of air entering the side opening of the carbonizer, (kg h^{-1});
Q_b = flow rate of air entering the bottom opening of the carbonizer, (kg h^{-1});

Then the matrix of input quantities (kg) of RH components, and side and bottom air may be derived from:

$$R_w = Q_r \times R_p \times t \times 100^{-1} \quad (13)$$
$$A_{ws} = Q_a \times A_p \times t \times 100^{-1} \quad (14)$$
$$A_{wb} = Q_b \times A_p \times t \times 100^{-1} \quad (15)$$

where:

R_w = matrix of the kg of RH components entering the carbonizer;
A_{ws} = matrix of the kg of air components entering the side opening;
A_{wb} = matrix of the kg of air components entering the bottom opening;
t = accounting period considered (one hour).

A portion of the C present in RH remains in the charcoal/biochar after the carbonization process, defined as equal to C_{crh} in the model. Therefore, the amount of gasified C (C_a) that took part in reactions one to nine is the difference in the amount of C present in the RH, and in the CRH:

$$C_a = R_w(1) - (Q_r \times y \times t \times C_{crh} \times 100^{-1}) \quad (16)$$

where:

$R_w(1)$ = the first element of the R_w matrix (%C);
y = yield of CRH (the ratio of the weight of CRH recovered after the carbonization process, and the total weight of the input RH, in decimal form).

The amounts of gasified C were allocated among the six C-consuming reactions (reactions one, two, four, five, six, and seven). The program itself generated C-allocation cases (up to 3,040,000) for use in the calculations.

Each case was a vector of nine elements, representing the nine reactions occurring at the reduction zone. Three of these nine reactions (reactions three, eight and nine) had a zero C-allocation throughout the generated cases. With the substitution of values for C, the other reactant and product amounts were calculable, following the procedure discussed by Joel [22]. Vectors of the same size as the C-allocation matrix were created for each other participating element or compound. In cases where a particular element or compound was both a reactant in one reaction and a product in another, two unique vectors were created. The elements or compounds that proceeded to the ignition chamber in the model were those that remained after subtraction of all the reactants from the products. For combustible gases, such as H_2, CO, and CH_4 entering the ignition chamber, the introduction of air results in H_2O and CO_2 as described in reactions twelve to fourteen. This finally consolidated the calculated amounts of all the emitted gases from the chimney.

For an allocation case to be accepted (and be considered realistic), the difference between the calculated emissions for O_2, CO, and CO_2 and those gathered from the carbonizer tests, should be equal to, or less than that the set margin of error. Any allocation case that yields an error equal to or less than the set value was displayed by the program. In cases where the program displayed more than one solution, the margin of error was lowered until only one solution was displayed.

MODEL OUTPUT

Table 3 shows output from the MATLAB® program for the carbonization process. With the accounting time of one hour, the gasified C from an input 40 kg RH was 8.4 kg. This was allocated to the six C-consuming reactions in the proportions shown in Table 3. This was the only solution out of the 90,000 C allocation cases considered within the set margin of error of 7.51879%.

The Table shows that 37% of the total gasified C was utilized in reaction one, which demonstrates that the reduction zone reactions are mostly partial combustion of RH. Conversely, each of the reactions six and seven consumed 19% of the total available C while reactions two, four and five consumed 9%, 15% and 1%, respectively. The results also show that 99.2% of the total amount (by weight) of CO was burned at the ignition chamber, meaning the remaining 0.8 % escaped as an emission at the chimney.

All reactions are associated with either by an absorption or evolution of energy, which usually manifest themselves as heat [23]. The operation of the rice husk carbonizer generates a net heat output, as shown by the sum of reactions one to twelve. Since the generated heat is planned to be utilized for productive rice farming purposes, it is necessary to analyze the heat loss during operation to evaluate the potential of heat recovery unit retrofits. The heat balance model was calculated using the material balance conditions described in Table 3.

Figure 5 shows the temperatures at different zones in the carbonizer based on data taken during performance testing. These heat balance equations are expressed as the enthalpies (H) of the materials entering and leaving the carbonizer:

$$H_{air} + H_{rh} = H_{crh} + H_{fg} + \text{Heat loss,} \qquad (17)$$

where:

H_{air} = enthalpy of the air;
H_{rh} = enthalpy of the RH;
H_{crh} = enthalpy of the CRH, and;
H_{fg} = enthalpy of the flue gas.

Since the total enthalpy of a material consists of three thermodynamic properties, heating value (HV), sensible heat (SH), and latent heat (LH). Equation 17 can also be rewritten to calculate heat loss:

$$\text{Heat loss} = (\text{HV, SH, LH})_{rh} + (\text{HV, SH, LH})_{air} - (\text{HV, SH, LH})_{crh} - (\text{HV, SH, LH})_{fg} \quad (18)$$

Table 3. Output of the computer program for material balance

```
Possible allocation of C
 Reaction 1: 3.11 kg (37%)
 Reaction 2: 0.76 kg (9 %)
 Reaction 4: 1.26 kg (15%)
 Reaction 5: 0.08 kg (1 %)
 Reaction 6: 1.60 kg (19%)
 Reaction 7: 1.60 kg (19%)
 TOTAL: 8.40 kg (100%)
 Amount of gases entering the combustion chamber
 H2: 0.00 kg
 O2: 46.95 kg
 CO: 14.18 kg
 CO2: 1.23 kg
 H2O: 15.47 kg
 CH4: 3.55 kg
 N2: 167.15 kg
 CO burned: 99.20% of total
 Emitted gases
 H2: 0.00 kg (0.00%wt) (0.00%vol)
 O2: 24.74 kg (10.05%wt) (8.84%vol)
 CO: 0.11 kg (0.05%wt) (0.05%vol)
 CO2: 30.61 kg (12.44%wt) (7.96%vol)
 H2O: 23.44 kg (9.53%wt) (14.89%vol)
 N2: 167.15 kg (67.93%wt) (68.26%vol)
 Difference between calculated and actual data
 for O2: 0.68%
 for CO: 7.46%
 for CO2: 7.52%
 ======================================
Number of possible solution(s) found: 1
>>
```

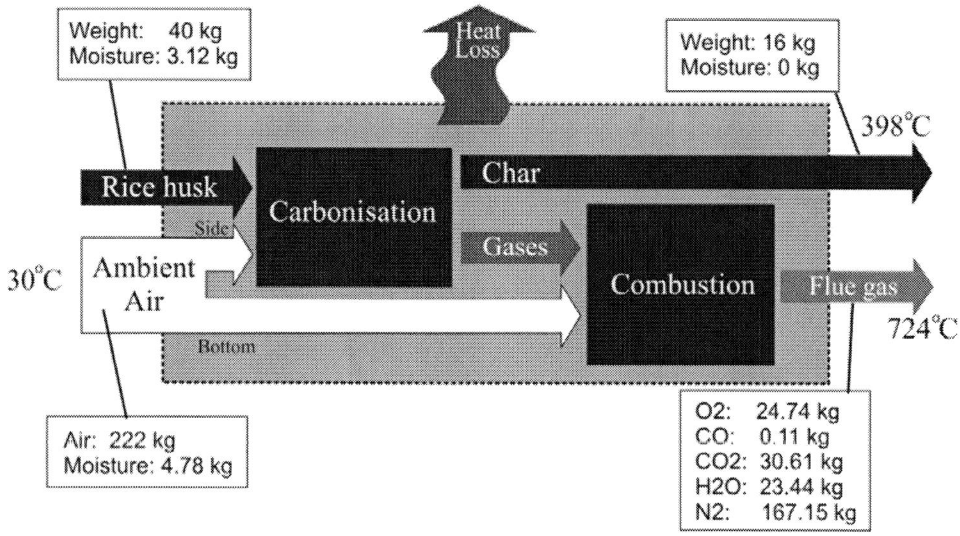

Figure 5. Parameters used in the heat balance analysis.

For materials or substances that are composed of several constituents, the total enthalpy is taken as the sum of the enthalpies of the individual constituents (elements or compounds) of which it is composed of. For sensible heat in particular, the following equation applies:

$$\text{SH} = mCp(\Delta T) \quad (19)$$

where:

m = mass of the material or substance,
Cp = specific heat capacity, and
ΔT = difference between reference and final temperature.

For the gases in the ideal state, the specific heat capacity at constant pressure is computed as follows [24]:

$$Cp = a + b \times T + c \times T^2 + d \times T^3 \quad (20)$$

where:

Cp = in J mole^{-1} K^{-1},
T = temperature in °K, and where
a, b, c, and *b* = are constants with values for selected gases shown in Table 4.

Table 5 shows the calculated results for HV, SH, and LH, at a reference temperature of 30°C (303°K) using values in Table 6. The analysis results show the calculated heat loss is 150,047 kJ, equivalent to 25.8% of the total enthalpy of the system.

Table 4. Constants for computing the specific heat capacity of selected gases

Constant	H_2	O_2	CO	CO_2	H_2O	CH_4	N_2
a	27.143	28.106	30.869	19.795	32.243	19.251	31.15
b	0.0092738	-3.68E-06	-1.29E-02	7.34E-02	1.92E-03	5.21E-02	-1.36E-02
c	-1.38E-05	1.75E-05	2.79E-05	-5.60E-05	1.06E-05	1.20E-05	2.68E-05
d	7.65E-09	-1.07E-08	-1.27E-08	1.72E-08	-3.60E-09	-1.13E-08	-1.17E-08

Source: Towler and Sinnot [24].

Table 5. Summary of heat balance computations

Material	Amount, kg	HV, kJ	SH, kJ	LH, kJ	Total
IN					
Rice husk	40	571,080	0	0	571,080
Moisture in RH	3.12	0	0	0	0
Air	222	0	0	0	0
Moisture in air	4.78	0	0	10,786	10,786
Sub-total		571,080	0	10,786	581,866
OUT					
Char	16	200,960	5,888	0	206,848
Flue gas:					
O_2	24.74	0	17,255	0	17,255
CO	0.11	1,115	85	0	1,200
CO_2	30.61	0	22135	0	22,135
N_2	167.15	0	128071	0	128,071
H_2O	23.44	0	3406	52904	56,310
Sub-total		202,075	176,840	52,904	431,819
Heat loss					150,047

Table 6. Rice husk data used in the material and heat balance analysis

Parameter	Value
Proximate Analysis	
Moisture, %	7.8
Ash, %	18.1
Volatile, %	55.5
Fixed Carbon, %	18.6
Ultimate Analysis	
Carbon, %	38.2
Hydrogen, %	5.5
Oxygen, %	35.2
Nitrogen, %	0.9
Ash, %	20.2

CONCLUSION

This chapter describes the use of modeling to optimize carbonizer performance and to obtain a greater understanding of the operational processes of converting RH into CRH. The material balance analyzed the flow and fate of the elements and compounds constituting the RH and the ambient air. A low (0.8%) level (weight basis) of CO emissions out of the chimney was achieved. Likewise, the energy balance model determined there is a good potential for the carbonizer to be retrofitted with heat capture technology to utilize the continuously generated heat on-site from RH conversion into CRH. The constructed PhilRice continuous carbonizer prototype was able to process 40 kg of RH in one hour at 40.3% CRH yield with a purity of 99.1%. This research demonstrates the operation of a much needed suitable technology where, unlike current carbonizers in use, it is able to produce a reliable supply of charcoal/biochar at the small-scale with a relatively clean and safe emission, and co-produce useful heat energy for potential farm applications [25-28].

REFERENCES

[1] Radovic, L.R., (2003) *Chemistry and Physics of Carbon, Vol. 28*. Marcel Dekker, New York, USA, p.283.

[2] Marsh, H., Reinoso, F.R., (2006). *Activated carbon*. Elsevier Ltd., Oxford, UK, p.479.

[3] Badger, P.C., (1999). *Solid fuels*, in *CIGR Handbook of Agricultural Engineering Vol. V – Energy and Biomass Engineering*. Kitani, O., Jungbluth, T., Peart, R., Ramdani, A., eds., American Society of Agricultural Engineers, St. Joseph, USA, pp. 253-254.

[4] Worasuwannarak, N.P., Tanthapanichakoon, W., (2004). *Carbonization characteristics of Thai Agricultural Residues*. Joint International Conference on Sustainable Energy and Environment (SEE), Hua Hin, Thailand, 1-3 December.

[5] Rezaiyan, J., Cheremisinoff, N.P., (2005). *Gasification technologies: a primer for engineers and scientists*. Taylor and Francis Group, LLC, Boca Raton, USA, p.145.

[6] Klass, D.L., (1988). *Biomass for Renewable Energy, Fuels, and Chemicals*, Academic Press, San Diego, USA, p.651.

[7] Demirbas, A., (2001). Biomass resource facilities and biomass conversion processing for fuels and chemicals, *Energ. Convers. Manage*. 42, 1357-1378.

[8] McKendry, P., (2002). Energy production from biomass (part 2): conversion technologies, *Bioresource Technol*. 83, 47-54.

[9] Gross, R., Leach. M., Bauen, A., (2003). Progress in renewable energy, *Environ. Int.* 29, 105-122.

[10] Maiti, S., Dey, S., Purakayastha, S., Ghosh, B., (2006). Physical and thermochemical characterization of rice husk char as a potential biomass energy source, *Bioresource Technol*. 97, 2065-2070.

[11] Prakash, N., Karunanithi, T., (2009). Advances in modeling and simulation of biomass pyrolysis. *Asian J Sci. Res*. 2(1), 1-27.

[12] Reed, T.B., Das, A., (1988). *Handbook of biomass downdraft gasifier-engine systems*. Solar Energy research Institute, The Biomass Energy Foundation Press, Colarado, USA.

[13] Kim, J., Lee, W.I, Lafdi, K., (2003). Numerical modeling of the carbonization process in the manufacture of carbon/carbon composites. *Carbon* 41(13), 2625-2634.

[14] Luo, Z., Wang, S., Cen, K., (2005). A model of wood flash pyrolysis in fluidized bed reactor. *Renew. Energ*. 30, 377-392.

[15] Díaz, C.J.D., (2006). *Understanding biomass pyrolysis kinetics: improved modeling based on comprehensive thermokinetic analysis*. Ph.D. Dissertation. Universitat Politècnica de Catalunya, Barcelona, Spain.

[16] Di Blasi, C., (2008). Modeling chemical and physical processes of wood and biomass pyrolysis. *Prog. Energ. Combust.* 34, 47–90.

[17] Hien, P.H., (1993). *Rice hull combustion system for crop drying*. Ph.D. dissertation. University of the Philippines, Diliman, Quezon City, Philippines.

[18] Vinluan, Jr. F.D., (1988). *Design, performance analysis, and modeling of a combined updraft-stratified downdraft (CUSD) gasifier*, Ph.D. dissertation, University of the Philippines, Diliman, Quezon City, Philippines.

[19] Basu, P., (2006). *Combustion and gasification in fluidized beds*, Taylor and Francis Group, LLC, Boca Raton, USA, p.86.

[20] Jaluria, Y., (1988). *Design and Optimization of Thermal Systems*. McGraw-Hill, New York, USA, p.626.

[21] Bartok, W., Sarofim, A.F., (1991). *Fossil fuel combustion: A source book*, Wiley, New York, USA, p.866.

[22] Joel, R., (1996). *Basic engineering thermodynamics, 5th edition*, Addison Wesley Longman Limited, Edinburgh Gate, UK, p.647.

[23] Glassman, I., (1996). *Combustion, 3rd edition*, Academic Press, Inc., San Diego, USA, p.632.

[24] Towler, G., Sinnot, R., (2008). *Chemical engineering design-principles, practice and economics of plant and process design. Elsevier* Ltd., Oxford, UK, p.1245.

[25] Orge, R.F., (2010) Design and development of the PhilRice continuous-type rice husk carbonizer. Unpublished PhD thesis, University of the Philippines – Diliman, Quezon City, Philippines.

[26] McHenry M.P., (2011). Soil organic carbon, biochar, and applicable research results for increasing farm productivity under Australian agricultural conditions. *Commun. Soil Sci. Plan.* 42, 1187-1199.

[27] McHenry M.P., (2009). Agricultural bio-char production, renewable energy generation and farm carbon sequestration in Western Australia: Certainty, uncertainty & risk. *Agr. Ecosyst. Environ.* 129, 1-7.

[28] McHenry M.P., (2012). *Sensitive variables for applying biochar as a fertiliser substitute and a method to sequester carbon in soils: a wheat crop scenario*, in *Carbon sequestration: technology, measurement technologies and environmental effects*. Ryan, B.J., Anderson, D.E., eds., Nova Science Publishers, Hauppauge, New York, USA.

INDEX

A

B

D

E

F

G

H

I

J

K

L

M

P

Q

R

S

X

Y

Z